🌹 🌹 🌹 🌹 🌹 🌹 🌹 🌹 🌹 🌹 🌹 MICHIGAN FLORA

Frontispiece

Iris lacustris (Dwarf Lake Iris)
First discovered by Thomas Nuttall at Mackinac
Island in 1810, and endemic to the northern shores of
Lakes Michigan and Huron (see map 667).

(Photo by Gary R. Williams in Emmet County,
June 14, 1970.)

MICHIGAN FLORA

*A guide to the identification and occurrence of
the native and naturalized seed-plants of the state*

Part I

GYMNOSPERMS AND MONOCOTS

Edward G. Voss

CRANBROOK INSTITUTE OF SCIENCE

and

UNIVERSITY OF MICHIGAN HERBARIUM

1972

BULLETIN 55

Second printing 1980, with corrections of text on pages xiv,
68, 79, 94, 113, 251, 276, 321, 322, and 344.
Third printing 1992, with addition to Preface on page x.

Designed by William A. Bostick
Composition by Cranbrook Press, Braun-Brumfield, Inc., and Michigan Typesetting Co.
Color separations by Lakeland Litho-Plate
Printed and bound in the United States of America by Kingsport Press
Edited by Margaret C. Fletcher

Dedicated to the memory of
DALE J. HAGENAH
1908-1971

Preface

THIS PORTION of a Flora planned to deal ultimately with all of the seed-plants of the state of Michigan covers somewhat over 700 species, probably slightly less than a third of the total. In addition to such familiar groups as the conifers, lilies, and orchids, it includes the grasses, sedges, and rushes, as well as most of the true aquatics. It is intended to make possible the identification of all species of gymnosperms and monocots known to grow (or to have grown) outside of cultivation in Michigan, whether originally native or not, and to indicate their general distribution and habitat in the state. By implication it will also show what is *not* known about our flora — what species do not grow here and where those which do occur have not yet been collected.

W. J. Beal's annotated list of the higher plants of Michigan was published in 1905. It reviewed previous listings and collections of the flora of the state. However, because of changes in nomenclature, in taxonomic understanding, and in standards for inclusion in an authentic accounting of the plants of the state, as well as a vast amount of subsequent field work, Beal's list is conspicuously out of date. In the meantime, a number of local lists have been published, most notably the excellent *Flora of Kalamazoo County* by Mr. and Mrs. Clarence R. Hanes (1947).

Although there have been both state and local lists, and treatments for certain families and genera in the state, never before has there been a guide for the identification of all the higher plants of Michigan alone or of any part of it.

Preparation of the present identification manual was begun in 1956 under a five-year grant from the Faculty Research Fund of the Horace H. Rackham School of Graduate Studies of The University of Michigan, which supported the work full-time through the 1960–61 academic year. Subsequently the project was included in the regular research program of the University Herbarium. Clerical and other assistance in completing this first part was made possible by another grant (1969–70) from the Faculty Research Fund. Major support in producing the manuscript and subsidizing its publication, including the eight color plates, came from the Clarence R. and Florence N. Hanes Fund, established by the bequest of Mrs. Hanes in 1966. All users of this Flora will be grateful — as am I — to the trustees of the Hanes Fund, and to Mr. and Mrs.

Hanes, for making possible a more thoroughly documented and illustrated work than would otherwise have been possible — and at a more reasonable cost. The Michigan Botanical Club subsidized the color frontispiece, which depicts one of the favorite and most distinctive wildflowers of the northern Great Lakes region.

Treatment of the dicots is expected to follow in two parts and should be completed more rapidly than the present part on gymnosperms and monocots, which goes to press 15 years after the project was optimistically begun. If a one-volume edition for handier use in the field can also be produced at the completion of the work, corrections should be made in the earlier parts at that time. It will not be feasible to search again exhaustively all of the herbaria examined in the preparation of Part 1. Persons collecting additional state or county records are urged to submit specimens to the University of Michigan Herbarium or at least to supply records of what is deposited elsewhere.

> *"I long to be delivered from the pressure of the engagements that have consumed so much of my time for the last year or two, and finish the 'Flora of North America.' "*
> —Asa Gray to A. DeCandolle, April 5, 1845

NOTE TO THE THIRD PRINTING

Updating the distribution maps is not feasible at this time. However, the list below includes those species of gymnosperms and monocots to be added to the Michigan list; their occurrence outside of cultivation in the state has been documented in the University of Michigan Herbarium since 1971. Species marked with an asterisk are not indigenous (or only questionably so) in Michigan.

*Picea abies**
*Pinus sylvestris**
*Alopecurus geniculatus**
*Alopecurus myosuroides**
*Apera interrupta**
*Bouteloua gracilis**
*Corynephorus canescens**
Danthonia compressa
*Elymus racemosus**
*Eragrostis trichodes**
*Festuca gigantea**
*Festuca myuros**
*Miscanthus sacchariflorus**

*Muhlenbergia asperifolia**
Panicum gattingeri
Panicum longifolium
Panicum polyanthes
Panicum tuckermanii
*Poa glaucifolia**
*Sporobolus indicus**
Carex assiniboinensis
Carex bushii(?)*
Carex heleonastes
Carex novae-angliae
Carex wiegandii
Eleocharis atropurpurea

Eleocharis microcarpa
Eleocharis nitida
Scleria reticularis
Wolffia papulifera
*Juncus compressus**
*Scilla siberica**
Sisyrinchium campestre(?)*
Spiranthes magnicamporum
Spiranthes ochroleuca
Spiranthes ovalis
Spiranthes tuberosa

Acknowledgments

In a work such as this, one is inevitably indebted to the entire botanical community of the past two centuries or more. Writers of other floras, monographs, and notes; collectors of the specimens examined; past annotaters of specimens in herbaria — all these have produced a basic resource freely drawn upon.

The curators of the 33 institutional herbaria cited on pp. 9–10 made the task as easy as possible by their unfailing cooperation in lending specimens and/or facilitating my examination of them on visits. In particular, massive loans from the herbaria of Michigan State University and Cranbrook Institute of Science made it possible to examine thoroughly the specimens of these two large Michigan collections. The care taken by their curators to set aside new Michigan accessions for my examination before inserting them into the collections made their work more complicated and mine more simple. Without exception, all those in charge of collections both in Michigan and elsewhere, at colleges and universities and at other institutions, were generous in their cooperation, and it was a pleasure to work with them. Today very few private herbaria exist in Michigan, but I am likewise grateful for the opportunity to examine material from those mentioned on p. 10. If there are others of which I am unaware, I hope to be informed before completion of the Flora.

Approximately 20,000 of the specimens examined have been in the genus *Carex* alone — our largest genus and one of the most difficult. A majority of these were examined by Dr. F. J. Hermann prior to this project or during the course of it. His patience and willingness to check questionable specimens and to explain to me the subtle distinctions between species are warmly appreciated, as is his examination of an early draft of the manuscript on that genus. Until the very last minute I was calling on him for help with *Carex,* which he knows so well, and also with *Juncus,* on which he is likewise a recognized authority. A devoted supporter of the idea of a flora of Michigan, he has been a source of inspiration and encouragement.

Other specialists, too, checked some of the specimens in their respective groups and offered helpful comments, although in all cases I take responsibility for the final results. Those to whom I have sent the knottiest problems and who have commented on early drafts of the manuscript include Dr. Eugene C. Ogden (*Potamogeton*), Dr. George L. Church (*Elymus*), and Dr. H. K. Svenson (*Eleocharis*). Many others have examined Michigan specimens in the course of their monographic work, and while I have not burdened them with my own specific requests I have benefited from their labors. The careful examination of Michigan grasses by N. W. Katz in a number of herbaria some 20 years ago has made my work much easier in that family. The late Dr. L. H. Shinners aided with *Panicum* and with some of the dicot groups to be covered later; his general encouragement and influence on my thought have been great, and I regret that he did not live to see the publication of this work. Frederick W. Case, Jr., who has more intimate knowledge than anyone else of the orchids of the state, has given much help in that family, including many distributional records.

Several persons offered suggestions and supplied information for the Introductory Section and general keys, including Drs. W. S. Benninghoff, Howard Crum, Thomas R. Detwyler, Jack L. Hough, E. L. McWilliams, Ronald L. Stuckey, John W. Thieret, and W. H. Wagner, Jr.

The Michigan Botanical Club, in establishing a new journal, *The Michigan Botanist,* in 1962, affected this work in perhaps unforeseen ways. My editing of the journal doubtless delayed production of this Flora a little − but the numerous manuscripts that crossed my desk dealing with our plants more than compensated in their contribution to the final product.

My thanks go to numerous citizens of Michigan who allowed collecting on their property, who called attention to significant collecting sites, who submitted specimens, and who otherwise demonstrated their interest. My students on the Ann Arbor campus and at the Biological Station on Douglas Lake contributed more than probably either they or I realize, through their demand for "good key characters" and their exasperation at many existing keys − including those of mine that were inflicted upon them for trial purposes.

Many students and other part-time assistants during the past 15 years have aided immeasurably in the assembling, recording, and organizing of data, both in the field and in the herbarium. My special appreciation goes to three undergraduate botany majors whose intensive work, intelligent understanding, and sympathetic good will during the final year of the project were essential to completion of maps, illustrations, and manuscript: Jacquelyn Kallunki, Kerry Walter, and Nancy K. White.

The present project was successfully launched in 1956 as a result of the efforts of Dr. Kenneth L. Jones, then chairman of the Department of Botany at The University of Michigan, with the active support of Dr. W. H. Wagner, Jr., and other botanists at the University. My colleague in the University Herbarium, Dr. Rogers McVaugh, has repeatedly been called upon for advice and help, including matters editorial and nomenclatural as well as strictly botanical. Dr. A. H. Smith, director of the Herbarium, has made possible the sustained effort which a project of this kind requires. My thanks for all of the support received from my associates in Ann Arbor cannot be adequately stated.

I must also express my gratitude to Cranbrook Institute of Science and the Cranbrook Press for undertaking to produce this volume so expeditiously. I take their zeal in this regard to be a deserved tribute to the late curator of the Cranbrook herbarium, Dale J. Hagenah.

ILLUSTRATIONS

The illustrations aim (with few exceptions) to show at least one species from each *family* in color and at least one species of each *genus* in either a line drawing or color. In the more diverse genera, an effort has been made to show representatives of the major groups. Selection of illustrations has been dependent not only on limitations of space but particularly on the availability of good depictions of the species in the sources used. (Often suitable figures were not readily available to show certain key characters.)

In the color photographs, preference has usually been given to selecting a general view showing something of the characteristic habit of the plant, rather than a detailed closeup. The color illustrations are all made from 35 mm transparencies of plants in nature and in Michigan (only the *Lemna* was brought indoors for photography). The county where the plant was growing is indicated with each photograph, along with the last name of the photographer. Those who responded to my plea for their best slides in certain families to supplement my own, and from whose offerings slides were − often with great difficulty − selected for use, are as follows: Frederick W. Case, Jr. (Saginaw, Michigan), Carmen and Bernard Horne (Jackson, Michigan), John S. Russell (Ann Arbor, Michigan), Marvin and Rudolf Schmid (Ann Arbor), Warren P. Stoutamire (Akron, Ohio), Richard Tetley (Ann Arbor), James R. Wells (Bloomfield Hills, Michigan), and Gary R. Williams (Glen Ellyn, Illinois). To all of these − my friends as well as photographers − I am especially grateful.

The line drawings were mostly not made from Michigan plants. When it is known that a figure is based at least in part on Michigan specimens, an asterisk follows the figure number in the list of acknowledgments below. All of the line drawings have been published previously, and it has been a pleasure to have them made available for use here.

Figure 1, the map of Michigan counties, is used through the courtesy of the University of Michigan Museum of Zoology.

Figures 2−10* were drawn for me by Edward M. Barrows and previously appeared in *The Michigan Botanist* 6: 46, 49. 1967.

Figures 11−19 are from Otis' *Michigan Trees,* published by the University of Michigan Press.

Figures 24−27, 31, 33, 34, 38, 41, 42, 46, 73, 126, 129, 138, 226, 245, 246, 253, 257, 271, 272a, 272b, 280, 291, and 294 originally appeared in Mason's *A Flora of the Marshes of California,* and are used by arrangement with the University of California Press.

Figures 47, 48, 50, 51, 53, 55, 56, 57, 58, 59, 60*, 61*, 62, 67*, 70, 71, 72, 74, 75, 76, 84, 86, 87, 89, 91, 93, 94, 95, 96, 97, 98, 99, 103, 104, 105, 108, 110, 111, 112, 113, 115, 117, 118*, 119, 120, 121, 122, 131, 133, 135, 136, 137, 138, 141, 142, 143, 144, 145, 148, 149, 150, 151, 152, 154, 155, 156, 157, 159, 160, 161, 162, 163, 164, and 167 appeared in the *Manual of Grasses of the United States* and other U. S. Department of Agriculture publications; the original drawings were lent from the collection of the Hunt Botanical Library, Pittsburgh, Pennsylvania.

Figures 20, 21, 23, 29, 40, 54, 77, 78, 79, 80, 81, 82, 90, 92, 123, 172, 207, 208, 210, 211, 229, 230, 233, 236, 252, 273, 281, 288, 295, 296*, 298, 299, 303, 305, 329, 336, 337, and 338 by Elizabeth Dalvé and Elizabeth King are reproduced from *The Monocotyledoneae: Cat-Tails to Orchids,* by E. Lucy Braun (The Vascular Flora of Ohio, Vol. I), and are Copyright © 1967 by the Ohio State University Press. They are used by permission of the late Dr. Braun, the Ohio State University Press, and the Ohio Academy of Science, which kindly lent the original drawings for reproduction.

Figures 283–285 are reproduced by permission of Dr. John W. Thieret from *The Michigan Botanist* 7: 70, 72. 1968.

Figures 318, 319, 320, 323, 328, 331, 334, 339, 340, 341, 342, 343, 344, and 345 were supplied by the Orchid Herbarium of Oakes Ames, Harvard University.

All of the remaining line illustrations are reproduced by permission of the New York Botanical Garden, which generously lent original drawings used for H. A. Gleason's *New Britton and Brown Illustrated Flora of the Northeastern United States and Adjacent Canada* and for K. K. Mackenzie's *North American Cariceae*. Of these, the following figures are based at least in part on Michigan specimens in the herbarium of the New York Botanical Garden: 85*, 170*, 171*, 178*, 183*, 185*, 186*, 191*, 197*, 198*, 199*, 202*, 216*, 217*, 232*, 234*, 235*, 238*, 258*, 260*, 300*, 309*, 311*, 322*, 330*, and 332*.

E. G. V.

Ann Arbor   University of Michigan Herbarium
January 1, 1972

Contents

Introductory Section

Introduction

Si quaeris peninsulam amoenam circumspice.
("If you seek a pleasant peninsula, look about you.")

THE OFFICIAL MOTTO of the state of Michigan recognizes the importance of plant life, for our thousands of lakes and miles of streams are made pleasant and beautiful by the vegetation which borders them. Our forests and fields have contributed vastly to the economy of the state. The tourist industry is based on the fact that both the Upper Peninsula and the Lower Peninsula include so many pleasant places one can enjoy. Nature study and conservation groups abound. Over 100,000 acres in the state have been dedicated as natural areas. Six million acres of State and National Parks, Forests, and Recreation Areas in Michigan attract millions of people annually. Even more than many regions, Michigan depends on its green plants!

This would be an uninteresting as well as barren world without a diversity of plant life. For many people, it becomes even more interesting when they are able to recognize some of the birds, mammals, insects, trees, wildflowers, and other creatures with which they share this planet. From the cold rock shores of Isle Royale, in northern Lake Superior, to the remnant prairies of the southernmost Lower Peninsula, Michigan includes a tremendous diversity of plant life and of the other organisms dependent on it. Some kinds of plants are found nearly throughout the state, but most have some definite geographic — and certainly habitat — limitations. The orchid family, for example, includes some of our showiest and best known wildflowers, a few of which are widespread — and it includes some of the most elusive and rare ones. The 51 species of orchids in the state are a particularly challenging group to many amateur naturalists and professional botanists.

This Flora is intended to help any interested person, whether a visitor in one of our State Parks or a botanist studying problems of plant geography, in an understanding of the plants around him: what is known about their kinds, their identification, and their distribution; and how much more there is to be learned by inquisitive souls who are fascinated by plants, as men always have been.

The Introductory Section defines the scope of the *Michigan Flora* in more detail, and gives necessary background material on how to interpret the Taxonomic Section, which then takes for granted some knowledge of Michigan and its plant habitats, taxonomy and nomenclature, and the use of keys for identification.

The Taxonomic Section begins with an explanation of the style and abbreviations adopted to present information concisely; and it concludes with a glossary.

THE SCOPE AND BASIS OF THIS FLORA

There are many ways in which the scope of a local flora may be defined and executed. A succinct statement has been made by F. R. Fosberg: "This is the true function of a local flora, to give local information." (Taxon 11: 205. 1962) But how does one select the information to be included, and where does he obtain it? Another botanist experienced in floristic work has rightly asserted: "it is the amount of original critical observation rather than the repetition of the errors of others, which is decisive for the standing of Floras. . . . The main aim of any Flora is to have well-defined trustworthy species and *ditto* genera." (Steenis, 1954, pp. 61–63)

A major purpose of the present Flora is to facilitate *identification* of the plants of Michigan. That is, of course, not the main goal of plant taxonomy, but it is the chief function of this work, which thus serves the citizens of the state and any others who may observe plants in Michigan or possess specimens collected here and who wish to know what they are. Identification is simplified when one does not have to take into account the kinds of plants that are not known from the area concerned, and when the identification aids, such as keys, are based on the range of variation existing in that area and not another (or on "repetition of the errors of others").

We have in Michigan almost exactly 50% of the monocot species included in *Gray's Manual* (Fernald, 1950) for all of the northeastern United States and adjacent Canada. The figure is high because of the large number of species characteristic of aquatic and bog habitats which are so abundant in this glaciated region: we have 85% of the pondweeds (*Potamogeton* and *Najas*), 65% of the *Carex,* and 70% of the orchids in *Gray's Manual.* We have less than half of the grasses and presumably will have less than half, on the average, of the dicots. Eliminating even half of the species, however, should make it possible to develop less complex identification keys than required for the flora of the larger area.

The second major purpose of this Flora is to provide condensed information on the distribution and habitats of the plants of the state. This kind of information is often a supplementary aid in identification. And it is precisely the kind of data for which one ought to be able to turn confidently to a local flora.

The distribution maps and the habitat statements, together with the keys and other descriptive information, are based on a fairly exhaustive examination of approximately 80,000 Michigan specimens in herbaria in the state plus several major herbaria elsewhere (see pp. 9–10) – along with my own field experience which has included some collecting in every county of Michigan and a dozen of the islands in the Great Lakes.

One should expect to turn to this Flora to find out where in Michigan a plant grows – or once grew (if it has occurred here at all), under what conditions, and how it may here be distinguished from other plants. Following such exemplary state floras as Deam's for Indiana (1940) and Steyermark's for Missouri (1963), descriptions of the taxa are not included; the keys usually contain a great deal of descriptive information, and further comments are often given in the text. Statements of the total range of every species; information on their chromosomes, possible evolution, or physiology; and documentation as to the history of the names (including synonyms) applied to them, important as these matters are, could not be verified and recorded for all our plants in the time available to produce a flora, even if there were space to include the additional volumes of material. Information on these points would have to be compiled largely from other sources; the reader who is so inclined can consult and evaluate such sources for himself. With the exception of the line drawings, which are included as an aid to identification for the beginner, this Flora does very little in the way of digesting and publishing once again what has already been published elsewhere on plants of other regions. Statements of range, for instance, if resulting from as deplorable a state of knowledge in some other areas as exists in the previously published literature for Michigan, would quite unnecessarily repeat countless errors. Instead, this Flora seeks to provide accurate data which should make Michigan less a "terra incognita" than it seems to be in much botanical literature, the goal of which *is* to synthesize available information from many sources.

This work may therefore be thought of as an annotated checklist, with keys and distribution maps, together with some illustrations. It is not an encyclopedia of information about the kinds of plants that happen to grow in Michigan. In its checklist function, it is intended to include *all* species authentically known from the state, including not only native plants but also those which have established themselves as escapes from cultivation or as unwanted "weeds." It is, however, not always easy to say whether a commonly cultivated species of plant is "established" (i. e., growing and reproducing where it was not planted) even if one sees it in the field, and much less if it is represented by a single specimen on a herbarium sheet and the collector neglected to record on the label an indication as to its apparent status. Some judgment has had to be used as to which taxa and which county records should be accepted, especially in such widely cultivated groups as gymnosperms and Liliaceae. Hence, figures on the number of species in the local flora are necessarily somewhat arbitrary. There is no list of "excluded species" provided, although unauthenticated reports are occasionally mentioned in the text. If a herbarium specimen has been seen, apparently made from a plant growing where it was not planted, the species is included.

Casual waifs collected once or twice may represent the first adventive specimens from cultivation or from a native range farther away. The first specimen of many a now-common weed must once have been considered a waif in the area. It is impossible to prophesy now which of such plants will one day become established and which will die out and never again be collected in the state. Likewise, it is impossible to designate a time limit after which a waif, if never again collected, should be expunged from the list of the state's flora. Some rare native species, not collected for many years, may also be extinct in Michigan. But conditions were once right for their survival, and in some unexplored corner these species might be discovered again. A few garden ornamentals which may persist after cultivation or spread slightly have been mentioned but are not included in the keys.

Dates of flowering are not routinely given, although seasons are mentioned for a number of species. Michigan extends more than 400 miles from the southern border to the tip of Isle Royale — extending over more than six degrees of latitude. Seasons vary tremendously from one end of the state to the other in a single year, not to mention the differences between years. Only a very broad range of dates could be given for the state as a whole for most species, even if the effort were to be made. And even then, there are always the precocious or late-blooming individuals that flower when they are not "supposed to."

Habitat information is condensed from data on labels, field experience, and statements in the literature on Michigan plants, as discussed on pp. 17–21. Distributional information is condensed by placing on the map for each species one dot in each county (or island unit) from which a herbarium specimen has been examined, as described below. This method is the only practical one for a map of the small scale necessarily used, and the fairly regular gridlike pattern of the counties of the state (at least in the Lower Peninsula) produces as fair a representation of the actual distribution of the species as could be expected from the very uneven distribution of collectors.

MAPS AND THE SOURCES OF DATA

There are 83 counties in Michigan; in addition, seven islands or island groups in the Great Lakes have been mapped separately as they are distinct enough geographically and/or phytogeographically from the mainland counties to which they are politically attached. These are shown in the map in Fig. 1, and are listed below:

1. Map of Michigan, showing counties and major islands in the Great Lakes.

Charity Islands (Arenac County)

Beaver Islands (Charlevoix County) [The entire group.]

Drummond Island (Chippewa County) [Including a few small adjacent islands, but not islands of the St. Mary's River.]

Isle Royale (Keweenaw County) [The entire archipelago comprising the National Park.]

Fox Islands (Leelanau County)

Manitou Islands (Leelanau County)

Mackinac, Round, and Bois Blanc Islands (Mackinac County) [But all other islands in the Straits of Mackinac have been included with the mainland.]

It must be stressed that the maps are based only on *specimens examined.* Of course, *Arisaema triphyllum* and *Trillium grandiflorum* occur in every county in the state, even if collectors seem to have neglected a few. But the only consistent standard for a work on this scale is to exclude all published, sight, and other records not vouched for by specimens which have been examined and can be rechecked by anyone who wishes to do so. For most species, enough records exist for the maps to indicate quite well what the general outlines of the range may be and whether the species becomes, for example, less frequent as one goes north or south. Absence of a dot from a county (or island) means only that a specimen has not been seen; the farther away the nearest county with a record is, the more likely it is that absence of a dot also means absence of the species in that area. If *every* specimen examined had been mapped, it would have taken years to track down the obscure localities (old lumbering towns, railroad stops, farmsteads, and woodlots) mentioned on labels, a larger scale map would be required, and the concentration of collections from certain well known areas would have been exaggerated.

Complete data from the labels of close to 80,000 specimens examined have, however, been recorded and are on file by species and county in the University of Michigan Herbarium, where they can be examined by anyone interested in the basis for any record mapped: Is there a single specimen or are there many? When were they collected? By whom? In what precise localities or habitats? Where are they now located? What specialists, if any, have checked them? Questions such as these can be answered from the files. All herbarium specimens from which complete data were recorded have been rubber-stamped "Noted, 19—, Michigan Flora Project". Several additional thousands of specimens, which would not provide further county records or habitat data, have not been thus stamped nor completely recorded; as time became critical in examination of herbaria outside of Michigan, these records have merely been noted by county and herbarium where the specimen is located.

As mentioned before, no list of species excluded from the state flora is provided, nor are there such lists for excluded county records. An example of what may be involved in preparing such a list can be seen in a floristic account of the region around the University of Michigan Biological Station (Voss, 1956).

There are many published maps and other data in botanical literature which, if they were to be accepted, would add county records to the maps. But the status of such records may reflect any of several situations:

(1) Specimens may never have existed to substantiate a report or may have been lost or discarded.
(2) Specimens may exist somewhere in a herbarium not examined (or may have been overlooked or misfiled in one which has been checked).
(3) Specimens intended to vouch for a report may be misidentified — or may be classified according to the concepts of this Flora in a manner other than they were originally.
(4) Published maps may not have included as separate units the island groups here mapped separately. (Thus records from Beaver Island or Drummond Island, for example, have often appeared on maps as if on the mainland of Charlevoix or Chippewa county, respectively.)
(5) Specimens may exist on which previous maps or citations were based, but they were incorrectly attributed to a Michigan county.

The problem listed last above is acute and is a major reason for my reluctance to use any symbol on maps for published records, even from taxonomically reliable works. For example, many specimens collected by Dennis Cooley at "Shelby" have erroneously been attributed to Oceana County, where there is a town of that name, instead of Shelby Township, adjacent to Cooley's home in Washington Township, Macomb County. Many early collectors used township names without specifying "township." Daniel Clarke's specimens from "Burton" are undoubtedly from Burton Township, on the southeast side of Flint, where he practiced medicine, not from the town of Burton in Shiawassee County. Grand Rapids collectors whose specimens are labeled "Paris" were certainly in the township of that name in Kent County, not in Mecosta County. The Detroit Zoo has too often been assumed to be in the same county as Detroit. "Grand Rapids" specimens from the Lower Wabash valley of Illinois have been erroneously attributed to Kent County, Michigan (see Voss, 1967).

E. J. Hill collected at Bear Lake in Manistee County in 1880 and at a Bear Lake in Van Buren County in 1872; his labels sometimes lack a county designation, and he also collected in Muskegon County, where there is another of Michigan's nearly 40 "Bear Lakes." There are many "Black Rivers," "Mud Lakes," "Round Lakes," and such localities in Michigan. For this reason, knowledge of the habits and itineraries of collectors is often required to assign specimens properly. Occasionally a collector may slip and actually cite the wrong county on his label. Rarely, the county border has been changed since a specimen was collected. C. A. Davis often labeled his material as collected "about Alma" (without county) but in some cases we know that it came from Montcalm or Isabella rather than Gratiot County (e. g., *Eriocaulon septangulare*, *Cypripedium arietinum*). Davis' annotated copies of floras in the University of

Michigan Herbarium have been helpful in these cases. Labels of Hill specimens from "Hanbury Lake," which is in Dickinson County in the Upper Peninsula, have been misread as "Hamburg Lake," and the records attributed to Livingston County in the southern Lower Peninsula, on the assumption there was a lake of that name near the community of Hamburg. Specimens from Oscoda, a town on the Lake Huron shore in Iosco County, have been mapped in the inland county of Oscoda.

Problems of label interpretation could be cited indefinitely, even when there is no question as to the identity of a specimen. One of the major efforts in the present project has been to assign specimens correctly to the county in which they were collected. Some mistakes have doubtless occurred, but extensive consideration of collectors' notes, annotated manuals, old maps, and other such documents has, it is hoped, greatly improved the accuracy of the maps here presented — and demonstrated the need to consider with a critical eye a number of published records for these species from additional counties. If distributional data are ever to be entered from herbarium specimens into the memory of a computer, some experienced human being must first evaluate the data!

Considerable effort has been made to track down the basis for many published records, and the cooperation of authors and curators has been greatly appreciated in this detective work.

It is also true that the locality cited on labels may be in a different county from the place where the collections were actually made. Specimens of bog species from "Olivet" quite possibly came from neighboring Calhoun County, for example. But labels have had to be taken at face value when there was no contrary indication. Some localities such as Hubbardston [Ionia County] and Fruitport [Muskegon County] occur so close to the county line (their limits now even closer than when collections were made decades ago) that in the absence of a county assignment by the collector the specimens might better have been treated as borderline, although they have generally been accepted for the one county in which the community barely lies. It has been necessary, however, to treat some localities as unquestionably "borderline," particularly lakes which extend across county lines (most commonly, for botanical records, Magician Lake in Cass and Van Buren counties). Records from localities which could not be assigned to either of two counties have not been mapped unless there were no other records from *either* of the counties; then the dot has been placed on the county line.

Thousands of unexamined Michigan specimens undoubtedly exist in herbaria around the world. Some of these would add dots to the maps, but somewhere the line must be drawn and the point of diminishing returns recognized. The size of a herbarium, the percentage of Michigan records expected, and whether they are organized geographically to ease the location of them are factors considered in selecting herbaria for examination. I have tried to cover Michigan herbaria quite thoroughly so that their specimens might be annotated, insofar as possible, in accordance with the Flora; the project has thus, I hope, been of some service to these institutions in return for their making specimens available for study.

I believe that this Flora has been based on a more thorough coverage of existing herbarium material than any previous work on Michigan plants. All specimens from the state in the following Michigan herbaria have been examined. Abbreviations in parentheses are the standard symbols listed in (or previously proposed for) the *Index Herbariorum* (ed. 5, Regnum Vegetabile Vol. 31. 1964).

Adrian College, Adrian (ADR)
Albion College, Albion (ALBC)
Alma College, Alma (ALM)
Aquinas College, Grand Rapids (AQC)
Central Michigan University, Mount Pleasant (CMC)
Cranbrook Institute of Science, Bloomfield Hills (BLH)
Cusino Wildlife Research Station, Shingleton (CUS)
Edwin S. George Reserve, University of Michigan, Pinckney (MGR)
Ford Forestry Center, Michigan Technological University, L'Anse (MCT-F)
Isle Royale National Park, Houghton (IRP)
Michigan State University, East Lansing (MSC)
Muskegon Community College, Muskegon (MUSK)
Northern Michigan University, Marquette (NM)
Olivet College, Olivet (OLV)
Seney National Wildlife Refuge, Seney (SEN)
University of Michigan, Ann Arbor (MICH)
University of Michigan Biological Station, Pellston (UMBS)
Wayne State University, Detroit (WUD)
Western Michigan University, Kalamazoo (WMU)

Michigan specimens in the following herbaria outside the state have been examined more or less completely, chiefly on visits or (when asterisked) entirely through loans:

Harvard University, Cambridge, Massachusetts (GH, Gray Herbaium; A, Arnold Arboretum; AMES, Orchid Herbarium of Oakes Ames)
Missouri Botanical Garden, St. Louis, Missouri (MO)
New York Botanical Garden, Bronx, New York (NY)
*Ohio State University, Columbus, Ohio (OS)
University of Illinois, Urbana, Illinois (ILL)
University of Notre Dame, Notre Dame, Indiana (ND)
*University of Wisconsin, Madison, Wisconsin (WIS)

Selected specimens, based on expectations for certain species or groups, have been examined from several additional herbaria (all sent on loan in response to specific requests except for US, where a number of grasses and a few others were checked on a brief visit):

Academy of Natural Sciences, Philadelphia, Pennsylvania (PH)
College of the Pacific, Stockton, California (CPH)
United States National Arboretum, Washington, D. C. (NA)
United States National Herbarium, Washington, D. C. (US)
University of Minnesota, Minneapolis, Minnesota (MIN)
University of Pennsylvania, Philadelphia, Pennsylvania (PENN)
University of Southern California, Los Angeles, California (USC)

At least selected specimens have been made available from the private herbaria of Mabel Demorest, Erna Eisendrath, Leroy H. Harvey, F. J. Hermann, Louis K. Ludwig, and E. G. Voss. Many former private herbaria are now in institutional hands and the collectors are mentioned in the following sketch of floristic work in Michigan. Notable among the private herbaria which have gone to institutions during the past 20 years are those of John A. Churchill (MSC), R. R. Dreisbach (MICH), Mr. and Mrs. Clarence R. Hanes (WMU), and Fred W. Rapp (WMU).

FLORISTIC WORK IN MICHIGAN

The earliest explorers in Michigan mentioned few plants. Charlevoix, for example, reporting on his 1721 voyage, discussed poison-ivy at Detroit and the importance of ginseng around the St. Joseph River. But he dismissed Mackinac Island as "only a quite barren rock, and scarcely covered with a little moss and herbs." The first person with a professional interest in natural history to collect and study plants in Michigan was apparently English-born Thomas Nuttall (1786–1859), a naturalist who set out from Philadelphia in 1810 on an expedition to the Northwest which brought him to Michigan. Travelling on behalf of Benjamin Smith Barton, professor of Materia Medica, Natural History, and Botany at the University of Pennsylvania, Nuttall travelled by stage from Philadelphia to Pittsburgh. He then went on foot to Erie and walked along the south shore of Lake Erie to near the site of the present town of Sandusky. Instead of hiking through or around the Great Black Swamp at the western end of Lake Erie, he realized that it was better to take a ship to Detroit. Becoming aware of the immensity of the journey he had undertaken, he also became aware of the fact that travel in this region was best by water, not overland. Consequently, he abandoned Barton's instructions to hike from Detroit to Chicago.

Nuttall's diary, which was not discovered and published until 1951 (see Graustein, 1951, 1967; Stuckey, 1967), records for July 29, 1810: "I left Detroit for Michilimackinak in a birch bark canoe . . ." What he did, actually, was to ride along with the surveyor of the Territory of Michigan, Aaron Greeley, who was going to the island to survey the town lots. No more entries are in Nuttall's diary until he arrived in Wisconsin in September. But we know that the surveyor arrived at Mackinac August 12. There, Nuttall met the Astoria party of the Pacific Fur Company, which was headed for the Columbia River, and he subsequently accompanied portions of that expedition, eventually returning to St. Louis and sailing from New Orleans for England in early 1812.

However, Nuttall returned to the United States in 1815, and in 1818 he published the first comprehensive treatment of American plants to be printed in this country: *The Genera of North American Plants, and a Catalogue of the Species, to the Year 1817.* Although Nuttall's diary is blank for the days he spent in Michigan, we know from his *Genera* that they were botanically profitable. He mentions about 60 species as occurring around the Great Lakes, and about a third of these were described as new to science. Three of these were described from Mackinac Island or nearby shores and they are to this day plants of considerable local interest:

Iris lacustris: "On the gravelly shores of the calcareous islands of lake Huron, near Michilimakinak."
Rubus parviflorus: "On the island of Michilimackinak, lake Huron."
Tanacetum huronense: "With *Artemisia canadensis* on the sandy shores of Lake Huron, near Michilimakinak; abundant."

Nuttall referred to the birdseye primrose (*Primula mistassinica*) on the "calcareous gravelly shores of the islands of Lake Huron; around Michilimakinak, Bois Blanc, and St. Helena, in the outlet of Lake Michigan . . ."; he mentioned *Calypso bulbosa* from St. Helena Island, described *Carex aurea* as new from the "shores of Lake Michigan" and *Orchis huronensis* [*Habenaria hyperborea* var. *huronensis*] as new from the "islands of Lakes Huron and Michigan." He noted such characteristic shore plants as sea-rocket (*Cakile edentula*) and beach pea (*Lathyrus japonicus*); fringed polygala (*Polygala paucifolia*) he noted as forming "almost exclusive carpets of great extent in the Pine forests of Lake Huron," and twinflower (*Linnaea borealis*) was said to be "abundant in the shady pine forests of Lake Huron." Surely Thomas Nuttall deserves credit as being the first botanist to have seen and recognized in his published work many of the distinctive plants of our northern shores and coniferous forests, plants which even today bring botanists into northern Michigan.

As Nuttall discovered, travel was predominantly by water in the earliest days in this region. By a happy coincidence, therefore, our most interesting plants of shores were among the first to be noted in the state.

Nuttall's *Genera* was published only two years before the first organized scientific expedition through Michigan, the Cass Expedition of 1820. Territorial Governor Lewis Cass, appointed by President Madison in 1813, conceived the idea of a scientific expedition to the Lake Superior—Upper Mississippi region, an expedition which would, as he wrote Secretary of War John C. Calhoun, "well accord with that zeal for inquiries of this nature which has recently marked the administration of the War Department." He added: "I am not competent to speculate upon the natural history of the country through which we may pass. Should this object be deemed important, I request that some person acquainted with zoology, botany, and mineralogy, may be sent to join me." Calhoun authorized the expedition and appointed Henry Rowe Schoolcraft, then aged 27, as its geologist and mineralogist. Capt. David Bates Douglass, of the Corps of Engineers and professor at West Point, was to prepare a map and serve as botanist.

Douglass sent the plants collected on this 4,000-mile canoe trip to the leading botanist of the day, John Torrey of New York, who published an account of them (1822). Several species are noted from Lake Huron — some, including the yellow lady-slipper (*Cypripedium calceolus*), specifically from Presque Isle (where a strong wind detained them on June 5). A specimen of *Iris lacustris* from Presque Isle, collected by Douglass, is in the herbarium of the New York Botanical Garden. Among Douglass' observations was that lousewort (*Pedicularis canadensis*), found near Detroit, was "Said by the Indians to cure the bite of a rattle-snake." Several species were also reported from Lake Superior. June 25–26 they crossed the Keweenaw Peninsula where the Portage Ship Canal now makes transit easier, and noted several bog plants, including grass-pink (*Calopogon tuberosus*) and showy lady-slipper (*Cypripedium reginae*). The bastard-toadflax (*Comandra umbellata*) was found at "Point Keeweenah," where it was "Used by the Indians and traders in fevers."

The Cass Expedition never reached its goal, the source of the Mississippi River. In 1830 Governor Cass requested Schoolcraft to head an expedition back into that region, and this left Sault Ste. Marie on June 25, 1831. The physician and botanist of the party was a young man, not yet 22, destined to become one of Michigan's best known citizens, Douglass Houghton. Many of Houghton's specimens are in the University of Michigan Herbarium. Others, sent to Torrey, are at the New York Botanical Garden. An excellent botanist, Houghton collected copiously — at Sault Ste. Marie, the Grand Sable Dunes, the Pictured Rocks, Presque Isle (Marquette), the Keweenaw Peninsula, and the Ontonagon River. This expedition failed to reach the source of the Mississippi and Schoolcraft persuaded the Office of Indian Affairs to send him again in 1832, when his party again included Douglass Houghton. They left the Sault June 7 and succeeded July 13 in reaching Lake Itasca. Houghton's 1832 collections in Michigan are from various places in Chippewa County, the Pictured Rocks, Presque Isle, the Yellow Dog River, Point Abbaye, the Keweenaw Peninsula, and the Ontonagon River (Rittenhouse & Voss, 1962).

When Michigan became a state in 1837, the first act of the legislature was to establish a geological survey, which was to include botanical and zoological studies. Douglass Houghton was appointed the first State Geologist and served until he drowned in Lake Superior in 1845. The botanical results of the First

Survey have been described in some detail by McVaugh (1970). About 800 specimens from this survey, largely collected under the supervision of John Wright, are in the University Herbarium in Ann Arbor and duplicates were distributed rather widely. A few species, such as *Echinodorus tenellus, Digitaria filiformis,* and *Commelina erecta,* have not been collected in Michigan since the First Survey.

Other botanical visitors did some collecting in the earliest days of the territory and state, but after the principal expeditions cited above, the bulk of botanical exploration was done by various resident botanists, mostly "amateurs" in various parts of the state, and by expeditions from the colleges and universities. It would be impossible here to chronicle all of the collectors whose specimens from almost a century and a half are the basis of this Flora. Most of the literature prior to 1945 is cited by Darlington (1945). More by way of acknowledgment than historical completeness, some of the major collectors are mentioned here, with herbaria where most of their specimens are located indicated by the abbreviations explained on pp. 9–10.

Our knowledge of the flora of Macomb County is largely due to the collections (MSC) of Dennis Cooley from 1827 to the mid 1850's. Daniel Clarke's collections (MSC) from the Flint area, made just over a century ago, supply the majority of our Genesee County records. G. M. Bradford in Bay County around the turn of the century (MSC), C. A. Davis in Gratiot County (MICH), C. K. Dodge in St. Clair County (MICH), and Emma J. Cole in Kent County (AQC) are among the other early botanists who concentrated (though not necessarily limited) their considerable collecting in one county. Other more recent collectors who have produced a large majority of the records for certain counties include R. R. Dreisbach in Midland County (MICH), Carl O. Grassl in Menominee County (MICH), Paul W. Thompson in Leelanau County (BLH), Jennie V. A. Dieterle in Grand Traverse County (MICH), Clarence R. and Florence N. Hanes and also Fred W. Rapp in Kalamazoo County (WMU), F. H. Test, E. U. Clover, and Larry Wolf in the Fox Islands (UMBS), Eric A. Bourdo, Jr., in Baraga County (MCT-F), and Jarl K. Hiltunen in Chippewa County (WUD).

Many collectors have contributed to the records from such places as Ingham County (Michigan State University), Washtenaw County (The University of Michigan), Emmet and Cheboygan counties (University of Michigan Biological Station, especially by John H. Ehlers, and earlier collections by C. W. Fallass (ALBC) and others; see Voss, 1956), the Detroit area (Wayne and neighboring counties), Marquette County, Muskegon County, and other comparatively well studied areas. Many collectors were associated with Emma Cole in the Grand Rapids area (see Cole, 1901); and later C. W. Bazuin (MSC, MICH, BLH) collected extensively there and elsewhere on the west side of the Lower Peninsula. Mackinac, Round, and Bois Blanc islands have been explored by Marjorie T. Bingham (BLH), C. K. Dodge (MICH), J. E. Potzger (ND), and others. Extensive collections have been made on Beaver Island by Margaret C.

Reis (BLH) and by Matthew Hohn and others from Central Michigan University (CMC). Isle Royale has long been attractive to botanists, with major collections by, among others, Harold and Virginia Bailey (MICH, IRP), Clair A. Brown (MICH), W. S. Cooper (MIN, GH), A. E. Foote (MICH, OS), Robert A. Janke (IRP), Philip C. Shelton (IRP), C. S. Stuntz and C. E. Allen (WIS), and Clifford M. Wetmore (MSC, MICH).

Many of the collectors cited above have not restricted their activities to the places mentioned, and others have been even more broad in their collecting. The herbarium at Wayne State University (WUD) is especially rich in county records because of the diverse collecting of C. M. Rogers and Jarl K. Hiltunen. C. L. Gilly, George W. Parmelee, and associates from Michigan State University (MSC) collected extensively 1947-1953, particularly in the south-central part of the state. John A. Churchill (MSC) built up a fine collection of specimens, including many *Carex*, from counties too much overlooked by others throughout the state. The collections of Frederick W. Case (MICH) and R. R. Dreisbach (MICH) in the east-central counties as well as elsewhere, and of H. T. Darlington (MSC) throughout the state, especially Gogebic and Ontonagon counties, are considerable. C. K. Dodge and C. A. Davis both collected at many places in Michigan besides St. Clair and Gratiot counties. The southwestern counties of the Lower Peninsula have received attention from many Indiana and Illinois collectors. Specimens from this part of the state by P. E. Hebert (ND) and J. A. Nieuwland (ND) are extensive. C. F. Wheeler and W. J. Beal (MSC) accomplished an extraordinary amount of collecting around the state, mostly just before the turn of the century.

One of the most prolific collectors of all in Michigan was Oliver A. Farwell (see McVaugh, Cain, & Hagenah, 1953). His collections (BLH, MICH) are chiefly from the several counties in the southeastern Lower Peninsula (including St. Clair, Macomb, Oakland, Wayne, Washtenaw, and Monroe) and from Houghton and Keweenaw counties in the Upper Peninsula. Operating in the same southeastern counties, often with Farwell, were Cecil Billington (BLH, WUD, MICH) and Branson A. Walpole (BLH).

Aquatic plants are often overlooked by general collectors. In Michigan their distribution is as well known as that of other plants, thanks in large part to a survey of aquatics conducted in the 1930's and early 1940's by staff of the Institute for Fisheries Research of the Michigan Department of Conservation in collaboration with the Botanical Gardens of The University of Michigan, where the botanical work of this survey was coordinated by Betty Robertson Clarke (MICH).

This itemization of some of the major collectors whose labors have contributed most to the distribution maps demonstrates how the work begun by such pioneers as Thomas Nuttall and Douglass Houghton toward inventorying our flora has grown to tremendous proportions. Yet an examination of the distribution maps will show how much more there is to do! My own collections, like those of several others, number about 10,000 in the state. I have tried to concentrate on regions and species thought to need particular attention, while doing some work in every county and 12 of the islands in the Great Lakes. The obvious omissions revealed by an examination of the maps are depressing to contemplate.

THE OCCURRENCE OF PLANTS IN MICHIGAN

The presence, abundance, and distribution of a given plant species in Michigan, as in any area, depend upon a number of factors. When the final portion of the *Michigan Flora* is completed, it is hoped to include a more comprehensive essay on aspects of Michigan plant geography, drawing upon the data assembled for the entire flora. Excellent works have been published on the state's geology (Dorr & Eschman, 1970) and soils (Veatch, 1953; Whiteside, Schneider, & Cook, 1968), and much information is available, although rather scattered, on climate (see Davis, 1964, pp. 39–53, for a summary). All these factors are of fundamental importance in governing the occurrence of plants. A number of general articles on geography and natural resources, including forest exploitation in Michigan, have been assembled by Davis (1964).

Exclusive of the extensive waters of the Great Lakes, the state covers 58,000 square miles of land and water, the latter including 35,000 lakes and ponds of all sizes, of which some 8,600 are named (Humphrys, 1965). Altogether, these lakes and ponds total about 1,300 square miles of water. They range in size up to Houghton Lake (31 square miles). Eleven inland lakes are each at least 15 square miles in area. Michigan is said to have over 36,000 miles of streams, but if this figure is as great an underestimate as the usual "11,000" figure quoted for lakes, the total length must be considerably more.

The state of Michigan has a shoreline of 3,222 miles on the Great Lakes, second only to Alaska in length of shoreline. This figure includes 948 miles of island shore, about a fifth of which is on Isle Royale. Because of the normal fluctuations in the level of the Great Lakes, this shoreline is far from stable, being subjected constantly to various processes of erosion, deposition, and disturbance. Fluctuations in water level control the shoreline marshes, beach pools, and dunes which are so rich in interesting plant species. The mean monthly levels of Lakes Michigan and Huron, which are treated as one for hydrographic purposes, have differed as much as 6 feet over the past century. Lake Superior, which is less variable due in part to the engineering structures at Sault Ste. Marie, varies less than 4 feet in monthly mean level. In these three lakes, the average change in mean levels in a single year, from winter low levels to summer high levels, is 1.1 feet. Michigan has a relatively short shoreline on shallower Lake Erie, where monthly mean levels have ranged over 5 feet (average of 1.5 feet in a single year).

The average elevations of the Great Lakes, as calculated by the Corps of

Engineers over the period 1860–1970, are 570.39 feet for Lake Erie, 578.68 for Lakes Huron and Michigan, and 600.38 for Lake Superior. Terrestrial elevations in Michigan range from the Great Lakes shores to the highest point, at about 1,980 feet. This is in Baraga County, but Summit Peak, in the Porcupine Mountains of Ontonagon County, comes close at 1,958 feet. Altitude does not exert the influence on plant distribution in Michigan that it does in areas with relief much exceeding our range of some 1,400 feet. More important – if such comparisons can be made – are the overall tempering and moisture-yielding effects of the Great Lakes on the climate some distance from their shores and the local severe effects of the lakes, with their fogs, ice, and waves, in modifying the microclimates and landscapes in their immediate vicinity.

Postglacial History

The story of Michigan's developing landscape may be found in a number of texts and articles, and new information on the details appears regularly. It seems obvious, however, that all of the plant life of the state is of necessity derived ultimately from propagules which moved into the region since the final retreat of glacial ice – about 13,000 years ago in portions of the southernmost Lower Peninsula and about 10,000 years ago in the Upper Peninsula. During and following the last glacial retreat, the present Great Lakes developed in basins which had been shaped by earlier glaciations. Lake Algonquin, in the Michigan–Huron basin (not Superior) before about 11,000 years ago, was succeeded by a series of lower stages until about 9,500–10,000 years ago, when the waters drained to very low levels (e. g., 350 feet below present level in Lake Michigan) – the result of retreat of the ice in the North Bay region of Ontario, opening up lower outlets at a time when the earth's crust was still depressed from the weight of the ice. As the crust subsequently rebounded, raising these outlets, the lakes returned to the former Algonquin elevation (605 feet) in the Michigan, Huron and Superior basins – the Nipissing Great Lakes. But the land had risen in the interval (to increasing elevations north of a "hinge line" across the middle of the Lower Peninsula), so that the old Algonquin shoreline was then at an even higher elevation. Gradual alterations in outlets led to the present configuration of the Great Lakes about 2,500 years ago. The relative stability of the Great Lakes during these past 2,500 years has been remarkable. Previous stages in their history lasted at most only a few hundred to perhaps a thousand years.

During the low-water stage preceding the Nipissing Great Lakes, the deep channel in the Straits of Mackinac passed north of Mackinac Island, leaving it, together with Round and Bois Blanc islands, attached to the Lower Peninsula. In the northern portion of Lake Michigan, the Manitou and Beaver island groups were connected with the mainland during the low-water stage, but the Fox islands remained insulated. At least portions of all of these island groups were high enough to be above the waters of Lake Algonquin, although some of the smaller ones in the Beaver group, as well as Round and Bois Blanc, were completely submerged at that time.

From this brief summary it can be seen that the entire vegetation of Michigan is of quite recent origin geologically speaking, and that the history of the Great Lakes following glaciation is likely to have been of key importance in influencing the distribution patterns of many species. The maps in this Flora show county (and certain island) records only. More detailed studies of distribution, based on the records already available plus additional critical field work, together with analysis of lake history, connecting channels, low and high levels, history of islands, and such features, should provide many a historical plant geographer or geobotanist in this region with a lifetime of problems to work on. Furthermore, it is clear that the thousands of bogs, marshes, swamps, and streams in this glaciated region are closely related to the history of the land — the depressions left by melting ice blocks and the progressive development of drainage patterns on the deglaciated surface.

Michigan Plant Habitats

The information on habitats given for the species in this Flora is compiled primarily from data on the labels of specimens examined, interpreted as seemed desirable by my own field experience and occasionally by statements in published works. A certain amount of common sense is needed in reading labels even if it was not used in writing them (*Lemna* from a "beech-maple forest" is presumably from a *pool* in such a forest), and some interpretation of labels has also been necessary to achieve a degree of uniformity in describing the common habitats — even admitting that they grade into one another. It would obviously be impossible to visit every collecting site in person and evaluate the situation against any set of standard ecological definitions. The habitats are therefore described in rather broad and overlapping terms except for certain species of great rarity or very specific habitats. The idea is to give an *impression* as to the kinds of places where each species grows. The few references cited in this section, and the bibliographies they contain, can help to direct the reader to some of the relevant literature about these habitats.

For more detailed and precise characterization of plant communities, for example, one can consult a work like Curtis (1959), although that has not been a conscious basis for the general characterizations presented here. Braun (1950, pp. 305–376) described the forest areas of Michigan, particularly the deciduous forests. The mixed conifer-hardwood forests of the northern part of the state have been included in an analysis by Maycock and Curtis (1960). Much information about habitats and associated species at the southern end of Lake Michigan, including Berrien County, can be found in Swink (1969).

The chief factors that have been used in describing habitats are moisture, soil, and plant cover. Rivers, lakes, ponds, and streams should need little explanation. "Wet places" and "low ground" are regularly moist under foot, if not wetter. "Marshes" are wet *treeless* areas. "Swamp forests" (or "swamps") are wet *wooded* areas, often subject to periodic flooding, as along many river flood-plains. "Bogs" are typically zoned from an open marsh-like area surrounding a pond, through low and high shrubs and tree invaders, to a surrounding swamp forest, usually coniferous. At the edge of a bog pond, a characteristic floating (and hence "quaking") mat borders the open water and is formed largely of interlacing rhizomes and roots, usually with much sphagnum moss which renders the mat waters very acid. Ericaceous shrubs, such as leatherleaf (*Chamaedaphne calyculata*), invade and become dominant. Bog ponds thus "fill from the top" as it were, as peat accumulates beneath the mat.* Some of the most interesting bogs have at least portions which are calcareous or marly, rather than acid, because of the presence of deposits of calcium carbonate or seepage of calcareous waters. In some bogs, the pond has been completely filled in by growth of the mat, which may remain treeless at the center for a long time. The older (outer) parts of bogs, commonly dominated by such trees as tamarack (*Larix laricina*), black spruce (*Picea mariana*), and/or white-cedar (*Thuja occidentalis*), could as well be described as "coniferous swamps" or "bog forests" and the terms are used interchangeably here. The younger (inner) portions of bogs, however, because of the floating nature of the mat, are generally not subject to flooding as is a true swamp. Many bog associations are described by Gates (1942), Dansereau and Segadas-Vianna (1952), and in a large additional literature on the subject of bogs.

Habitats described as "woods" or "forests" are clearly places dominated by trees. Conifers predominate in "coniferous forests," including the older portions of bogs and other stands of similar composition, generally on the acid side and in wet ground; other coniferous forests include stands of pine, especially jack pine (*Pinus banksiana*) on the driest and usually sandy sites (although jack pine does also occur in bogs in the northern part of the state). "Mixed woods" include substantial proportions of both hardwoods and conifers.

"Hardwoods," "deciduous forest," or "hemlock-hardwoods" may include some conifers, particularly hemlock (*Tsuga canadensis*), with beech (*Fagus grandifolia*), sugar maple (*Acer saccharum*), and associated trees. Beech does not range into the western part of the Upper Peninsula, but if a plant is said to grow in "beech-maple" woods it is understood that if the distribution map shows it extending into the western Upper Peninsula it may be expected in similar woods which lack beech. In addition to some hemlock, yellow birch (*Betula alleghaniensis*) particularly in the western Upper Peninsula, and basswood (*Tilia americana*), a number of other trees are often included in the deciduous forest,

---

*Among the usual conditions for bog formation are (a) a relatively deep depression (as in the numerous glaciated kettleholes in the northern part of the state where a block of ice melted after being surrounded by glacial drift), so that filling from the bottom does not overtake the growth of a bog mat; (b) a relatively small lake or pond, so that wave and ice action do not completely break up the mat; and (c) absence of significant inflowing or outflowing streams, resulting in poor oxygenation and the accumulation of peat rather than decayed organic matter. The term "boggy" is used rather loosely but is intended here to include sites which at least in part are *bog-like* in their peaty soil and floristic composition but which are not structured like a true bog.

especially in the more diverse hardwoods toward the southern Lower Peninsula. These stands constitute a major forest habitat of the state, with a rich spring flora before the trees are in full leaf. On drier sites in the southern Lower Peninsula, species of oak (*Quercus*) and hickory (*Carya*) may dominate. In the north-central Lower Peninsula and locally in the Upper Peninsula, jack pine is dominant on many dry and rather barren sites, seldom forming a densely stocked forest. Thus the common phrases, "jack pine plains" or "jack pine barrens." These situations may occur where formerly there were good stands of red pine (*Pinus resinosa*) and/or white pine (*P. strobus*), but repeated fires following the initial lumbering of the area have had drastic effects — sometimes resulting in extensive stands of jack pine and sometimes in aspen (*Populus grandidentata* or, particularly on moister sites, *P. tremuloides*).

The term "thicket" is admittedly ill-defined, used for usually small areas (or narrow ones, as along a stream) characterized by dense shrubs or small second-growth trees, thus offering some shade without qualifying as a forest. In a "woodland" the trees are less dense than in a forest or woods.

The sandy beaches of much of our Great Lakes shoreline are a well known habitat. Because of the normally fluctuating water levels in the lakes, resulting in periodic drying out of large amounts of sand, and the sweep of prevailing westerly winds across the lakes, sand dunes often develop along these shores. Some fixed (inactive) dunes occur inland, chiefly as relic features of former stages in the history of the Great Lakes when water levels were higher. Between the dunes fringing a shore there may be low, more or less moist, interdunal flats, beach pools, or swales. Sometimes sphagnum moss and other characteristic bog plants may be established in such places. Boggy shores may also occur where there are no dunes, and other shores may be described as beach meadows or simply as marshy (wet and treeless, with extensive grass and/or sedge growth). Where gravel or marl deposits prevail, a shore may be described in such terms (many of the shores on or near limestone or dolomite bedrock in the northern portion of the state are typically calcareous or marly). Since a combination of crustal uplift and lowering water levels has characterized the northern shorelines of the Great Lakes over the past several thousand years, one can easily imagine the extensive and complex range of habitats available among shores, dunes, and interdunal areas of varying age and distance from the present shorelines. Botanically these are among the most interesting areas in the state.

Moisture-laden winds from across the Great Lakes contribute to the richness of the forests often found on wooded dunes and islands, where the soil may be poor but the vegetation luxuriant.

Bedrock crops out through the glacial drift in some areas of Michigan, and may provide a special habitat for plants. Along the shore of the western part of Lake Superior, there are extensive rock outcroppings, including the famous "Pictured Rocks" east of Munising in Alger County, where the sandstone cliffs in places rise 200 feet above the lake; the capping layer on top is often quite calcareous. Sedimentary rock (sandstones and conglomerates) are exposed westward from the Pictured Rocks, as on some shores of the Keweenaw Peninsula and the scenic West Bluff (Brockway Mountain) in northern Keweenaw County. But the rocks exposed west of Marquette are more often igneous or metamorphic in origin. On rock shores at the tip of the Keweenaw Peninsula and on Isle Royale, many species very local in Michigan may grow in pools and rock crevices. Some of these plants are characteristic of similar habitats along the north shore of Lake Superior and of the Hudson Bay region northward. Outcrops of rock are also numerous inland in the western portion of the Upper Peninsula, a region noted for its mining history. Except for the Pictured Rocks sandstones, which outcrop to the east at Tahquamenon Falls, and a few other sandstones, as at Pointe Aux Barques in Huron County at the tip of the "Thumb" east of Saginaw Bay, most of the comparatively few rock exposures in the eastern Upper Peninsula and the Lower Peninsula are limestone or dolomite. "Rocky woods" are woods in which outcrops of bedrock, whatever it may be in the region, or at least large boulders, are frequent; such areas are generally well drained and fairly dry compared to otherwise similar woods on deep soils.

A mixture of peat or muck and sand occurs around many small lakes and ponds, especially ones with greatly fluctuating water levels, and this combination provides a favorable situation for a number of plants which are often considered very local in their distribution, if not rare. In the northern part of the state, at least, such shores appear to result from old bogs that have been so severely burned in the post-lumbering forest fires that most of the peat was consumed and no mat remains. The waters of such lakes may be very "soft" (extremely low alkalinity) or acid (see, for example, description of Cusino Lake, Voss, 1965a). Certain aquatic plants of restricted distribution may be found in these places, such as *Sparganium fluctuans, Potamogeton confervoides, Eriocaulon septangulare,* and (among the dicots) *Myriophyllum tenellum, Elatine minima,* and *Littorella uniflora.* In the southwestern Lower Peninsula, somewhat similar situations (in part old interdunal areas) may, in years of low water, display moist shores covered with plants of species characteristic of the Atlantic Coastal Plain and very rarely found inland, such as *Psilocarya scirpoides, Fuirena squarrosa* var. *pumila,* and *Juncus scirpoides.* Associated with these may be other plants of somewhat wider range but very local distribution in this region, such as *Hemicarpha micrantha, Scirpus hallii,* and *Xyris* spp.

Prairies have never been extensive in Michigan and the term is often broadly used in this region. Naturally treeless areas (often maintained as such by fires), true mesic prairies were dominated by grasses that formed a thick sod on rich black loamy soil. Such areas are now almost all intensively cultivated, although they were scattered in the southwestern and south-central Lower Peninsula (Veatch, 1928). So-called "dry prairies" (not in the sense of Veatch), with many of the same dominant species, including the bluestem grasses (*Andropogon* spp.). but on more or less sandy soils, are found scattered in the southern half of the Lower Peninsula, such as in considerable areas of Newaygo County. They grade into what are sometimes called "sand barrens." "Wet prairies" again have some of the same dominants but are periodically flooded and include a number of marsh or even bog species (Hayes, 1964; Brewer, 1965).

While scarcely any portion of Michigan has escaped the destructive hand of man, some habitats are very clearly associated with his activities, and the meanings of terms used to describe these are largely self-evident. "Ditches" are channels dug by man and at least seasonally wet (whereas "swales" are natural elongate depressions, such as interdunal troughs). "Fields" are areas that are, or have been, cultivated or pastured. The word "meadow" is vague but has been used so often on herbarium labels that it can hardly be avoided here as a term applying to a diversity of treeless areas such as grassy fields, pastures (grazed areas), and other grassy (or sedgy) places which are usually less level (and therefore less wet) than are generally termed "marshes." "Waste places" and "waste ground" refer to any seriously disturbed area that is not cultivated, such as roadsides, vacant lots, dumps, and construction sites.

Distribution Patterns

The preceding brief survey of postglacial history and habitats in the state should suggest how productive an endeavor it can be to analyze the distribution patterns of the species in our flora and attempt to relate them to factors of the environment past and present. However, only a few of the more interesting or striking patterns of species which do not occur throughout the entire state are mentioned here, as examples of the elements represented in the local flora.

I know of no familiar and well accepted species of vascular plant whose range is entirely restricted to Michigan. (Certain unusual Sisyrinchiums described from the state have not been found elsewhere, and there are numerous local segregates in *Rubus* and *Crataegus.*) The species nearest to being a recognized endemic that we have is the dwarf lake iris, *Iris lacustris.* All of the present range of this species is near the northern shores of Lakes Michigan and Huron, including the Door Peninsula of Wisconsin and the Bruce Peninsula of Ontario as well as extensive areas in northern Michigan. Pitcher's thistle (*Cirsium pitcheri,* a dicot and hence not included in this part of the Flora) is endemic but more widespread on the sandy shores and dunes of Lakes Michigan, Huron, and Superior — again with stations in Wisconsin and Ontario (also Illinois and Indiana), although the bulk of its range is in Michigan.

We have a number of other plants of shores and dunes that occur not only in the Great Lakes region but also on ocean shores and sometimes on other large inland lakes. The beach grass (*Ammophila breviligulata*) is the best monocot example of such a range, and dicots like sea-rocket (*Cakile edentula*), beach pea (*Lathyrus japonicus*), and seaside spurge (*Euphorbia polygonifolia*) can also be mentioned. Some northern plants come south only as far as the shores of the Great Lakes, sometimes quite disjunct from the Hudson Bay region. These include species of sandy habitats, like the dune grass (*Elymus mollis*) or the Lake Huron tansy (*Tanacetum huronense*), and species of the rock crevices along Lake

Superior and occasionally elsewhere — often circumpolar in total range — such as *Trisetum spicatum, Carex media, Tofieldia pusilla,* and (among the dicots) *Sagina nodosa, Polygonum viviparum,* and *Saxifraga aizoön.* The sandy shores of the Great Lakes are also the home of varieties of some western plains plants, e. g., the grasses *Calamovilfa longifolia* var. *magna* and *Agropyron dasystachyum* var. *psammophilum.* So although there is an interesting assemblage of plants found in this region nowhere except along the shores of the Great Lakes, the species have no single geographic affinity otherwise (Guire & Voss, 1963).

As mentioned under the preceding discussion of habitats, some plants characteristic of the Atlantic Coastal Plain and rarely found elsewhere do occur at disjunct stations, usually in the southwestern Lower Peninsula and the Indiana Dunes. A more northern Coastal Plain disjunct is *Juncus militaris,* not yet found in southwestern Michigan although formerly in the Indiana Dunes.

One of the most conspicuous phytogeographic features seen as one travels northward through the state, particularly in the center, is a striking change in the vegetation (e. g., a few miles north of Clare), where the "north begins." At this general latitude, westward or northwestward from Saginaw Bay, the predominantly deciduous forest of the southern portion of the Lower Peninsula gives way to the conifers of the north — with many pockets of hardwoods, to be sure. The former great pine forests of the northern half of the Lower Peninsula, usually growing on sandy soils, meant much to the state's economy in the lumbering days but now — except for a small stand in Hartwick Pines State Park — are largely reduced to charred stumps, blueberries, and second-growth forest or woodland.

Across the middle of the Lower Peninsula, then, we have the equivalent of the "tension zone" that has received much attention in Wisconsin but considerably less in Michigan (Elliott, 1953). It can be easily seen from the distribution maps that while many common forest species of eastern North America are found throughout the state in suitable habitats, a number of species are found rarely if at all north of this zone, and other, boreal, species rarely if at all south of it. Still others, such as the white trout-lily (*Erythronium albidum*), Michigan lily (*Lilum michiganense*), the sedges *Carex grayi, C. sprengelii,* and *C. woodii,* and dicots like box-elder (*Acer negundo*) and hornbeam (*Carpinus caroliniana*), occur chiefly south of the tension zone but also in the western Upper Peninsula. The tension zone seems to be much more significant a feature than the Straits of Mackinac in relation to plant distribution north and south. However, a few species (such as the sedge *Fimbristylis autumnalis* and white fringed orchid, *Habenaria blephariglottis*) are apparently not found north of the Straits and a few (such as the bur-reed *Sparganium fluctuans* and a rush, *Juncus filiformis*) not south of it, although fairly widespread in their respective peninsulas.

Some species of distinctly southern range barely reach Michigan (and some southern ones may be only adventive this far north). These include, among many, a beak-rush (*Rhynchospora globularis*), a dayflower (*Commelina erecta*), and toadshade (*Trillium sessile*) – paralleling the distribution of such dicots as buckeye (*Aesculus glabra*) and redbud (*Cercis canadensis*).

A number of plants common in the western part of North America have disjunct stations in the northern Great Lakes region, often with intermediate stations in the Black Hills of South Dakota. Some of these occur again in the Ottawa district of Ontario or near the lower end of the St. Lawrence valley. Among the plants in this "Cordilleran element" are the grasses *Festuca occidentalis*, *F. scabrella*, and *Melica smithii*; fairy bells (*Disporum hookeri* var. *oreganum*); the orchids *Corallorhiza striata*, *Goodyera oblongifolia*, and *Habenaria unalascensis*; and many dicots, including the thimbleberry (*Rubus parviflorus*), a hawthorn with purplish-black fruits (*Crataegus douglasii*), a bilberry (*Vaccinium ovalifolium*), and trail plant (*Adenocaulon bicolor*).

A few prairie plants such as the cream wild indigo (*Baptisia leucophaea*), are apparently restricted to remnants of mesic prairie left in cemeteries, roadsides, or railroad rights-of-way in Michigan. Most characteristic prairie plants, however, range into similar moist or dry habitats, sometimes northward from any prairie area – particularly in the jack pine plains and even as far as the Upper Peninsula: e. g., the bluestems (*Andropogon gerardii* and *A. scoparius*), Indian grass (*Sorghastrum nutans*), and among the dicots the blazing-stars (*Liatris* spp.).

The above few examples of some of the more clearcut and distinctive distribution patterns of native plants represented in our flora may suggest why the maps presented here can provide in most cases only a small piece of the total floristic picture. Yet they can be used to make that local piece more accurate and complete than it has been before. I readily admit that in my own field work I have been more interested in concentrating on the species that do show some distinctive pattern than on those that are more or less widespread throughout the state, where a few more county records would add little of significance to the know ranges.

Abundance

Terms relating to abundance are always subjective when used on the scale of a flora like this. They are not based on precise counts. But they may give some idea of what to expect, at least in some years and in the appropriate season. The careful field observer knows well that abundance of any plant (notably mushrooms!) can vary a great deal from year to year in a given spot, depending on changes in such factors as moisture, temperature, competition from other plants, plant succession, and disturbance by man or other animals. Many plants appear to be cyclic in their abundance without a clear relationship to possible causal factors.

When statements regarding abundance are made in this Flora, the terms are used approximately as follows:

Rare: occurring in small numbers wherever found and often absent in the expected habitats.

Local: occurring in a very restricted habitat or region, but may be common when found.

Occasional: scattered or infrequent throughout the habitat(s) or area(s) in which it grows.

Frequent: intermediate in abundance between occasional and common: likely to be found if one looks for it.

Common: widespread and reasonably plentiful throughout its range in appropriate habitats, where easily found.

Abundant: very common, likely to form dense stands.

TAXONOMY AND NOMENCLATURE

The sequence of families in this Flora — and, indeed, the fact that the monocots have not been left for the final part — follows the long-traditional sequence based on the system of Engler and Prantl. This is the sequence used in most large American herbaria (that of Michigan State University is a notable exception); it is the same sequence as in other works covering this area (e.g., Fernald, 1950; Gleason, 1952; Gleason & Cronquist, 1963; Smith, 1966), and its use here facilitates comparison with them. As an expression of current knowledge regarding directions of plant evolution, this system is of course as obsolete for our times as the Linnaean system was a century or so ago. For teaching purposes and as a synthesis of modern knowledge, a system derived from the principles developed in this country by C. E. Bessey, such as the thorough one presented by Cronquist (1968), is strongly to be recommended. The long-familiar Englerian sequence is rather like the alphabet — this is the way people have learned to use it for years, even though it may seem rather arbitrary. (Linotype operators and typewriter manufacturers have used a more "natural" sequence of letters for their purposes, but we still learn the alphabet "ABCDEF . . . " rather than "ETAOIN . . . " or "QWERTY . . . ")

Within each family the sequence of genera is simply the same as the sequence in which they fall in the key to genera; and within each genus, the species are likewise in the same sequence as they appear in the key. Since all taxa are numbered consecutively, this method makes it easy to find the treatment of them once a name has been reached in the keys, and it simplifies further comparison, beyond the characters mentioned in the keys. The only important exceptions are that the tribes of grasses and the groups (or sections) of *Carex* are left in their customary sequence, as are the families.

The keys to species and any further comments there may be about the species should make clear the sense in which the names are being used — to what each name is being applied and therefore what meaning can be attached to the distribution maps. Anyone who prefers to recognize genera defined either more broadly or more narrowly in any particular case is of course free to do so, as discussed, for example, under the Cyperaceae on p. 244. The information here presented on habitat and distribution in Michigan and on identifying characters is in no way affected whether one chooses to call the ragged fringed orchid *Habenaria lacera* (Michaux) Loddiges or *Blephariglottis lacera* (Michaux) Farwell — or for that matter, to take up Michaux's original *Orchis lacera.* These are matters of taxonomic judgment that cannot be legislated and about which there may be legitimate differences of opinion.

Similar options for judgment exist in defining species, but here there may be problems interpreting the maps. If one chooses to "lump" species that are here treated as distinct, he can easily combine the information provided about them. And if he prefers to place *all* of our plants in a particular variety or subspecies, it is easy to do so. If, however, he prefers to define his taxa more narrowly in any case, the map may be disappointing. Sometimes clues are given in the text as to variations in habitat or distribution at an infraspecific level. More often, varieties are merely mentioned whose ranges in the state are even less distinct than their morphology. In a few instances, named varieties may be distinct morphologically and geographically, as shown by Stephenson (1971) in *Brachyelytrum.* In some cases, as in *Maianthemum canadense,* the varieties are morphologically distinct but show no geographic orientation in Michigan. In the great majority of cases, I feel that the named varieties represented in our flora have neither morphological nor geographic significance sufficient to warrant recognition except under circumstances when relatively trivial variation or variation over a much larger geographic area is considered important.

Standards for mentioning varieties in this Flora may seem erratic. Those included in current manuals for the area are generally mentioned if either or both of two conditions apply:
(a) they appear to be reasonably distinct taxa in this area, worthy of note or perhaps likely to be thought of as different species;
(b) the type locality is in Michigan, i. e., the locality at which a type specimen (see below) was collected.
A few other varieties, not in current manuals, are also mentioned when they seem important to note. But numerous taxa for which the type locality is in this state are not mentioned if they are neither significantly worthy of note nor treated in current manuals where one is likely to find them placed as part of a broader geographic range (note, for example, the many Farwell names listed by McVaugh, Cain, & Hagenah, 1953).

As a result of the effort to check key characters against all the material collected in Michigan, I am more convinced than ever of the great variability of plant species — just as in the human species (where taxonomic refinements are seldom useful). Many persons thoroughly familiar with plants in the field will have more intimate knowledge of variation and hybrids in this state than is indicated by information in the Flora. Oliver A. Farwell, one of Michigan's most active field botanists during the half-century preceding his death in 1944, had a particularly keen eye for aberrations and unusual expressions of species, and gave names to many of these. The present keys will include more of these unusual forms within the range of variation allowed for many species than the average key in manuals, but they still cannot hope to be complete. Albino forms, as suggested by Asa Gray, may be expected in almost any plant with the flowers normally colored: "Expect a white-flowered state of every colored species. They are sure to turn up sooner or later. And I find it no good therefore to say var. *alba* over and over." (Asa Gray in a letter dated June 1, 1886, quoted in Asa Gray Bull. 1(7): 42. 1894.) Other unusual color forms, as well as hybrids and variants of all sorts, have been and will continue to be discovered by alert field workers in the state, who can contribute to botanical knowledge at a more refined level by a thorough study of them, with deposition of representative specimens in herbaria.

I have thought it more important to study carefully the normal forms of plants and their occurrence in the state than to attempt to catalog all of the hybrids and variants that may occur, interesting as these are. Hybrids are not included in the keys, but are sometimes mentioned in the text. In general, they may be expected to be intermediate between the parents. Sometimes, after first-generation hybrids have back-crossed with their parent species, a host of intermediate forms may occur, more closely resembling one parent than the other. Such *introgression* of characters from one species into another through hybridization and back-crossing may be of considerable importance in accounting for the variability in certain groups of plants, but certainly not all normal variability of species should be attributed to hybridization.

Relatively little work has been done by any botanist on studying the chromosome numbers of flowering plants in Michigan. It can only be assumed that in most cases they are the same as for the same species elsewhere on the North American continent, but this assumption has not led to routine citing of chromosome numbers as if they were known to apply to our plants. In a number of species, including such monocots as *Poa compressa, Tradescantia ohiensis,* and *Polygonatum biflorum,* more than one chromosome number is known to exist in a species; although of these three, only *Tradescantia* has been studied carefully in Michigan, where we have both diploid and tetraploid races (Dean, 1954).

Nomenclature

As suggested above, there is considerable room for differences of opinion in plant taxonomy. The widely held view that a plant can bear only one correct scientific name is not true without some qualification. The three names cited above for the ragged fringed orchid are all correct, depending upon one's taxonomic judgment as to generic limits in the Orchidaceae. Whether a particular water-plantain in Michigan is to be called *Alisma plantago-aquatica* L., *A. plantago-aquatica* var. *americanum* Schultes & Schultes, or *A. triviale* Pursh similarly depends, in large part, on the classification accepted. A plant may bear only one correct scientific name *when classified in a particular manner.* No botanist is obligated to accept the opinion of another as to how a plant is to be classified if he is convinced by the evidence that a different classification is preferable. The most recent classification is not automatically the best. The rules governing scientific names, as set forth in the *International Code of Botanical Nomenclature* (Stafleu, Voss, et al., 1972), can be applied once one has settled upon the taxonomic status of the plants to be named.

The scientific name of a species consists of two words: the first is the name of the *genus* to which the plant is assigned and the second is the *specific epithet.* The generic name in such a binomial is always a singular noun. The specific epithet may be another noun in apposition (as in *Cypripedium calceolus*); a noun in the Latin genitive (possessive) case, either singular or plural (e. g., *Cypripedium reginae, Carex grayi, Carex scopulorum*), or – most frequently – an adjective, in which case it must agree with the generic name according to the rules for Latin declension, since all scientific names are treated as Latin (e. g., *Streptopus roseus, Carex rosea, Cypripedium candidum, Cypripedium acaule, Fissipes acaulis*). Generic names, used as such, always begin with a capital letter; specific epithets are recommended always to begin with a lower-case letter although one is permitted to capitalize those which are old vernacular or generic names and those directly derived from the names of persons (e. g., *Cypripedium Calceolus, Carex Grayi*).

The names of higher categories (families, orders, etc.) have standardized endings except for the authorized alternative names for certain families (e. g., Poaceae vs. Gramineae for the grass family), and all such names are *plural* in form. (It is thus grammatically incorrect to say, for example, "The Orchidaceae is . . ." or "The Liliopsida has . . .")

The names of categories lower than species exist in three ranks, which, in descending order, are subspecies, variety, and form; they follow the same conventions as do specific epithets. One is not obligated to recognize infra-specific taxa, but if they are used they must be kept in the established hierarchy: e. g., subspecies may be divided into two or more varieties, but varieties may not be divided into subspecies. Many botanists find this amount of refinement useful in describing variable species; many others maintain that use of both subspecies and variety is unnecessary in most if not all cases and also that most "forms" are not worth official recognition. (Forms are generally trivial variations, as in flower color or stature of the plant, which may occur anywhere in the range of a species.) In this Flora, if any infraspecific categories are mentioned they are nearly always treated as varieties, although attention is called to a few striking forms. (Note: The word *species* is the same in both Latin and English and in both singular and plural; the Latin for variety is *varietas* and for form, *forma.*)

The basic principles of the Code of Nomenclature require that each kind of plant, when classified in a particular way, may bear only one correct scientific name; that this must be the *earliest* name (beginning with Linnaeus for most plants) for which a description was published under certain regulations regarding the formation and proper ("valid") publication of names; that identical names may not be used for different kinds of plants; and that each name is to be standardized in its application by designation of a particular representative or *type* to which the name always applies.

Requirements for designation of types are fairly recent and there are regulations for determining the types of older names when the author did not establish them. For the name of a species or infraspecific taxon, the type is a *specimen*. The name always applies to that specimen — plus all other specimens which one's judgment includes in the same taxon. The type of a generic name is a *species*; a genus includes its type species, plus all other species deemed to belong in the same genus. The type of a family name is a genus, and so forth. If one prefers not to accept a properly published name of a genus, it is basically because he considers its type species to belong to some earlier described genus and hence the names to be taxonomic synonyms. If two or more names of species are treated as synonyms, it is because the types of these names are judged to belong to the same species. Conversely, if a taxon is divided into two or more elements, the name remains with that element which includes the type and a new name is applied to the other element(s). No one is required to use a name which he considers to be a synonym, but if a name *is* used it cannot be used in a sense which excludes its type.

This Flora is not the place for a further elaboration of the Code of Nomenclature, which includes about 150 Articles and Recommendations, but it should be pointed out that such rules exist and are modified (seldom in a major way) at intervals of several years during International Botanical Congresses. Names of plants may change as a result of changes in the Code. More often, name changes follow inevitably upon changes in taxonomic judgment (about which a nomenclatural code can say nothing) or upon discovery that a certain name has been incorrectly applied or misinterpreted (as determined by its type), that it should be superseded by a prior name for the same taxon, or that some other rule was violated by a particular usage. Validly published names which are contrary to one or more Articles of the Code are said to be *illegitimate* and must be rejected. Ultimately, nomenclatural changes resulting from factors other than changes in taxonomic opinion should become increasingly reduced in number as stability is reached in application of the Code and rejection of illegitimate names. Indeed, present floristic manuals are much more in agreement regarding the names that they use than are many of their predecessors.

Since 1950 the Code has specified a clear way to deal with the names of infraspecific taxa. The variety of a species which includes the type of that species is designated by the repetition of the specific epithet, without citation of an authority (e. g., *Maianthemum canadense* Desf. var. *canadense*); this can then be easily contrasted with one or more óther varieties (e. g., *Maianthemum canadense* var. *interius* Fernald) which together comprise the species. It is not logical, as formerly was often done, to contrast a variety with "the species" for this is contrasting a part with the whole rather than with other parts. This system means, too, that when no variety is indicated, typical or otherwise, no restriction to any one variety is necessarily implied in the use of a binomial. It is just as correct to refer to our plants of a certain grass as *Milium effusum* L. as it is to refer to European plants in the same way. However, if it seems necessary or desirable to do so, ours may be designated as *M. effusum* var. *cisatlanticum* Fernald, and European ones as *M. effusum* var. *effusum*. Note that it is not acceptable style to give the name of a subspecies, variety, or form (in all of which ranks these conventions apply) as an "undesignated trinomial": e. g., *Maianthemum canadense interius* or *Habenaria psycodes albiflora*; these must be designated as *M. canadense* var. *interius*; *H. psycodes* f. *albiflora*. Note, too, that in this context reference to the "typical variety" (or other rank) in a species refers only to the variety which includes the type specimen of the name of the entire species; it does not necessarily apply to what is the most common or representative variety in any given area.

Scientific names may be followed by the name (or an abbreviation thereof) of the author who first validly published a description of the taxon; his name is placed in parentheses if the position (e. g., classification in one genus or another) or rank (level in the taxonomic hierarchy, e. g., species vs. variety) has been changed, and the name of the author responsible for publishing the change is added as a bibliographic aid. It hardly seems necessary to abbreviate names of authors by dropping only one or two letters and replacing them by a period. Most names which are often thus abbreviated are written in full here. No list is provided explaining the abbreviations of other names (e. g., L. = Linnaeus). Those who do not recognize them but are sufficiently curious to investigate will find such lists in Fernald (1950) and Gleason (1952).

A number of names and/or author citations in this Flora will be seen to differ from those in other manuals. The reasons for the use of unfamiliar names are usually mentioned briefly if they are not easily gained from a realization of a differing taxonomic treatment. They have been discussed more fully, for those who want nomenclatural details, in other places (Voss, 1965b, 1966, 1972). Corresponding names from the current manuals are usually indicated when there would be any confusion as to correlation with those works.

Realizing that there may easily be differences of opinion as to the proper disposition of some of our taxa, I have often indicated taxonomic and nomenclatural alternatives, especially when they seem to have almost or quite as great merit as the ones adopted here. But it has not been desirable or possible to call attention to *all* of the recent taxonomic and nomenclatural innovations proposed for our plants. Keeping a flora in manuscript up to date and also within bounds could be a never-ending task of re-evaluation; somewhere the work must stop and be prepared for publication. When differing treatments for our plants exist in reliable manuals and monographs, I have pragmatically based the disposition of our plants on the treatment which seems to deal most satisfactorily with specimens from this region.

Common Names

True common or vernacular names are a part of folklore and are often of interest. "Duckweed," "arrowhead," "timothy," "lady-slipper" are indeed in common use among non-botanists. "Wheat," "corn," and such names are familiar to all. Manufactured common names, generally derived by a literal translation of the scientific name, are much less interesting or useful. Why develop a long list of names like "porcupine sedge," "beaked sedge," "Bebb's sedge," etc., for *Carex hystericina, C. rostrata, C. bebbii,* and our other 166 species of *Carex?* I refuse to believe that "sessile-leaved twisted-stalk" is any easier to pronounce or remember than the euphonious *Streptopus roseus.* Common names have been indicated in this Flora for species or genera when they seem to be in general use or apply to well known plants like weeds and familiar wildflowers, but seldom otherwise if they appear to have been invented simply because someone thought that every plant *ought* to have a common name (as if one name were not enough to learn). Some obviously misapplied or manufactured common names (such as "Muhly" for *Muhlenbergia*) are placed in quotation marks. Many familiar names are both "common" and "scientific," such as Asparagus, Trillium, Iris, and Calypso, and many others can just as easily serve both purposes.

Common names which are consistent with taxonomic correctness are written as separate words; e. g., three-way sedge (in the sedge family, Cyperaceae), canary grass and sweet grass (in the grass family, Gramineae), wood rush (in the rush family, Juncaceae). (Some names, such as bluegrass, are such natural generic common names that they are retained as one word.) Common names which are not taxonomically correct are hyphenated; e. g., skunk-cabbage (not a cabbage), water-plantain (not a *Plantago),* nut-grass (in the Cyperaceae, not the Gramineae), blue-eyed-grass (in the Iridaceae), grass-pink (in the Orchidaceae, not the Caryophyllaceae), wool-grass (in the Cyperaceae), broom-sedge (in the Gramineae).

When a common name applies to all of our species in a genus, it is given after the generic name except when there is only one species in the genus. Common names appled to a single species in our flora are given after the scientific name of the species.

PLANT IDENTIFICATION AND THE USE OF KEYS

Even a minimal understanding of plant taxonomy and geography makes clear that plant identification is not an exercise that takes place in a vacuum. We are inevitably dealing with living organisms (or specimens that once were living) — organisms that vary from one another in countless ways and for countless reasons. Some of them may hybridize with members of other kinds; some appear in unexpected habitats or places; some develop — or certain of their parts develop — to unexpectedly large (or small) sizes or vary in other ways, because of conditions under which they grow, differences in their genetic constitution, or for unknown reasons. The biologist is interested in the causes and expressions of all these differences. But we are all taxonomists at heart; classification and identification are the very basis of language. The child learns that objects with four legs, a seat, and a back may be called "chairs"; and that plants with three-lobed leaves and flowers in a peculiar erect pencil-shaped structure covered by a hood are called "jack-in-the-pulpit." We are always more comfortable when we know the names of objects around us — or at least it makes our life more interesting.

Determining the names of plants is often necessary for the botanist and for scientists in related fields; and it is often the source of much pleasure for many others who find study of the flora an agreeable hobby. Any intelligent person who would like to know about the kinds and diversity of flowering plants can master the minimum of vocabulary and understanding required. Countless so-called amateurs have done it through the years and have contributed to science, although as Shinners (1958) has concisely stated: "Nature gives away few secrets to the lazy, and none to the incompetent." Inexperience should not, however, be confused with incompetence. Practice with keys and with the terminology required will soon ease the process of identification insofar as this can be done without neglecting the fact that plants can not always be placed into neat, easily recognized pigeon-holes when the diverse processes of their evolution have failed to do so. It is easier to write (and use) a simple key to species of *Uvularia* or *Erythronium* than it is to the species of *Panicum*. It is impossible to write a simple key for many genera, because they are not simple genera.

The species is not a precise measurement, like degrees Centigrade, or acre, or pound. Species vary greatly in their distinctness and closeness of relationship. They are not all equivalent. Some, such as *Calypso bulbosa* or *Calla palustris*, are absolutely distinctive in our flora — the only member of their genus. Other species seem to be in a more active state of evolution and the lines are not clearly drawn. *Potamogeton gramineus* and *P. illinoensis* persistently cause problems. Are *Polygonatum biflorum* and *P. commutatum* distinct? How many taxa of yellow (and white) lady-slippers are there? How much weight should be given to the fact that intermediates between the clearcut extremes of *Habenaria dilatata* and *H. hyperborea* are rather common? Which are the important, stable characters, if any, for defining a species?

Problems such as these should not discourage the botanist, whether amateur or professional, but should challenge him. One of the important functions of a flora like this is to call attention to problems requiring study. As soon as one realizes that a specimen he has found need not be pigeon-holed, for example, as definitely *Typha angustifolia* or definitely *Typha latifolia,* but may be a hybrid, *T.* ×*glauca,* he knows an interesting fact and something about evolution and variation in the cat-tails.

Identification very often involves judgment on a number of characters. One must select the choice in a key (see below) which *best* fits his specimen, not always one which fits perfectly, although this would be ideal. The word "usually" appears often in the keys, not to discourage the user but to remind him that exceptions may occur in some of the characters used to describe a species. When one realizes that a plant specimen is an individual, just as he is himself, he should not feel worried when it fails to fit completely the compartment to which, all things considered, it apparently does belong.

The use of good keys is the most dependable and straightforward way to identify most unknown plants — short of having a specialist near at hand whom you can ask. A key is simply an efficient way of rapidly narrowing down the possibilities. It is like a highway system with no crossroads but only forks. The keys in this Flora are always strictly dichotomous; that is, with only two choices at any given point. Each of the two is called a *lead* and the pair is called a *couplet*. The two leads of a couplet always bear the same number, which is never repeated in the same key. (One never runs out of numbers, which are the easiest means of identifying contrasting choices; some keys rely solely on equal indentation or employ letters of the alphabet, sometimes in different type faces, assorted symbols, and other complicated devices.)

Start in a key with the two choices numbered 1. The two will bear this same number and will be at the left margin, though they may be some distance apart if the key is long (the smaller of two choices is consistently given first in these keys, to bring the alternative lead closer). Select the lead which best fits the unknown to be identified. If it does not direct you to a plant name, simply drop down to the next couplet below it, and choose between the two leads there offered. Again, the one which fits best will give you a name or else you move to the couplet immediately below it. **Always read leads completely, and compare both of them in a couplet, before selecting one of them!**

Each successive couplet is indented slightly from the lead by which it is reached, for easy following. The two leads of a couplet are always indented the same distance, in addition to bearing the same number, and start with the same word. (To help avoid confusion, adjacent couplets so far as possible begin with another word.) Occasionally a lead will direct you backward or forward to a couplet other than the next one, by stating which number you should move to. In this way, account has been taken of some ambiguous situations where one might try an unknown under either of two leads. Sometimes a taxon may be entered at more than one place in a key, thus accounting more satisfactorily for its variability or making it possible to use simpler key characters than would otherwise be necessary.

The keys in this Flora usually mention more than a single character at each point — as many, in fact, as seem to be good and useful for identification, including characters of both vegetative and reproductive parts. (The ones that are most reliable or easiest to determine are generally mentioned first.) Nevertheless, one will sometimes be confronted with a plant that does not run well through a key. Even after repeated attempts, selecting the alternative lead where there may have been some doubt, it still does not seem to fit. There may be any number of reasons for the problem, among them the following:

(1) The plant may not be in this book at all. It may be a dicot, or even a monocot previously unknown in Michigan, or a cultivated plant. Try a manual covering a larger portion of the flora or a larger area.
(2) The key may require characters not well displayed on your specimen. If possible, try another specimen, avoiding ones that are over-mature, or too young, or abnormal in their population. Of course, this may mean waiting for another season!
(3) There may be an error in the key. The author is human, and so are proofreaders. There is no easy way to overcome this problem.
(4) There may be an error in reading the key. The user is also human. Did you skip a couplet? Was it pedicels or peduncles or petioles which were to be examined? Did the key move from characters of sepals to those of petals? Did it ask about the length or the width of a structure? *Always read a key with extreme care and take nothing for granted*; it is intended to mean exactly what it says. Do not guess at measurements, proportions, number of parts, ovary position, etc. Measure, count, examine!
(5) The terminology used is vague or unclear. Check the glossary to be sure you understand the meaning of a word. Compare the alternative statement carefully to understand the meaning of a contrast. Every author has his own idiosyncracies in the use of certain words; get used to this Flora and do not assume that every word is necessarily used here in precisely the same sense as it is used in some other work.

The use of seemingly vague and ambiguous terminology is not always an evidence of lack of precision in a key or description. In fact, it is usually the result of an attempt to deal *more* precisely with the variability of nature, which does not produce individuals by die-cutting them on a machine. Leaves, for example, may be intermediate between the arbitrary shapes defined by botanists as "linear" and "lanceolate," or they may vary between the two shapes; so the only honest way to describe them concisely is with a hyphen: "linear-lanceolate." Structures are seldom truly spherical in a geometric sense, so we add prefixes or modifiers and say "subglobose," "nearly or quite globose," or use similar expressions.

Once a name has been reached in the key, do not automatically assume that you have correctly identified the plant. Check all other aids available. Look at the figure if there is one. Read the discussion, if any, in the text. If there is any doubt at all, check a more complete description and picture in another manual, or if possible compare with authoritatively identified specimens in a herbarium — that's what herbaria are for. Remember, too, to look at the distribution maps and statement of habitat. If you have identified a specimen as *Sparganium fluctuans* and it came from Washtenaw County, or from dry sand under jack pines, it is almost certainly misidentified (see Map 18). The farther you are from the range indicated on the maps, the more likely it is that further checking is called for. If, to continue with this example, you identify a plant from a soft-water lake in Schoolcraft or Dickinson county as *Sparganium fluctuans,* it is much more likely to be correct even though the map has no dots in those counties. Always use common sense and draw upon your own experience, which as it increases will help in the interpretation of the variability that plants display. The more plants you know, the easier it becomes to fit additional species into the "system." Becoming familiar with the first ones is always the most difficult.

Once you become familiar with a species, it will often be possible to recognize it without resort to the "key characters." Almost everyone can tell a tulip from a lily at a glance, without inspecting the stamens to see whether the filament is attached to the end or the middle of the anther. This is a precise character, easily stated in a key, to separate the genera *Tulipa* and *Lilium.* To use more superficial "recognition" characters would require lengthy and involved statements which the person who did not already know these plants (or have both of them at hand simultaneously for comparison) would rightly deplore.

The keys in this Flora are designed for identification — the true function of a key. They use the characters that I believe are the most reliable for this purpose (avoiding as far as possible those requiring microscopic examination or elaborate dissection). They try never to use such characters as "annual" vs. "perennial" or the nature of underground parts as the sole character at any point, for the user is often dealing with a specimen that lacks basal parts. The keys do not intend to imply anything about the characters which may be most important in the assumed evolution of the taxa or in describing their natural relationships with other taxa. They are not, in short, synopses of a classification.

Specialized terminology has been kept to a minimum, but some is essential. One cannot describe plants without an appropriate vocabulary, any more than he can talk about rockets, hi-fi equipment, an automobile, or a modern American kitchen in the vocabulary of Elizabethan England. Anyone who can manage words like riboflavin, mononucleosis, stereophonic sound, carburetor, or sodium bicarbonate can conquer words like panicle, lemma, perigynium, scape, or hispidulous. Unfamiliar words not in the glossary (and most that are) can be found in any good dictionary. See also the statements on style at the beginning of the Taxonomic Section.

It may be helpful to state that for the most part the details mentioned in the keys have been determined from examination of specimens under a magnification of about 12x and with *good illumination.* Insufficient lighting will make many details obscure. Measurements have been made (estimated) with a scale graduated in half-millimeters. Excellent 10-cm plastic rules thus marked are made in Germany (e. g., Nestler No. 1020 or Aristo No. 1317) and can be obtained for less than a dollar.

While formal descriptions are not provided here, a concise description of a taxon may generally be obtained by "reading the key backward." Start with the name and work back through all of the choices that led to it, adding up the characters mentioned. Caution in this regard is necessary whenever a taxon appears more than once in a key, and such taxa should be read back from all points.* Another caution in regard to the keys should be mentioned, to clear up what may appear to be contradictions. For example, the first lead of couplet 7 in the key to species of *Elymus* (p. 152) calls for paleas 7.5–10.5 mm long, whereas this couplet appears under the second lead of couplet 3, which calls for paleas 5.5–8.5 mm [not as large as 10.5]. This apparent inconsistency is clarified by noting that couplet 7 can *also* be reached from the first lead of couplet 3 [paleas 8.6–12.7 mm] and couplet 4. Certain species, in other words, may be reached via both halves of couplet 3, and neither provides a complete description of these species.

REFERENCES

Often the user of this Flora will want to consult other works for additional information, confirmation (hopefully) of his identifications, or possible alternatives when a key here fails to give satisfaction.

Under many of the families and genera, references are cited which give some further local information or are useful in identification for those taxa. These citations are by no means complete bibliographies; they merely call attention to some publications especially helpful in this region or which have been most often drawn upon for useful key characters or taxonomic clarity. Some good treatments of families and genera are omitted, particularly if they add little to the usual taxonomic treatment of species in this area, or omit a significant number of our species, or differ greatly from the taxonomy here adopted.

For general identification of vascular plants in the northeastern United States and adjacent Canada, the standard manuals, with keys, descriptions, and statements of range (to be interpreted with caution), are Fernald's eighth edition of *Gray's Manual* (1950), Gleason's three-volume *New Britton and Brown Illustrated Flora* (1952), and Cronquist's unillustrated condensation of the latter, as Gleason and Cronquist's *Manual of Vascular Plants* (1963).

Poa compressa, for example, comes in couplet 15 of the key to species of *Poa*, pp. 124–125. To reach couplet 15, one must take the second lead of couplet 7, "Margins of lemma ± hairy," and the second lead of couplet 1, "Callus . . . with a small or large cobwebby tuft . . ." However, some specimens of *P. compressa* completely lack the cobwebby tuft; hence they are keyed under the first lead of couplet 1: "Callus . . . without a cobwebby tuft . . ." This much is fairly obvious from a glance at the key: *Poa compressa* may or may not have a cobwebby tuft on the callus. What is not evident from the key is that most of the specimens which lack the cobwebby tuft also lack hairs on the margin of the lemma. The choice, therefore, made in couplet 7 for specimens with the tuft is not valid for all specimens of the species. The key should work satisfactorily for identifying an unknown specimen of this variable species, but one may not safely read back from couplet 15 and conclude that *all* specimens have hairy margins on the lemma.

On the other hand, those who wish to approach what appears to be a common wild plant using a *less* thorough or complete work may find valuable the concise keys of Gleason's *Plants of Michigan* (1939) or the well illustrated *Michigan Wildflowers* by Helen Smith (1966), both of which omit most aquatics, grasses, and sedges — a majority of the monocots. Excellent works for neighboring states, which should prove useful in southern Michigan at least, are Deam's classic *Flora of Indiana* (1940) and E. Lucy Braun's beautifully illustrated *Monocotyledoneae* [of Ohio] (1967). A complete *Illustrated Flora of Illinois,* in many volumes, edited by Robert H. Mohlenbrock, is currently in the process of publication.

These works are cited below, along with all other references cited in this Introductory Section of the Flora. In the Taxonomic Section that follows, references are given at the beginning of a genus or family or, if they are not found there, are in the general list below.

Beal, W. J. 1905 ["1904"]. Michigan Flora. Rep. Mich. Acad. 5: 1–147.
Billington, Cecil. 1952. Ferns of Michigan. Cranbrook Inst. Sci. Bull. 32. 240 pp.
Braun, E. Lucy. 1950. Deciduous Forests of Eastern North America. Blakiston, Philadelphia. 596 pp. + map.
Braun, E. Lucy. 1967. The Monocotyledoneae [of Ohio] Cat-tails to Orchids. Ohio State Univ. Press, Columbus. 464 pp.
Brewer, Richard. 1965. Vegetational Features of a Wet Prairie in Southwestern Michigan. Occ. Pap. Adams Cent. Ecol. Stud. 13. 16 pp.
Cole, Emma J. 1901. Grand Rapids Flora. A. Van Dort, Grand Rapids. 170 pp.
Cronquist, Arthur. 1968. The Evolution and Classification of Flowering Plants. Houghton Mifflin, Boston. 396 pp.
Curtis, John T. 1959. The Vegetation of Wisconsin. Univ. Wisconsin Press, Madison. 657 pp.
Dansereau, Pierre, & Fernando Segadas-Vianna. 1952. Ecological Study of the Peat Bogs of Eastern North America I. Structure and Evolution of Vegetation. Canad. Jour. Bot. 30: 490–520.
Darlington, H. T. 1945. Taxonomic and Ecological Work on the Higher Plants of Michigan. Mich. Agr. Exp. Sta. Tech. Bull. 201. 59 pp.
Davis, Charles M. 1964. Readings in the Geography of Michigan. Ann Arbor Publ., Ann Arbor. 321 pp.
Deam, Charles C. 1940. Flora of Indiana. Dep. Conservation, Indianapolis. 1236 pp.
Dean, Donald S. 1954 ["1953"]. A Study of Tradescantia ohiensis in Michigan. Asa Gray Bull., N. S. 2: 379–388.
Dorr, John A., & Donald F. Eschman. 1970. Geology of Michigan. Univ. Michigan Press, Ann Arbor. 476 pp.
Elliott, Jack C. 1953. Composition of Upland Second Growth Hardwood Stands in the Tension Zone of Michigan as Affected by Soils and Man. Ecol. Monogr. 23: 271–288.

Fernald, Merritt Lyndon. 1950. Gray's Manual of Botany. Ed. 8. Am. Book Co., New York. lxiv + 1632 pp.

Gates, Frank C. 1942. The Bogs of Northern Lower Michigan. Ecol. Monogr. 12: 213–254.

Gleason, Henry Allan. 1939. The Plants of Michigan. Ed. 3. George Wahr, Ann Arbor. 158 pp.

Gleason, Henry A. 1952. The New Britton and Brown Illustrated Flora of the Northeastern United States and Adjacent Canada. N. Y. Bot. Gard., New York. 3 vol. [Minor corrections in later printings.]

Gleason, Henry A., & Arthur Cronquist. 1963. Manual of Vascular Plants of Northeastern United States and Adjacent Canada. Van Nostrand, Princeton. li + 810 pp.

Graustein, Jeannette E. 1951. Nuttall's Travels into the Old Northwest. An Unpublished 1810 Diary. Chron. Bot. 14: 1–88.

Graustein, Jeannette E. 1967. Thomas Nuttall Naturalist. Explorations in America 1808–1841. Harvard Univ. Press, Cambridge. 481 pp.

Guire, Kenneth E., & Edward G. Voss. 1963. Distributions of Distinctive Shoreline Plants in the Great Lakes Region. Mich. Bot. 2: 99–114.

Hanes, Clarence R., & Florence N. Hanes. 1947. Flora of Kalamazoo County, Michigan. Vascular Plants. [Authors], Schoolcraft, Mich. 295 pp.

Hayes, Bruce N. 1964. An Ecological Study of a Wet Prairie on Harsens Island, Michigan. Mich. Bot. 3: 71-82.

Humphrys, C. R., et al. 1965. Michigan Lakes and Ponds. Mich. Agr. Exp. Sta., Dep. Resource Development. 2 + 6 + 15 + [215] + 33 + 14 pp.

Maycock, P. F., & J. T. Curtis. 1960. The Phytosociology of Boreal Conifer-Hardwood Forests of the Great Lakes Region. Ecol. Monogr. 30: 1–35.

McVaugh, Rogers. 1970. Botanical Results of the Michigan Geological Survey under the Direction of Douglass Houghton, 1837–1840. Mich. Bot. 9: 213–243.

McVaugh, Rogers, Stanley A. Cain, & Dale J. Hagenah. 1953. Farwelliana: An Account of the Life and Botanical Work of Oliver Atkins Farwell, 1867–1944. Cranbrook Inst. Sci. Bull. 34. 101 pp.

Mohlenbrock, Robert H. 1967– . The Illustrated Flora of Illinois. Southern Illinois Univ. Press, Carbondale.

Nuttall, Thomas. 1818. The Genera of North American Plants, and a Catalogue of the Species, to the Year 1817. Author, Philadelphia. 2 vol.

Rittenhouse, Janice L., & Edward G. Voss. 1962. Douglass Houghton's Botanical Collections in Michigan, Wisconsin, and Minnesota on the Schoolcraft Expedition of 1832. Mich. Bot. 1: 61–70.

Schwarten, Lazella, & Harold William Rickett. 1958. Abbreviations of Titles of Serials Cited by Botanists. Bull. Torrey Bot. Club 85: 277–300.

Shinners, Lloyd H. 1958. Spring Flora of the Dallas-Fort Worth Area, Texas. [Author] 514 pp.

Smith, Helen V. 1966. Michigan Wildflowers. Cranbrook Inst. Sci. Bull. 42, revised. 468 pp.

Stafleu, F. A., E. G. Voss, et al. 1972. International Code of Botanical Nomenclature Adopted by the Eleventh International Botanical Congress, Seattle, August 1969. Regnum Vegetabile Vol. 82. 426 pp.

Steenis, C. G. G. J. van. 1954. General Principles in the Design of Floras. VIII Congr. Internatl. Bot. Paris, Rapp. Comm. Sect. 2,4,6: 59–66.

Stephenson, Stephen N. 1971. The Biosystematics and Ecology of the Genus Brachyelytrum (Gramineae) in Michigan. Mich. Bot. 10: 19–33.

Steyermark, Julian A. [1963]. Flora of Missouri. Iowa State Univ. Press, Ames. lxxxiii + 1725 pp.

Stuckey, Ronald L. 1967. The "Lost" Plants of Thomas Nuttall's 1810 Expedition into the Old Northwest. Mich. Bot. 6: 81–94.

Swink, Floyd. 1969. Plants of the Chicago Region. Morton Arboretum. 445 pp.

Torrey, John. 1822. Notice of the Plants Collected by Professor D. B. Douglass, of West Point, in the Expedition under Governour Cass, During the Summer of 1820, Around the Great Lakes and the Upper Waters of the Mississippi . . . Am. Jour. Sci. 4: 56–69.

Veatch, Jethro Otto. 1928. The Dry Prairies of Michigan. Pap. Mich. Acad. 8: 269–278.

Veatch, J. O. 1953. Soils and Land of Michigan. Michigan State Coll. Press, East Lansing. 241 pp. + map.

Voss, Edward G. 1956. A History of Floristics in the Douglas Lake Region (Emmet and Cheboygan Counties), Michigan, with an Account of Rejected Records. Jour. Sci. Lab. Denison Univ. 44: 16–75.

Voss, Edward G. 1965a. Some Rare and Interesting Aquatic Vascular Plants of Northern Michigan, with Special Reference to Cusino Lake (Schoolcraft Co.). Mich. Bot. 4:11–25.

Voss, Edward G. 1965b. On Citing the Names of Publishing Authors. Taxon 14: 154–160.

Voss, Edward G. 1966. Nomenclatural Notes on Monocots. Rhodora 68: 435–463.

Voss, Edward G. 1967. The Status of Some Reports of Vascular Plants from Michigan. Mich. Bot. 6: 13–24.

Voss, Edward G. 1972. Additional Nomenclatural and Other Notes on Michigan Monocots and Gymnosperms. Mich. Bot. 11: 26–37.

Whiteside, E. P., I. F. Schneider, & R. L. Cook. 1968. Soils of Michigan. Mich. Agr. Exp. Sta. Ext. Bull. E-630 (former Spec. Bull. 402). 52 pp. + map.

Taxonomic Section

B. Larix laricina
Russell   Emmet Co.

C. Juniperus horizontalis
Voss   Emmet Co.

PLATE 1

A. Taxus sp. (cultivated)
R. Schmid   Washtenaw Co.

E. Sparganium fluctuans
Voss   Schoolcraft Co.

D. **Typha latifolia**
M. Schmid   Oakland Co.

F. Potamogeton epihydrus
Voss   Cheboygan Co.

A. Zannichellia palustris
Voss   Emmet Co.

F. Elodea canadensis
Voss   Cheboygan Co.

PLATE 2

D. Alisma plantago-aquatica
Voss   Cheboygan Co.

B. Najas flexilis
Voss   Emmet Co.

C. Triglochin maritimum
M. Schmid
Washtenaw Co.

E. Butomus umbellatus
M. Schmid   Washtenaw Co.

G. Phragmites australis
Voss   Emmet Co.

A. Elymus mollis
Voss   Chippewa Co.

F. Eriophorum virginicum
Horne   Emmet Co.

D. Cladium mariscoides
Voss   Cheboygan Co.

PLATE 3

E. Scirpus hudsonianus
Voss   Emmet Co.

B. Hordeum jubatum
Horne   Jackson Co.

C. Carex comosa
Voss   Emmet Co.

E. **Xyris difformis**
Voss   Lake Co.

A (above) and B (below)
Arisaema triphyllum
Horne   Jackson Co.

PLATE 4

C. **Lemna minor**
Stoutamire
Clinton Co.

D. **Eriocaulon septangulare**
Voss   Otsego Co.

F. **Tradescantia ohiensis**
Horne   Jackson Co.

D. **Clintonia borealis**
Wells   Ontonagon Co.

PLATE 5

A. **Pontederia cordata**
Voss   Emmet Co.

B. **Heteranthera dubia**
Voss   Washtenaw Co.

C. **Juncus militaris**
Voss   Cheboygan Co.

E. **Uvularia grandiflora**
Voss   Emmet Co.

D. Erythronium americanum
Russell   Washtenaw Co.

E. Lilium michiganense
Horne   Jackson Co.

PLATE 6

C. Trillium grandiflorum
Russell   Washtenaw Co.

A. Smilax illinoensis
Horne   Jackson Co.

B. Trillium sessile
Horne   Lenawee Co.

A. **Streptopus roseus**
Horne   Ogemaw Co.

B. **Dioscorea villosa**
Horne   Hillsdale Co.

PLATE 7

D. **Cypripedium calceolus**
Voss   Emmet Co.

C. **Hypoxis hirsuta**
Tetley
Washtenaw Co.

E. **Cypripedium arietinum** (with
  Cornus canadensis
  & *Iris lacustris*)
Horne   Emmet Co.

B. Isotria medeoloides
Case   Berrien Co.

PLATE 8

C. Arethusa bulbosa
Voss   Emmet Co.

E. Calopogon tuberosus
Voss
Beaver Island

D. Habenaria orbiculata
Williams   Emmet Co.

A. Calypso bulbosa
Williams   Emmet Co.

USING THIS FLORA: STYLE, CONVENTIONS, ABBREVIATIONS

Having first familiarized himself with the material on "Plant Identification and the Use of Keys" on pp. 30–35, the beginner with an unknown plant to identify should start with the General Keys and associated aids which begin on p. 44. Here he can determine whether he has a plant which is in this part of the Flora and, if it is, to what family it belongs. In some instances, the General Keys will lead directly to a genus rather than a family, or even to a single species. Unless the genus or species has been stated in the General Key, the next step is to turn to the family, where there will be a key to genera (if the family includes more than one genus in our flora). All local genera are included in these keys, even when certain ones are reached directly in the General Keys. Thus, one who already recognizes the family can go directly to these keys without encountering omissions.

Finally, for each genus with more than one local species, there is a key to species. When there is a single species in a genus, or a single genus in a family, the fact is obvious without requiring a statement explaining the absence of a key. In all instances, the keys and descriptive comments are intended to apply *in our flora,* not necessarily elsewhere in the range of the taxa.

Following the keys to species, each species is listed with the number of its distribution map and information on its habitat in Michigan (as explained on pp. 17–21). There follow such other comments as seem pertinent on variation, additional (often more subjective) identifying characters, alternative classifications or names, and other matters of interest.

No special typography distinguishes the names of species not considered indigenous in the local flora. Some species are apparently native in part of the state but adventive elsewhere. Others are completely introduced, and the status of some is doubtful. For species considered to be wholly or partly non-indigenous, there is a statement regarding their origin; others are considered to be native.

Records of special interest (the rarest species, unusual forms, doubtful records, etc.) are documented in the customary style, with collector's name (omitting initials if included in the sketch of Michigan floristics on pp. 10–15) and collection number (if any) in italics; the date, if cited, not in italics; and the herbaria where specimens have been seen indicated by the standard abbreviations as listed on pp. 9–10.

For the sake of uniformity and consistency, all measurements of plant parts are based on dry specimens. In some cases, especially with delicate structures like anthers, measurements of fresh material may run a trifle larger. However, it is manifestly impossible to check a full range of variability in all species on fresh material in the course of a few years, so that overall the broadest range of measurements was obtainable from herbarium specimens — which is not to deny that an even broader range may in fact exist for some species.

In describing plants and their parts, the standard scientific units of the metric system have been used. In describing locations and habitats (depth of water, distances, land descriptions, etc.) the English system has been employed, perhaps inconsistently but in the interest of clarity and familiarity.

The use of square brackets in giving locality data indicates information (usually the county) which was not included in the source cited (label or literature), but which has been added — I hope correctly.

Several other conventions and idioms used in the keys and descriptive notes should be mentioned:

In keys, nouns are given first for ready comparison, followed by the modifiers. It is intended that *all* characters mentioned in one lead of a couplet be contrasted in the alternate lead, if only to state that a condition is "various" (to make clear it is not *necessarily* different from the first lead). Positive statements are preferred in keys, rather than the negative and uninformative "not as above" which does not distinguish whether the alternative to red is pink or blue, or whether "not pilose" means "glabrous" or, more ambiguously, "hispid."

The commonest range of measurements or conditions is stated, with rare extremes at one or both ends added in parentheses. Thus, "Scales (3) 3.5—4.5 (5) mm long" means that they are *usually* 3.5—4.5 mm, but one or a few individuals have been seen in which they are as small as 3 mm or as large as 5 mm; this is a concise though not statistically precise way of saying that the great majority of specimens will fall in a certain range but that some individuals are known to be exceptional.

The word "mostly" generally means "in most instances on a specimen," e. g., most of the leaves, perigynia, tepals, or whatever is mentioned, on a plant. On the other hand, "usually" generally means "on most specimens," e. g., most plants in our area have the widest leaves, or largest fruit, of the stated size or description.

Comparatives in a key with the noun first, e. g., "leaves wider," are intended to contrast with the alternative lead; i. e., "leaves wider than stated in the alternative." Comparative adjectives preceding a noun, e. g., "wider leaves", simply allow for more than a single extreme measurement, i. e., in this example, not necessarily the one widest leaf but several of the wider leaves on a plant.

The symbol ±, meaning "more or less," is frequently used as a space-saving way of saying "rather," somewhat," "about," or "to a greater or lesser extent."

Illustrations

The scale of enlargement or reduction of a whole plant or inflorescence (or part of an inflorescence) is given first in the legends, followed by indication of any details and their scale. Separate figure numbers are assigned to details only when they are not clearly associated with the main drawing or if they have been taken from a different source. The scales are only approximate, and usually those stated in the sources from which the drawings were taken have been accepted (a major exception being in *Carex,* where Mackenzie gave no magnifications for details and those assigned by Gleason are in most instances about a fourth to a third off). On each page of figures, they are numbered in the same sequence as they appear in the text. Occasionally space has required that a species appear "out of order" on a different page, however, and it is then numbered in the sequence for that page.

When critical points mentioned in the keys, especially qualitative comparisons, are likely to be significantly clarified by an illustration, figure numbers have often been cited; but most figures are not cited in the keys. All figures and color plates are cited after the name of the species in the text.

Abbreviations and Symbols

Abbreviations of titles for scientific journals basically follow Schwarten and Rickett (1958). Herbarium abbreviations are listed on pp. 9–10; private herbaria are designated by the last name of the owner.

Generic names are abbreviated to the first letter when it is clear from the context what name is intended.

ca. about (Latin: circa)
cf. compare (Latin: confer)
cm centimeter (see scale below)
dm decimeter (see scale below)
e. g. for example (Latin: exempli gratia)
f. form (forma)
i. e. that is (Latin: id est)
m meter (= 10 decimeters, 100 centimeters, 1000 millimeters)
mm millimeter (see scale below)
sp. species (singular)
spp. species (plural)
ssp. subspecies
TL type locality; the locality from which a type specimen came (this is indicated by the abbreviation or by fuller discussion for all taxa mentioned when the type locality is in Michigan)
Tp. Township
var. variety (varietas)
± more or less
× in figure legends, scale of enlargement or reduction; otherwise, the sign of a hybrid: before a specific epithet, × indicates that the binomial applies to a hybrid; between two binomials, it indicates a hybrid (sometimes putative) of that parentage

INCHES   1   2   3   4

METRIC   1 2 3 4 5 6 7 8 9 10

GENERAL KEYS TO THE FAMILIES OF
MICHIGAN GYMNOSPERMS AND MONOCOTS

A question inherent in any partial flora, such as this one, is whether the plant one wishes to identify is in the book at all. It is premature at this time to offer a key to all families of Michigan seed-plants, including those dicots expected to be covered in future parts. Keys to the families of gymnosperms and monocots are presented here, together with some suggestions for recognizing whether one has a plant in either of these groups. It is also necessary to be sure that one does not have a pteridophyte.

PTERIDOPHYTES

This Flora does not include the diverse plants known as the "ferns and fern allies" (all of the non-seed-bearing vascular plants), in however many divisions of the plant kingdom one chooses to classify them. Billington's *Ferns of Michigan* (1952) has treated these plants, with distribution maps and illustrations; but a great deal of additional information has accumulated since his volume and many new taxa, including an extraordinary number of hybrids, have been discovered in the state. Many of these have been mentioned in articles in *The Michigan Botanist*. Most people are familiar with typical members of many of the pteridophyte groups: the ferns (with their leaves or fronds usually compound or dissected but occasionally simple), the club-mosses (*Lycopodium*), the spike-mosses (*Selaginella*), the aquatic quillworts (*Isoëtes*), and the horse-tails and scouring-rushes (*Equisetum*).

These plants all bear nothing resembling flowers or seeds. They reproduce (when not vegetatively) by one-celled spores which are produced in sporangia borne on the fronds (ferns), in leaf axils, or in cones. Sterile pteridophyte specimens without sporangia may at times be difficult to distinguish from sterile seed plants. But except for certain aquatics, this Flora is not designed for use with sterile plants (unless sometimes one knows at least the family or genus from the start), and the aquatic pteridophytes are included in Key A, p. 50.

GYMNOSPERMS

The gymnosperms represent a division, Pinophyta, of the plant kingdom, as placed in the system of Cronquist (1968). They differ, technically, from all of our other seed plants (i. e., from the angiosperms, division Magnoliophyta) in not bearing the seeds in a *closed* vessel (the ovary, ripening into a true fruit). Instead, the seeds are exposed ["gymnosperm" = "naked seed"] on the scale of a cone or, in *Taxus,* in a cup-shaped fleshy aril. In *Juniperus,* the cone-scales become more or less fleshy and coalescent, the resultant cone simulating a berry. The Pinophyta do not produce true flowers with petals or other perianth.

More superficially, our native and established gymnosperms may be recognized as *woody* plants (trees or shrubs) with leaves usually described as "scale-like" or "needle-like" (sometimes actually narrowly linear). Those with *scale-like* leaves have them *opposite* or *whorled*. Our only other seed plant which might be considered to be woody and to have opposite scale-like leaves is the tiny parasitic dwarf mistletoe, *Arceuthobium pusillum,* which grows on the branches of spruce (and perhaps other gymnosperms). In two other woody plants with scale-like leaves, these are alternate: *Hudsonia* (beach-heath), with densely pubescent leaves and bright yellow flowers; *Tamarix* (tamarisk), with glabrous leaves and pink flowers. Gymnosperms with *needle-like* leaves have them less than 2.7 mm broad, several times as long. They may be opposite, alternate, or clustered. Our only other native (or established) woody plant with needle-like leaves which might be confused with those of a gymnosperm is *Empetrum* (crowberry), a prostrate shrub of the Lake Superior region with leaves less than 7 mm long and the fruit a dark several-seeded berry.

KEY TO FAMILIES OF MICHIGAN GYMNOSPERMS

1. Leaves opposite or whorled, scale-like or needle-like . . . . . . CUPRESSACEAE (p. 66)
1. Leaves alternate or in clusters, needle-like or linear
  2. "Fruit" red and berry-like; leaves flattened, with strongly decurrent bases, persistent, appearing 2-ranked, all green on both sides (may be yellower beneath)
  . . . . TAXACEAE (p. 58)
  2. "Fruit" a dry woody cone; leaves flattened or not, but if so, then not as above (i.e., not decurrent, readily falling when dry, not 2-ranked, and/or with distinct white lines) . . . . PINACEAE (p. 59)

MONOCOTS

The monocotyledones, as placed by Cronquist, are a class of the division Magnoliophyta (flowering plants) and in this position may be called the Liliopsida. The dicotyledones (class Magnoliopsida) will be treated in other parts of this Flora. Except in certain obvious cases of vegetative or other asexual propagation of one kind or another, the Magnoliophyta reproduce by seeds developed in an enclosed ovary, as in the various familiar types of fruits. The Magnoliophyta produce true flowers.

The monocots and dicots have long been recognized as natural groups, although the technical differences separating them are each subject to occasional exceptions. The monocots have a single cotyledon in the embryo (whence the name) and the leaves are usually parallel-veined (i. e., all of the main veins running from base to apex of the leaf, without true branching or netted venation). There is usually no cambium and the vascular bundles are usually scattered. The flower parts (when not numerous) are typically in multiples of 3 (never 5, rarely 4). In the dicots, the cotyledons are normally 2 and the leaves with few exceptions have a branching or netted venation. A cambium is usually present, at least in the vascular bundles, which are typically in a ring. The flower parts are often in multiples of 5, sometimes of 4, and only rarely of 3 (in which case only *some* of the parts are in 3's except for *Floerkea*). Because of the exceptions that occur to the technical characters — some of which are cited above — and the comparative difficulty of determining, in many instances, the number of cotyledons in the ovule or seed and the arrangement of vascular tissue, some characters more readily seen and some general warnings are stressed here. These notes apply only to the species *in our flora*:

PRACTICAL HINTS FOR DISTINGUISHING MONOCOTS FROM DICOTS IN MICHIGAN

1. All trees and shrubs (except for gymnosperms, discussed above) are *dicots*.
2. All plants with one or more cycles of flower parts numbering 5 or a multiple of 5 are *dicots*.
3. Except for two submersed aquatics (*Potamogeton crispus,* fig. 23, and *Najas marina*), all plants with definitely *toothed* leaves are *dicots*. (Some monocots, however, have minutely spinulose, conspicuously ciliate, or strongly scabrous leaf margins.)
4. Except for the jack-in-the-pulpit genus (*Arisaema,* plate 4-A, B), with characteristic spadix and spathe, all plants with *compound* leaves are dicots.
5. Except for *Eryngium*, a rare member of the Umbelliferae with spiny-margined leaves, and *Tragopogon*, with typical Composite head and milky juice, all plants with parallel-veined grass-like leaves the base of which *sheathes the stem* are *monocots*.
6. Except for *Floerkea*, a weak-stemmed annual with pinnately compound leaves, plants with *all* flower parts in multiples of 3 are *monocots*. (Three carpels may be united into a single pistil.)
7. All plants with net-veined leaves are *dicots* except for certain vines with 3-merous flowers (covered in the key below); *Trillium* (with all parts, even leaves, normally in 3's); some Araceae (with characteristic spadix and spathe); and some Alismataceae (plants of wet ground, with 3 white to pinkish petals).
8. **Note**: A number of aquatic dicots (or aquatic forms of terrestrial or marsh species) are often collected without reproductive parts and might be confused with some of the numerous aquatic monocots. They are covered under the first choice in the key below, which leads directly to Key A, p. 50, for true aquatics and which is based primarily on vegetative characters.
9. **Note**: A number of terrestrial dicots (or terrestrial forms of aquatic ones) have leaves so small and/or with the venation so obscure that they might easily be thought to have the venation (if any) parallel. These include at least some species in the following genera: *Arenaria, Hippuris, Linum, Littorella,*

Lobelia (*L. dortmanna*), *Polygonella, Ranunculus* (e. g., *R. reptans*), *Stellaria.*
Certain larger and coarser-leaved dicots have a venation pattern that is
basically parallel: e. g., *Eryngium, Plantago, Tragopogon.* All of these plants
can be distinguished from monocots by their possession of either of the
following:
(a) A conspicuous perianth of 5 petals, sepals, or tepals (or a typical
    Composite head and milky juice in *Tragopogon*);
(b) A perianth inconspicuous and chaffy (or absent), the stamens solitary
    (*Hippuris*) or 4 (*Littorella, Plantago*). [Monocots with the perianth
    chaffy have definitely parallel-veined leaves and the stamens usually 3 or
    6.]

KEY TO FAMILIES OF MICHIGAN MONOCOTS
(Including some plants likely to be confused with monocots.)

This key is designed to direct the beginner who is unfamiliar with the families
of monocots to the name of the family (or in some cases, the genus) to which an
unknown plant suspected of being a monocot belongs. **This is not a general key
to flowering plant families** — plants which are clearly dicots (see hints 1–9
above) should not be sought in it. A few plants experience has shown are often
confused with monocots are included in the key and should run to a
determination as "Dicot" or, if more precision has been possible, to the name of
a dicot (or non-vascular) genus or family. Names of monocot families are in
CAPITALS. This is a strictly artificial key designed for *identification* on the
basis of the characters most useful for that purpose.

1. Plants strictly aquatic, the leaves or plant body *entirely* submersed or floating on
   the surface of the water (at most, the inflorescence and bracts, not leaves, held
   above the surface) . . . . [Key A, p. 50]
1. Plants with at least some leaves (or stem if plant apparently leafless) above the
   water, or plants strictly terrestrial
   2. Plants lacking green color and also apparently leafless at flowering time
      3. Stem buried in ground; flowers in late winter or earliest spring, crowded in a
         spadix with a nearly or partly buried hood-like brownish or mottled spathe
         (green leaves from buried rhizome appearing after flowering); stamens 4; plant
         with skunk-like odor . . . . *Symplocarpus* in ARACEAE (p. 367)
      3. Stem or an elongated flowering stalk above ground; flowers solitary or in a few-
         to many-flowered raceme or umbel, not a spadix; stamens 1, 2, or 6; plant with
         odor other than skunk-like
         4. Flowers regular, in an umbel on a naked scape; plant with odor of onion;
            stamens 6 . . . . *Allium tricoccum* in LILIACEAE (p. 410)
         4. Flowers bilaterally symmetrical, solitary to racemose; plant without odor
            except fragrance of flowers; stamens 1 or 2
            5. Sepals and petals 3, the lower petal a definite lip, the others little modified;
               ovary inferior; plants of woods or bogs, with perianth of various colors. . . .
               . . . . ORCHIDACEAE (p. 433)

5. Sepals apparently 2 and petals 5, but flower basically 2-lipped; ovary superior; plants of wet shores and bog pools, with perianth yellow or magenta
. . . . *Utricularia* in Lentibulariaceae
2. Plants with green color and the leaves usually developed
6. Inflorescences producing only bulblets or sterile tufts of small leaves or modified flower parts instead of flowers or fruit − although plants apparently fully mature otherwise
7. Leaves broad (or compound or dissected), neither grass-like nor terete . . . . . Dicots
7. Leaves elongate, linear and grass-like or terete
8. Plant with strong odor of garlic or onion . . . . . . *Allium* in LILIACEAE (p. 410)
8. Plant odorless
9. Leaves terete, septate (with hard cross-partitions easily seen in dry specimens or felt by gently pinching a leaf and drawing it across one's fingernail) . . . . *Juncus* in JUNCACEAE (p. 381)
9. Leaves neither terete nor septate
10. Stems terete and hollow (in internodes) . . . .
. . . . various grasses, particularly *Phragmites* in GRAMINEAE (p. 116)
10. Stem ± triangular and solid . . . . *Scirpus* in CYPERACEAE (p. 348)
6. Inflorescence normal, the flowers (if pistillate) ultimately producing seeds
11. Plant a climbing or twining vine, in most species with tendrils; flowers unisexual; leaves net-veined
12. Leaves several-lobed; sepals and petals each 6, the former green and minute, the latter petaloid, united . . . . *Echinocystis* in Cucurbitaceae
12. Leaves not lobed (at most cordate); sepals and petals each 3, similar (= tepals), separate
13. Inflorescence an umbel; plants with tendrils; ovary superior; fruit a berry
. . . . *Smilax* in LILIACEAE (p. 396)
13. Inflorescence spicate to paniculate; plants without tendrils; ovary inferior; fruit a capsule . . . . DIOSCOREACEAE (p. 423)
11. Plant not a vine, without tendrils; flowers perfect or unisexual; leaves parallel- or net-veined
14. Inflorescence a spadix, subtended by a spathe which may be broad and hood-like or elongate; leaves in some species compound or net-veined . . . .
. . . . ARACEAE (p. 365)
14. Inflorescence not a spadix (if flowers in a head, this with neither an elongate fleshy axis nor a conspicuous subtending spathe); leaves simple, rarely net-veined (in *Smilax ecirrata, Trillium,* and some Alismataceae)
15. Perianth much reduced: absent, or composed solely of bristles (these small and stiff or elongate and cottony), or of chaffy or scale-like parts − never conspicuously petaloid
16. Individual flowers subtended by 1 or 2 scales; leaves ± elongate, grass-like, usually with a sheath at the base surrounding the stem; inflorescence various but never a single terminal globose head on a scape; fruit a 1-seeded grain or nutlet (achene)
17. Each fertile flower subtended by a *single* scale (others may be at base of spikelet); sheaths of leaves closed (margins connate); stems frequently triangular (but 4- to several-angled or terete in many species), usually solid; leaves usually 3-ranked (especially so in a species with terete hollow stem); stamens with filament attached to end of anther; fruit a definitely 2- or 3-sided (rarely nearly terete) nutlet (achene) . . . .
. . . . CYPERACEAE (p. 244)

17. Each flower subtended by 2 scales (almost opposite each other, one rarely absent); sheaths often open; stems ± terete (sometimes flattened), never triangular in section; leaves not clearly 3-ranked (basically 2-ranked); stamens with filament attached near middle of anther (or apparently so because of sagittate anther); fruit usually a grain neither flattened (2-sided) nor triangular . . . . GRAMINEAE (p. 109)

16. Individual flowers subtended by no scales or only by bristles, or with a *regular* perianth of chaffy scales (or tepals); leaves, inflorescence, and fruit various

18. Inflorescence a single, very compact, almost spherical head (terminating an erect scape) less than 12 mm across

19. Surface of head (tips of receptacular bracts) white-woolly; flowers chaffy, not concealed by involucral bracts; roots with abundant conspicuous transverse markings (fig. 7) . . . . . ERIOCAULACEAE (p. 374)

19. Surface of head (bracts) glabrous; flowers yellow (properly running in alternate lead 15) or largely concealed by bracts; roots without transverse markings . . . . XYRIDACEAE (p. 372)

18. Inflorescence not a single terminal head and/or exceeding 12 mm

20. Inflorescence composed of separate staminate and pistillate portions, the former of conspicuous stamens, sooner or later withering, leaving only the pistillate portion conspicuous

21. Pistillate flowers in (1–) several globose heads; perianth of greenish sepals; leaves strongly keeled (3-angled in section) . . . .
. . . . SPARGANIACEAE (p. 71)

21. Pistillate flowers in an elongate densely flowered spike; perianth of white hairs; leaves flat-elliptical in section . . . . . . TYPHACEAE (p. 69)

20. Inflorescence composed of perfect flowers, without conspicuously separate staminate and pistillate portions

22. Flowers in a branched or umbellate inflorescence, solitary or, more often, clustered into small heads of 2 or more; fruit a 3- to many-seeded capsule . . . . JUNCACEAE (p. 379)

22. Flowers in a single elongate spike or ± zigzag raceme; fruit indehiscent or a 1–2-seeded follicle

23. Spike (truly a spadix) apparently lateral (fig. 277); fruit of each flower indehiscent . . . . *Acorus* in ARACEAE (p. 366)

23. Spike or raceme terminal (figs. 35, 37); fruit of each flower consisting of 3 or 6 1- or 2-seeded follicles . . JUNCAGINACEAE (p. 98)

15. Perianth at least in part of ± conspicuous white or colored petals

24. Flowers bilaterally symmetrical

25. Ovary inferior; fertile stamens 1 or 2, united with the pistil; flowers not blue (almost any other color) . . . . ORCHIDACEAE (p. 433)

25. Ovary superior; fertile stamens 3 or 6, free; flowers blue (except albinos), at least in part

26. Sepals colored like the petals; stamens 6, all fertile; flowers in a dense elongate inflorescence (plate 5-A). . . . .
. . . . *Pontederia* in PONTEDERIACEAE (p. 378)

26. Sepals greenish, unlike the petals; fertile stamens 6, 3 with imperfect anthers; flowers few. . . . . *Commelina* in COMMELINACEAE (p. 375)

24. Flowers regular (radially symmetrical)

27. Sepals and petals of quite different color and/or texture, the former green or brownish

28. Leaves in a single whorl of 3 on the stem . . *Trillium* in LILIACEAE (p. 400)

28. Leaves all basal or nearly so

29. Petals yellow; flowers in a single compact head less than 12 mm across
. . . . XYRIDACEAE (p. 372)
29. Petals blue, white, or pink; flowers in a more open or larger inflorescence
30. Pistils several in each flower, each developing into an achene; stamens 6–many; flowers unisexual or perfect; petals white or pinkish; leaves often broadly elliptical or sagittate, usually ± net-veined . . . . ALISMATACEAE (p. 100)
30. Pistil 1 in each flower, developing into a capsule; stamens 6; flowers perfect; petals blue, purple, or rose (except in occasional albinos); leaves elongate, definitely parallel-veined . . . .
. . . . *Tradescantia* in COMMELINACEAE (p. 376)
27. Sepals and petals both colored and petaloid, usually similar in shape (tepals) or the former (in *Iris*) of different size and shape
31. Ovary superior (or flowers unisexual)
32. Pistils 6, united only at the very base, ripening into follicles; stamens 9; flowers pink, in an umbel terminating a long scape . . . .
. . . . BUTOMACEAE (p. 105)
32. Pistil 1, sometimes the carpels slightly separate near the summit; stamens 3–6; flowers and inflorescence various
33. Stamens 3; tepals 6, yellow; plants creeping on wet shores (plate 5-B) . . . . *Heteranthera* in PONTEDERIACEAE (p. 378)
33. Stamens and tepals 4 or 6, the latter yellow or not; plants erect, of various habitats . . . . LILIACEAE (p. 392)
31. Ovary inferior (flowers perfect)
34. Stamens 3; leaves equitant . . . . IRIDACEAE (p. 424)
34. Stamens 6; leaves not equitant
35. Ovary clearly inferior, hairy . . . . AMARYLLIDACEAE (p. 423)
35. Ovary only "half-inferior," part of it adnate to the perianth, glabrous (at most granular-roughened) . . . . LILIACEAE (p. 392)

KEY A: FOR STRICTLY SUBMERSED OR FLOATING AQUATIC PLANTS*

Note: In the field, plants are usually readily recognized as being aquatic if one is not misled by a rise in water level to assume that a temporarily inundated plant belongs to a normally aquatic species. In the herbarium, a proper label should record the habitat, but most of the larger true aquatics, even without complete data, can be recognized as such by the delicate structure of submersed stem and leaves, which are often extremely limp and flexible; hence when dry they still convey the impression of having been supported by water. Of course, the presence of algae, other aquatic organisms, or marl encrustations is also a handy clue to an underwater source. Rush-like plants (grasses, sedges, rushes) with erect stems extending above the water should *not* be sought in this key unless they have definite limp aquatic foliage; go to couplet 6 in the key above for them.

*Modified from relevant portions of "A Vegetative Key to the Genera of Submersed and Floating Aquatic Vascular Plants of Michigan," E. G. Voss (Mich. Bot. 6: 35–50. 1967).

1. Plants without distinct stem and leaves, free-floating at or below the surface of the water (except when stranded by a drop in water level), the segments (internodes) small (up to 15 mm, but in most species much smaller), often remaining attached where budded from parent plant
  2. Plant body once to several times equally 2-lobed or 2-forked . . . .
  . . . . Ricciaceae (a family of bryophytes)
  2. Plant body not consistently dichotomous . . . . LEMNACEAE (p. 368)
1. Plants with distinct stem and/or leaves, usually anchored in substrate, mostly larger
  3. Plants with floating leaves present (blades, or at least their terminal portions, floating on the surface of the water, usually ± smooth and leathery in texture, especially compared with submersed leaves – or submersed leaves none)
    4. Blades of some or all floating leaves on a plant sagittate or deeply lobed at base, or compound, or peltate
      5. Floating blades compound (4-foliolate) . . . Marsileaceae (a family of pteridophytes)
      5. Floating blades simple
        6. Floating blades (at least some of them) sagittate (the apex and lobes acute) [Caution: Plants with sagittate leaves extending *above* the surface of the water do not belong in this key.] . . . . . . *Sagittaria* in ALISMATACEAE (p. 101)
        6. Floating (and any other) blades circular to ± elliptical in outline, peltate or rounded at apex with deep sinus at base . . . . Nymphaeaceae
    4. Blades of floating leaves all unlobed (at most subcordate at base), simple, the petiole marginal or (in ribbon-like leaves) absent
      7. Floating leaves small (less than 1 cm long), crowded in a terminal rosette; submersed leaves distinctly opposite; flowers solitary, axillary . . . . . Callitrichaceae
      7. Floating leaves larger, not in a rosette; submersed leaves alternate, basal, or absent; flowers mostly in a terminal inflorescence
        8. Leaves narrow and ribbon-like, the blades many times longer than broad, without distinct petioles (though in some species a sheath surrounds the stem)
          9. Leaves ± rounded at tip (even if tapered), the floating portion smooth and shiny, somewhat yellow-green to bright green when fresh, occasionally keeled but midvein scarcely if at all more prominent than others; leaf not differentiated into blade and sheath, the submersed portion similar to the floating but more evidently with a fine closely checked pattern; flowers and fruit in spherical heads (plate 1-E) . . . . SPARGANIACEAE (p. 71)
          9. Leaves sharply acute at tip, the floating portion rather dull, ± blue-green when fresh, with midrib; leaf including a sheath around the stem and a membranous ligule at junction of sheath and blade; flowers and fruit in paniculate spikelets*
            10. Floating blades minutely but densely pubescent or strongly papillose above; sheaths closed at least most of their length (margins connate to form tube around stem); plants rhizomatous; florets perfect, all alike . . .
            . . . . *Glyceria* in GRAMINEAE (p. 145)
            10. Floating blades glabrous; sheaths open (split down one side nearly or quite to the node); plants loosely rooted, not rhizomatous; florets unisexual, staminate toward base of panicle, pistillate toward summit . .
            . . . . *Zizania* in GRAMINEAE (p. 212)

*The grasses which most commonly produce elongate floating leaves in our area are manna grass, *Glyceria borealis* and *G. septentrionalis*, and young plants of wild-rice, *Zizania aquatica*. Identification is assured by comparison with normal plants in the vicinity or by checking flowers if present (as they often are, with floating leaves, in *Glyceria*).

8. Leaves (at least floating ones) with ± elliptical blades and distinct petioles
   11. Leaves all basal; petals 3, white . . . . ALISMATACEAE (p. 100)
   11. Leaves alternate or opposite; petals 4–6, pink or dull and inconspicuous
      12. Venation netted; flowers bright pink, in dense ovoid to cylindrical spike
      . . . . *Polygonum* in Polygonaceae
      12. Venation parallel; flowers dull, in narrow cylindrical spike . . . .
      . . . . POTAMOGETONACEAE (p. 75)
3. Plants without any floating leaves, entirely submersed (except sometimes for inflorescences and associated bracts)
   13. Leaves (or leaf-like structures) all basal and simple
     14. Leaves definitely flat, several times as broad as thick (widest about the middle or parallel-sided)
       15. Leaf blades not over twice as long as broad . . . . Nymphaeaceae
       15. Leaf blades more than twice as long as broad
         16. Leaves stiff and erect or somewhat outcurved, less than 20 cm long . . . .
         . . . . *Sagittaria* in ALISMATACEAE (p. 101)
         16. Leaves limp, more than 20 cm long (tape-like or ribbon-like)
           17. Midvein not evident: all veins of essentially equal prominence, with the tiny cross-veins giving a checkered appearance to the leaf, which is thus uniformly marked with minute rectangular cells ca. 1–2 mm long or smaller (fig. 3) . . . . SPARGANIACEAE (p. 71)
           17. Midvein (and usually some additional longitudinal veins) evident, the veins not all of equal prominence, not dividing the leaf into minute rectangular cells
             18. Leaves with the central third (or more) of distinctly different texture (more densely reticulate) than the two marginal zones (fig. 2); plants dioecious, the staminate flowers eventually liberated from a dense inflorescence submersed at base of plant, the pistillate solitary on a long ± spiraled peduncle which reaches the surface of the water (fig. 44); rhizomes without reduced leaves; plants without milky juice . . . .
             . . . . *Vallisneria* in HYDROCHARITACEAE (p. 105)
             18. Leaves ± uniform in venation pattern, not 3-zoned (fig. 4); plants monoecious, with emergent inflorescence of white-petaled flowers (but these scarce on plants with submersed tape-like leaves); rhizomes with remote reduced bracts; plants often with some milky juice . . . .
             . . . . *Sagittaria* in ALISMATACEAE (p. 101)
     14. Leaves* filiform (thread-like) or terete or only slightly flattened (especially basally), elongate and limp to short and quill-like, less than twice as broad as thick
      19. Major erect structures solitary, spaced along a simple or branched delicate rhizome, consisting either of rather yellowish stems bearing minute alternate bumps as leaves or of filiform leaves mostly buried in the substrate and with a few minute bladder-like organs**
        20. Leaves minute alternate bumps on stem; bladders not present; flowers sessile, inconspicuous, regular . . . . . *Myriophyllum tenellum* in Haloragidaceae

---

*The conspicuous structures here called "leaves" for ease of identification are in some instances actually stems or sterile culms.
**Occasional populations of *Utricularia cornuta* without buried bladders may be placed by the inrolled (circinate) tips of the young singly spaced leaves.

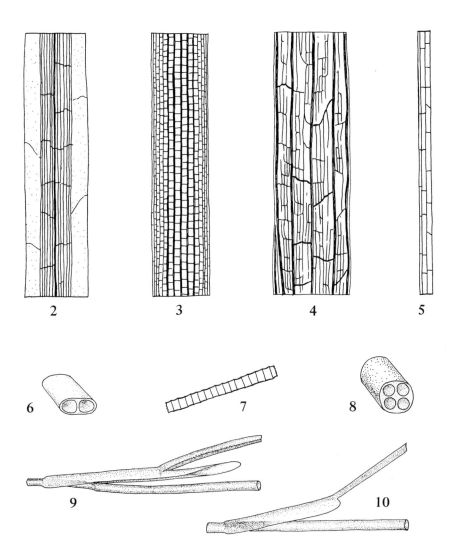

2. *Vallisneria americana,* segment of leaf ×1½
3. *Sparganium* sp., segment of leaf ×1½
4. *Sagittaria cuneata,* segment of leaf ×1½
5. *Scirpus subterminalis,* segment of leaf ×1½
6. *Lobelia dortmanna,* section of leaf ×2
7. *Eriocaulon septangulare,* segment of root ×2
8. *Isoëtes* sp., section of leaf ×2
9. *Potamogeton pectinatus,* sheath of leaf base and stipule ×2
10. *Ruppia maritima,* sheath of leaf base and stipule ×2

20. Leaves filiform, mostly buried in substrate (only the green tips, incurled when young, protruding); bladders (minute) usually present on the delicate branching rhizomes and buried leaf bases; flowers short-pedicelled, showy (yellow or purple), bilaterally symmetrical .... *Utricularia* in Lentibulariaceae
19. Major erect structures solitary to densely tufted, consisting of filiform or quill-like leaves or culms, with neither alternate bumps nor bladders
21. Leaves very limp (retaining no stiffness when removed from water and hence irregularly curved, sinuate, bent, or matted on herbarium specimens) – though a stiffer straight culm may also be present, mostly more than 20 cm long, ca. 0.2–1 mm in diameter
22. Leaves (actually sterile culms) terete their entire length, not expanded basally nor sheathing each other, but each separate and closely surrounded at base for ca. (0.6) 1 cm or more by a very delicate membranous tubular sheath (this sometimes requiring careful dissection to distinguish); rhizome less than 2 mm in diameter; inflorescence (rare on plants otherwise entirely submersed) a single strictly terminal spikelet
23. Rhizome reddish, at least on older portions; leaves mostly over 20 cm long, very limp; fertile culm triangular in section on emersed portion, much larger in diameter than the leaves, but spikelet no thicker than culm .... *Eleocharis robbinsii* in CYPERACEAE (p. 338)
23. Rhizome whitish throughout; leaves sometimes shorter, usually stiffer, than in preceding species; fertile culm terete, no larger than leaves, but spikelet distinctly thicker than culm ....
.... *Eleocharis acicularis* in CYPERACEAE (p. 340)
22. Leaves slightly expanded basally for ca. (0.7) 2–10 cm, sheathing the next inner leaf at least dorsally (usually the sheath continued ventrally as an almost invisible membrane), with tiny ligule or pair of auricles at the summit; rhizome various; inflorescence a lateral spikelet or terminal cyme
24. Leaf somewhat flattened or grooved ventrally for at least a few cm above the sheath (± crescent-shaped in section), with 1–5 longitudinal nerves evident, the tiny cross-veins connecting between nerves but not extending entirely across the leaf (fig. 5); sheath with a tiny ligule at summit; rhizome less than 2 mm in diameter; inflorescence a solitary lateral spikelet on a stiff wiry culm just above or near the surface of the water; flowers without petals and sepals; fruit an achene ....
.... *Scirpus subterminalis* in CYPERACEAE (p. 354)
24. Leaf terete above sheath, with no evident longitudinal nerves, but numerous definite septa extending entirely across the blade (which shrinks between septa on drying); sheath with a minute pair of auricles at summit; rhizome ca. 2–5 mm thick; inflorescence an open cyme of many several-flowered heads on a very stout culm (several mm in diameter, over 50 cm tall); flowers with 6 tepals; fruit a capsule .....
.... *Juncus militaris* in JUNCACEAE (p. 389)
21. Leaves usually firm (retaining stiffness when removed from water and hence straight or with an even curve in herbarium specimens), less (in most species much less) than 20 cm long, of various diameter
25. Leaves filiform throughout, not broader basally nor sheathing each other, solitary (rarely) or in small tufts along a filiform whitish rhizome, each leaf (actually a sterile culm) closely surrounded at its base for ca. 6 mm or more by a very delicate membranous tubular sheath (this sometimes requiring careful dissection to distinguish); inflorescence (rare on completely submersed plants) a single terminal spikelet ....
.... *Eleocharis acicularis* in CYPERACEAE (p. 340)

25. Leaves linear or tapered from base to apex, or if otherwise uniformly filiform then sheathing or expanded at base, without individual tubular sheaths as described above; inflorescence various
    26. Leaf in section appearing composed of 2 hollow tubes (fig. 6), linear (± parallel-sided), broadly rounded at tip; flowers bilaterally symmetrical, in a few-flowered raceme . . . . *Lobelia dortmanna* in Campanulaceae
    26. Leaf not (or rarely) of 2 hollow tubes, tapered and ± acute (or filiform); flowers regular and racemose, or solitary, or in a dense head or spike, or plant producing spores at base
        27. Roots with prominent cross-septate appearance (checkered with fine transverse lines, fig. 7); inflorescence a small whitish or gray head (flowering in shallow or rarely deep water or on wet shores) . . . . . .
        . . . . ERIOCAULACEAE (p. 374)
        27. Roots not distinctly septate or cross-lined; inflorescence not as above
            28. Leaves rather abruptly expanded at base to enclose sporangia, often dark green, composed of 4 hollow tubes (seen in section, fig. 8), surrounding a hard corm-like stem; plant always submersed (unless stranded), non-flowering . . . . . . Isoëtaceae (a family of pteridophytes)
            28. Leaves gradually and slightly expanded or grooved on one side at a somewhat sheathing base but not composed of 4 tubes nor enclosing sporangia and no corm-like stem present; plants (except *Subularia*) not flowering when submersed but only on wet shores
                29. Leaves somewhat flattened at least basally, widest at the base, gradually tapered to sharp apex; plants with buried rhizome or none
                    30. Plants connected by slender rhizomes (ca. 1 mm or narrower); sheathing basal portion of leaf (ca. 7 mm or more) with pale membranous borders abruptly terminating (or with minute auricles); leaves often 4 cm or more in length, somewhat flattened laterally below, with 2−3 conspicuous hollow tubes evident in section; inflorescence (not on wholly submersed plants) a spreading cyme of solitary to paired 3-merous flowers . . . .
                    . . . . *Juncus pelocarpus* in JUNCACEAE (p. 387)
                    30. Plants without rhizomes; sheath not abruptly auricled; leaves less than 4 cm long, somewhat flattened dorsoventrally (especially toward base), with numerous small hollow areas of irregular size; inflorescence (often submersed) a few- (often only 2-) flowered raceme of 4-merous flowers . . . . *Subularia aquatica* in Cruciferae
                29. Leaves ± terete, scarcely or no wider at base than at middle, of ± uniform width at least to middle (or even slightly thicker there before tapering to apex); plants with rhizomes or stolons at, near, or above surface of substrate
                    31. Plants with green stolons strongly arching above substrate; leaves filiform, ± uniform in diameter, ca. 0.5−1 mm thick, truncate at tip . . . . *Ranunculus reptans* in Ranunculaceae
                    31. Plants producing delicate horizontal white to green stolons at or near (above or below) surface of substrate (in addition to stouter short rhizome); leaves ca. 0.7−3 mm thick at middle, whence tapered to apex . . . . *Littorella uniflora* in Plantaginaceae
13. Leaves cauline, simple or compound
  32. Leaves compound, dissected, forked, or deeply lobed . . . . Dicots
  32. Leaves simple, unlobed, usually entire
    33. Leaves ± scale-like, not over 7 mm long, never distinctly opposite or whorled

34. Leaves minute, yellowish, merely widely spaced bumps on stem . . . . . .
. . . . *Myriophyllum tenellum* in Haloragidaceae
34. Leaves up to 7 mm long, green or brownish, loosely overlapping . . . . . .
. . . . (aquatic mosses & liverworts, bryophytes)
33. Leaves much longer or distinctly opposite or whorled (or both)
  35. Leaves alternate, with ligule-like stipules (these wholly adnate to leaves in
    Ruppia, fig. 10)
    36. Leaf blades ± filiform, terete or at least half as thick as broad, *and* the
      stipule adnate to leaf base for 10–30 mm or more, forming a sheath
      around the stem
      37. Sheath with no free stipular ligule at summit (the stipule wholly adnate
        to leaf blade, merely rounded at summit, fig. 10); leaf blade terete; fruit
        stalked in an umbel-like arrangement on a ± spiraled and elongating limp
        peduncle (fig. 30) . . . . RUPPIACEAE (p. 93)
      37. Sheath with a short ligule-like extension of free stipule at summit (the
        stipule only partly adnate, fig. 9); leaf blade often somewhat flattened;
        fruit sessile or subsessile in a spike with a straight ± stiff peduncle . . . .
        . . . . POTAMOGETONACEAE (p. 75)
    36. Leaf blades definitely flattened and several times broader than thick (even
      if narrow) *or* stipule little if at all adnate to blade (or both conditions)
      38. Blades flattened, ribbon-like (up to 5 or even 7.5 mm wide), with no
        definite midrib (no central vein more prominent than others except
        rarely toward base); flowers solitary, rare, cleistogamous in axils of
        submersed leaves or (these almost never on submersed plants) with 6
        bright yellow tepals . . . . . . *Heteranthera* in PONTEDERIACEAE (p. 378)
      38. Blades filiform or flattened with a definite midrib; flowers in globose or
        cylindrical spikes, neither cleistogamous nor with showy yellow perianth
        . . . . POTAMOGETONACEAE (p. 75)
  35. Leaves opposite or whorled, without stipules as described above (except
    very inconspicuously in *Zannichellia*)
    39. Leaves nearly filiform, not over 0.5 mm broad, very gradually tapered
      from base to apex but not abruptly expanded basally, perfectly smooth;
      plants perennial by slender rhizomes; flowers axillary, 1 staminate flower
      (a single stamen) and (1) 2–several carpels at a node; fruit slightly curved
      and ± minutely toothed on convex side (figs. 31, 32) . . . .
      . . . . ZANNICHELLIACEAE (p. 94)
    39. Leaves broader; or if filiform then abruptly expanded at base and with
      spiculate or toothed margins, the plants annual, and the fruit solitary and
      ellipsoid
    40. Leaves opposite (in some species, with bushy axillary tufts of leaves
      which may give a falsely whorled appearance)
      41. Leaves less than 6 times as long as wide; fruit various . . . . Dicots
      41. Leaves at least 6 times as long as wide; fruit solitary in axils
        42. Leaves filiform to linear-lanceolate (but ± expanded at very base),
          acute at apex; fruit ± ellipsoid, 1-seeded . . . . . NAJADACEAE (p. 94)
        42. Leaves linear and bidentate at apex when well submersed, often
          becoming obovate, ± weakly 3-nerved, and not necessarily bidentate
          toward summit of stem (or floating in rosettes); fruit somewhat
          heart-shaped, of two 2-seeded segments . . . . Callitrichaceae
    40. Leaves definitely whorled
      43. Whorled structures ("branches") cylindrical, elongate, usually stiff with
        calcium deposits; plants with distinctive musky odor . . . .
        . . . . Characeae (the "stonewort" algae)

43. Whorled structures (true leaves) flattened, short (not over 20 mm long) or elongate and very limp; plants without odor
    44. Leaves 6–12 (usually 9) in a whorl, not over 2.5 mm wide, ca. 12–25 times as long as wide; flowers perfect, without petals, sessile in axils of emersed leaves or bracts . . . . Hippuridaceae
    44. Leaves mostly 3–4 (rarely 6) in a whorl, 0.8–5 mm wide, at most 10–13 times as long as wide; flowers perfect or unisexual, but with petals
        45. Leaves mostly 3 (rarely 6) in a whorl, very thin (2 cell layers) and delicate; stem round (not angled) in section, smooth; flowers unisexual, with 3 often pink petals, at least the pistillate long-stalked from entirely submersed stem (fig. 46) . . . .
        . . . . *Elodea* in HYDROCHARITACEAE (p. 106)
        45. Leaves mostly 4 in a whorl, stiff and firm; stem square (4-angled) in section, often with minutely retrorse-scabrous angles; flowers perfect, with 3–4 white petals (usually not developed on wholly submersed plants) . . . . *Galium* in Rubiaceae

Gymnospermae (Gymnosperms)

Many gymnosperms are cultivated, both native and exotic species, but this treatment accounts for only the native ones. Among the interesting exotic species is the famed maidenhair tree, *Ginkgo biloba* L., which was first introduced into the United States from the Orient in 1784. The sexes are separate in *Ginkgo*, and the "fruit" is a smelly plum-like seed, so that staminate plants are preferred for cultivation.

The conifers include the oldest and the largest living things on the face of the earth (species of the western United States). Their importance as forest trees, particularly in this region, can scarcely be overestimated.

Unless otherwise specified, all references to cones are to the female cones. A preliminary version of this treatment appeared in 1968 (Mich. Bot. 7: 121–128).

TAXACEAE   Yew Family

1. Taxus   Plate 1-A

In our area we have only one genus of this family, with a single native species. Various cultivated yews may be recognized by their resemblance in fruit and needles (flat, ± uniform yellow-green beneath, with decurrent bases) to the native plant — even though some cultivated yews are upright, like small trees, rather than spreading shrubs.

1. T. canadensis Marsh.   Ground-hemlock; Yew

Map 1. Rich often swampy woods: deciduous, mixed, or coniferous (hemlock, white pine, fir, and especially cedar), often thriving on banks and in ravines; favored by the moist winds from Lake Michigan and often luxuriant on wooded dunes and in coniferous woods near the shore.

A spreading monoecious shrub (most other species are dioecious) with attractive red "fruit." The latter consists of a single seed surrounded by a fleshy cuplike structure (the aril). The sharp-pointed needles are persistent on dry branches (unlike the readily falling ones of hemlock and spruce). The bright red pulpy aril is edible, but the seed is reputed to be poisonous, as is the foliage, at least to some animals. (The seed, even if it contains a poisonous principle, may pass safely through the digestive tract if not chewed.) The foliage is reported to have been used by several Indian tribes in the Michigan region and Quebec in the making of a beverage. It is heavily browsed by deer and moose in Michigan.

The European yew (*T. baccata* L.) is sometimes cultivated for ornament and is apparently much more poisonous. *T. cuspidata* Sieb. & Zucc., an Asian species, is also cultivated, as are hybrids between these introduced species. Rarely, a cultivated yew may spread (presumably by animals) and appear spontaneously where not planted by man; such an occurrence is reported in a pine plantation in Washtenaw County.

PINACEAE   Pine Family

KEY TO THE GENERA

1. Leaves needle-like, all or mostly grouped in definite clusters on short shoots
  2. Leaves deciduous, crowded and numerous on short lateral shoots (alternate leaves
     on new twigs); cones less than 2 cm long . . . . 1. **Larix**
  2. Leaves evergreen, in clusters of (normally) 2 or 5; cones more than 2 cm long . . 2. **Pinus**
1. Leaves flattened or 4-sided, alternate (spiraled), not in definite clusters
  3. Leaves persistent on dry branches, sessile, separating cleanly from an orbicular
     leaf-scar without any raised projection, the twig hence basically smooth; cones
     erect, 3.5–6.5 cm long, the scales falling at maturity from the persistent central
     axis . . . . 3. **Abies**
  3. Leaves readily falling from dry branches, leaving persistent peg-like bases, the
     twig hence very rough; cones pendent, (1) 1.3–6 cm long, falling entire at
     maturity
    4. Leaves flattened, rounded at apex, distinctly short-stalked in addition to the
       persistent narrow base . . . . 4. **Tsuga**
    4. Leaves ± 4-sided, acute or sharp-pointed, sessile on the persistent peg-like base
       . . . . 5. **Picea**

1. Larix

1. **L. laricina** (DuRoi) K. Koch   Plate 1-B   Larch; Tamarack
   Map 2. In almost all sorts of low places, open or wooded, shores, and sometimes on drier ground where there is not too much competition; especially characteristic of the older stages in bog succession (and invading bog mats), sometimes with spruce and less often cedar; poison sumac (*Rhus vernix*) is often associated in tamarack bogs in southern Michigan; frequent in beach thickets and interdunal hollows.

The European *L. decidua* Miller is sometimes planted; it differs in its larger cones, with pubescent scales, and is rarely reported as reproducing where planted (as on the campus of Michigan Technological University in Houghton).

2. **Pinus**   Pine

KEY TO THE SPECIES

1. Needles usually 5 in a cluster, ± triangular in section; membranous sheath surrounding base of each cluster deciduous; cones cylindrical, at least twice as long as wide . . . . 1. **P. strobus**
1. Needles usually 2 in a cluster, ± semi-circular in section; membranous sheath surrounding base of each needle cluster ± persistent; cones short-ovoid, much less than twice as long as wide
    2. Needles ca. 8–15 cm long, straight; bark of older branches and trunk reddish; cones generally deciduous, subterminal on the branchlets . . . . 2. **P. resinosa**
    2. Needles ca. 1.5–3 (6) cm long, ± twisted; bark of older branches and trunk dark gray to black; cones generally long-persistent, lateral on the branchlets . . 3. **P. banksiana**

1. **P. strobus** L.   Figs. 12, 13   White Pine
   Map 3. Often in mixed woods (the "hemlock–white pine–northern hardwoods" of Braun, 1950) but also on sandy plains and dunes with red and sometimes jack pine, invading bogs with tamarack, in swampy woods (mixed or on banks, rather than deciduous swamp forest or floodplains), on rock ridges, and even in cedar swamps.

   The familiar and important white pine is the official state tree of Michigan. It is a fitting designation, for this was the backbone of the lumber industry, particularly in the last quarter of the 19th century, when Michigan led the nation in lumber production. Here were the finest stands of this species in the world: trees attaining at their best 5–7 feet in diameter and 150–200 feet in height (although usually smaller). Branches at the top of tall trees in exposed situations are generally bent away from the direction of the prevailing winds, producing a characteristic and picturesque shape which makes it possible to recognize white pine even at a distance or in silhouette along shores and dunes and woodlands.

   The needles of white pine are lighter in color, softer in texture, finer and less stiff than those of our other two species.

 1. Taxus canadensis   2. Larix laricina   3. Pinus strobus

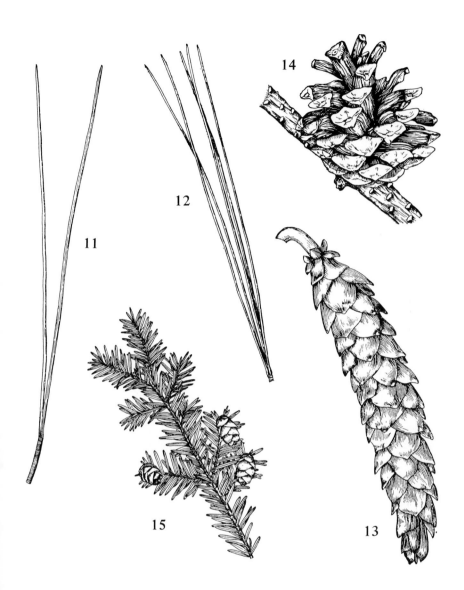

11. *Pinus resinosa,* pair of leaves ×1
12. *P. strobus,* cluster of leaves ×1
13. *P. strobus,* cone ×¾
14. *P. banksiana,* cone ×1
15. *Tsuga canadensis,* underside of branchlet ×½

2. **P. resinosa** Aiton   Fig. 11   Red Pine

Map 4. Usually a tall straight tree of striking appearance when solitary or in groves on sand dunes, ridges through boggy ground, or rock outcrops; common on well drained sandy plains, seldom on moist ground, sometimes with balsam fir, often with jack pine and oak. An important timber tree, in some places succeeding the jack pine which originally replaced it after lumbering and fire. In some ways red pine is ecologically intermediate between white pine and jack pine, growing with the former on better soils and with the latter on light sandy soils, although white and jack pines less often grow together. Reported as far south as Van Buren County. The largest red pines in Michigan — and in the nation — slightly exceed 3 feet in diameter.

This stately tree is frequently called by the misleading name "Norway Pine"; the species grows naturally only in northeastern North America. A somewhat similar European species, *P. nigra* Arnold (Austrian Pine), is occasionally planted. Its bark is gray or darker in contrast to the very distinctive reddish flaky bark of *P. resinosa*; and the stiff needles do not break cleanly upon bending, compared to those of *P. resinosa*, which snap sharply.

3. **P. banksiana** Lamb.   Fig. 14   Jack Pine

Map 5. Most often a rather scrubby tree, often mixed with some oak, in woodland or savana on dry sandy soil, as in the vast areas of jack pine plains in the north-central Lower Peninsula. Some dense stands of straight, mature trees occur, especially northward toward Lake Superior; on dunes as well as sandy outwash plains, and in the Upper Peninsula also in boggy situations with leatherleaf (*Chamaedaphne*) as well as on barren rocky sites (e. g., Isle Royale). One tree is known on Beaver Island and jack pine is frequent on low dunes on South Manitou Island; otherwise it is apparently absent from the islands of Lake Michigan. Occurs south along Lake Michigan into Indiana.

It has been estimated that jack pine predominates on over a million acres of forest in Michigan, over half of it in the northern Lower Peninsula, where certain restricted areas of young growth are the home of our most famous bird, the Kirtland's or "Jack Pine" Warbler, which breeds nowhere else in the world.

The needles of jack pine are ± abruptly obtuse to acute but blunt at the apex. In the similar *P. sylvestris* L. (Scots pine; Scotch pine), the needles average a bit longer than in *P. banksiana* and they often tend to taper to sharp tips as well as being more silvery in aspect. The curved cones of *P. banksiana* tend to point forward toward the ends of their branches and are long-persistent, while the more readily deciduous cones of *P. sylvestris* are ± horizontal or reflexed, pointing toward the base of their branches. The bark of *P. sylvestris* is a distinctive orange-brown, noticeable especially on the upper part of the trunk. While the latter species is widely planted, it does not often become established on its own. Old plantations seen from highways, as in Roscommon and Crawford counties, may give the impression of stands of a native tree.

3. Abies

1. A. balsamea (L.) Miller   Fig. 16   Balsam Fir

Map 6. Coniferous and mixed woods, often with aspen or paper birch (especially in a rather characteristic association near the Great Lakes shores); cedar swamps, bogs, and spruce—fir stands. The "National Champion" balsam fir is a tree 2 feet 2 inches in diameter in the Porcupine Mountains State Park. Our southernmost station is a sphagnum bog in Ingham County, just south of the Clinton-Shiawassee county line (*P. Cantino* in 1969, MICH, MSC).

The young twigs vary from densely to sparsely puberulent. The needles usually appear ± 2-ranked, in flattened sprays, and are usually rounded or slightly notched or 2-toothed at the apex, 0.8—2.5 (3.2) cm long. However, especially on older cone-bearing branches, the needles may curve upward, giving a bushy spruce-like appearance to the branches, and may be quite acute. Such needles tend to be shorter and thicker than the flatter, longer needles of sterile lower branches. But all are readily recognized by the smooth round leaf scars.

This aromatic plant is a popular Christmas tree since the needles do not fall readily as do those of the spruces. The blunt fragrant needles are used in balsam pillows, and the "Canada Balsam" of the microscopist is distilled from the bark and needles. Resin-filled pustules are characteristic of the bark of this species, and the young cones are often heavily resinous.

4. Tsuga

1. T. canadensis (L.) Carr.   Fig. 15   Hemlock

Map 7. Typically with beech, sugar maple, yellow birch, and often white pine, but also in coniferous swamps and on wooded dunes; often on knolls, in small groves, or in ravines.

The young twigs are ± densely pubescent. The short needles, mostly 6—13 (16) mm long, in flat sprays, and the small cones, mostly 13—22 mm long, are characteristic of this species. Even shorter needles occur on the upper side of the twigs, closely appressed and thus displaying conspicuously the prominent white

4. Pinus resinosa

5. Pinus banksiana

6. Abies balsamea

lines of the under surface. This characteristic is not easily seen on dried specimens, because the needles are so readily deciduous, but it is a useful field mark. The needles of hemlock are minutely toothed toward the rounded apex and have distinct petioles about 0.5 mm or a little longer.

Hemlock inner bark is purple, and the bark has been the main source of natural tannin for the leather industry. This genus is not native in Europe and should not be confused with the poison-hemlock of Socratic fame. The latter is an herb in the Umbelliferae, *Conium maculatum* L.

5. Picea   Spruce
Spruce needles fall readily from a prominent peg-like projection (or sterigma) on the twigs.

Black spruce is the usual host of dwarf mistletoe, *Arceuthobium pusillum* Peck, although the parasite has occasionally been found on white spruce in the vicinity of the Straits of Mackinac (Beaver Island; Wilderness State Park, Emmet Co.; Bois Blanc Island; Drummond Island).

KEY TO THE SPECIES

1. Young branchlets ± densely fine-pubescent; leaf bases below the sterigmata ± obscure on 1-year-old twigs; cones rather globose, (1) 1.5–2.8 (3) cm long, the scales with slightly irregular or erose margins, woody and rigid at maturity . . . .
. . . . **1. P. mariana**
1. Young branchlets glabrous; leaf bases prominent below the sterigmata on 1-year-old twigs, appearing to cover them with ridges and grooves; cones cylindrical, (2.5) 3–6 cm long, the scales nearly or quite entire, thinner and more leathery . . . . **2. P. glauca**

1. P. mariana (Miller) BSP.   Fig. 17   Black Spruce
Map 8. In the southern portion of its range in the state, almost entirely restricted to bogs; northward, found also in the same low woods, dune ridges, gravelly shores, etc., as *P. glauca* but most often in bogs, usually with *Larix*, sometimes with *Thuja* or *Abies*. True boreal spruce–fir forest is very little developed in Michigan although common north of Lake Superior.

7. Tsuga canadensis   8. Picea mariana   9. Picea glauca

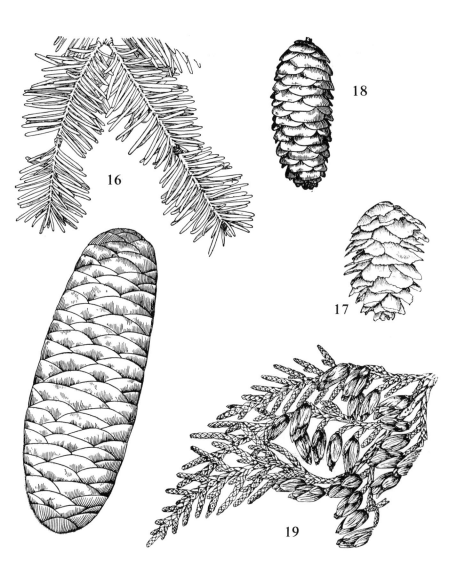

16. *Abies balsamea,* branchlet ×1 (above) & unopened cone ×1½ (below)
17. *Picea mariana,* cone ×1
18. *P. glauca,* cone ×1
19. *Thuja occidentalis,* branchlet with cones ×1

The sterigmata tend to be widely spreading in this species, even at right angles to the twig, while in the next they tend to be strongly ascending. The needles are quite variable in size and shape and sharpness. It has been reported that the needles of seedlings of white spruce are minutely denticulate, while those of black spruce are entire; there is no such distinction on mature needles, and it should be carefully checked on seedlings.

2. **P. glauca** (Moench) A. Voss   Fig. 18   White Spruce
    Map 9. Coniferous swamps, mixed woods, bogs, and stream borders, often seen in thickets and woods on dunes and gravelly shores along the Great Lakes. Reported from Newaygo, Mecosta, and Bay counties, which apparently form the southern limit of its range in Michigan except for a *Thuja* bog in Bruce Tp., Macomb County (*Parmelee 1902* in 1950, MSC). Michigan's largest white spruce is 3½ feet in diameter and grows about 15 miles north of Newberry, Luce County.
    The needles may be up to 18 mm long, usually averaging a little longer than those of black spruce, although they may be as short as 5–8 mm.
    Picea abies (L.) Karsten, the Norway spruce, with drooping branchlets and cones at least twice as large as white spruce, is often planted as an ornamental.

CUPRESSACEAE   Cypress Family

This family is sometimes included in the Pinaceae.

KEY TO THE GENERA

1. Mature cones ca. 8–15 mm long, brown and ± woody, the elongate scales distinct; leaves all opposite and appressed, scale-like, dimorphic (a pair of opposite strongly keeled leaves at right angles to adjacent pair of flat leaves); plant a tree . . . . . 1. **Thuja**
1. Mature cones up to 8 (10) mm long, green or bluish, fleshy and berry-like; leaves opposite or whorled, needle-like or if scale-like and appressed then quite uniform (none keeled); plant a tree with needle-like leaves or a shrub . . . . 2. **Juniperus**

1. Thuja

1. **T. occidentalis** L.   Fig. 19   Arbor Vitae; White-cedar; "Cedar"
    Map 10. This is the charateristic tree of "cedar swamps" which occupy hundreds of thousands of acres of low ground in northern Michigan, much of it nearly impenetrable except by the deer for which both shelter and a favorite food are provided. But cedar may be found at least sparsely in almost all kinds of woods except the driest; on sand dunes, shores, and rock outcrops (including limestone); along streams, in springy areas, and in bogs where it is often described as the major climax species; thrives on calcareous gravelly shores and ridges near Lakes Michigan and Huron in the northern part of the state. The

"National Champion" white-cedar is a tree almost 5½ feet in diameter on South Manitou Island.

The youngest seedlings have opposite or whorled flattened needle-like leaves, but the first branches have the characteristic scale-like leaves. The flat unkeeled leaves (i. e., those on top and bottom of twigs in contrast to the keeled leaves straddling the sides) have a ± prominent resin gland near the apex; this is especially evident on the lower leaf.

The only other native species of *Thuja* in North America is the western red-cedar, *T. plicata* D. Don, of the Pacific Northwest. The true cedar (*Cedrus*) of the Old World is quite different; it has foliage somewhat like that of *Larix*, but evergreen, and the cone scales fall separately like those of *Abies*.

2. Juniperus

The female cones are berry-like and dark bluish, often glaucous. The volatile oil they contain is the source of the distinctive flavor of gin. One often sees on *Juniperus* colorful orange galls caused by rust fungi of the genus *Gymnosporangium*, a widespread disease of apples and related plants.

KEY TO THE SPECIES

1. Leaves in whorls of 3, all awl-like, articulated at the base, not decurrent; cones on very short, scale-covered peduncles in the axils of awl-like leaves . . . . . 1. **J. communis**
1. Leaves mostly opposite, some or all scale-like (awl-like leaves when present often whorled but not articulated, decurrent at base); cones apparently terminal on short, scale-covered peduncles or branchlets like the branchlets from which they arise
  2. Plant an erect small tree with central trunk, growing in southern Michigan; fruit on ± straight, ascending peduncles (or branchlets) . . . . 2. **J. virginiana**
  2. Plant a very prostrate trailing shrub, growing in northern Michigan; fruit usually on ± arched or recurved peduncles . . . . 3. **J. horizontalis**

1. **J. communis** L.   Common or Ground Juniper

Map 11. Most widespread on or near sandy shores and dunes along the Great Lakes, often associated with pines; less common far inland except in the Upper Peninsula and southeastern Lower Peninsula, and occurring in a diversity of habitats: old fields and gravelly banks, usually with scattered red-cedar in characteristic "juniper savanas"; also occasionally in coniferous swamps (on hummocks in wet places); in oak—hickory woods and northward scattered under jack pine and aspens; and in the western Upper Peninsula in crevices of rock outcrops.

Most of our plants are var. *depressa* Pursh, a ± decumbent form whose large cup- or saucer-shaped mats are characteristic of dunes and sandy places. Some plants in the vicinity of the Sleeping Bear Dunes, Leelanau County, are unusually tall and robust, including the "National Champion" of this variety. The upper surface of the needles is slightly concave and is generally marked with a conspicuous white stripe the entire length. The needles may be as long as 2.6 cm.

Various forms of this species are cultivated for ornament.

2. J. virginiana L.   Red-cedar

Map 12. Stabilized sand dunes, lake shores, low deciduous woods, oak–hickory woods, and even swamps; but most characteristic, especially in the southeastern Lower Peninsula, of old fields and hillsides in open juniper savanas with *J. communis*. It is not certain whether any of the collections north of Newaygo County represent native plants (including one seen [no data] from Grand Traverse County but not mapped). Red-cedar is probably more common and widespread in the state now than it was before the clearing and exposure of the landscape and subsequent spread by seeds eaten by birds. It is often planted.

This species is said to have 3–5 seeds in each berry-like cone, while the next has only 1 or 2. Both species have sharp awl-like leaves on the young growth and seedlings (or sometimes after injury), while the leaves on old growth are scale-like and overlapping. See also notes under the next species.

3. J. horizontalis Moench   Plate 1-C   Creeping Juniper

Map 13. On rock at Isle Royale and in Marquette and Keweenaw counties (including a very blue-glaucous form on Mt. Bohemia) and on dolomite pavement on Summer Island (Delta County) and Drummond Island; elsewhere almost entirely confined to the sandy or gravelly shores and dunes of the Great Lakes, usually associated with pine and the common juniper, and relic on older shores and beach ridges among cedar and fir thickets; in Luce County, found by Hagenah inland in alder thickets and rocky woods. Grows in bogs (or muskegs) in Ontario and rarely in the Upper Peninsula.

I can see no consistent difference in leaf shape by which to distinguish this species from the preceding; both have a wide range of scale shapes and awl-like leaves, chiefly the former. Sterile specimens for which the collector neglected to include on the label any statement regarding the stature of the plant may usually be distinguished by the definitely one-sided aspect of the trailing branches of *J. horizontalis*, which furthermore grows in Michigan only north of the range of *J. virginiana*. Where the two species overlap (in Wisconsin and Maine), hybrids are reported.

10. Thuja occidentalis   11. Juniperus communis   12. Juniperus virginiana

Liliopsida (Monocots)

TYPHACEAE   Cat-tail Family

1. Typha   Cat-tail

This is the only genus in the entire family (unless one includes *Sparganium* here). The staminate flowers are in the terminal portion of a dense cylindrical spike, the pistillate flowers below. Spending more than 10 evenings on the task, Edwin Way Teale once counted the number of seeds in an average-sized cat-tail. The total was 147,265. Another study (Yeo, 1964) estimated an average of more than 220,000 seeds per spike. If a spike has fallen in a wet area, the seeds may germinate in place, producing a striking green-colored cat-tail. In a single season, growth from one seed may produce a rhizome system 10 feet in diameter, with a hundred shoots and an equal number of buds (Yeo).

REFERENCES

Hotchkiss, Neil, & Herbert L. Dozier. 1949. Taxonomy and Distribution of N. American Cat-tails. Am. Midl. Nat. 41: 237–254.

Fassett, Norman C., & Barbara Calhoun. 1952. Introgression Between Typha latifolia and T. angustifolia. Evolution 6: 367–379.

Smith, S. Galen. 1967. Experimental and Natural Hybrids in North American Typha (Typhaceae). Am. Midl. Nat. 78: 257–287.

Yeo, R. R. 1964. Life History of Common Cattail. Weeds 12: 284–288.

KEY TO THE SPECIES

1. Staminate and pistillate portions of the spike separated; stigmas slender and elongate; pistillate flowers each accompanied by a hair-like bract with a flat spatulate tip (resembling the stigma of the next species, but smaller); hairs of pistillate flowers usually very slightly enlarged and darkened at the tip; leaf blades 3–8 mm wide; summit of leaf sheath usually prominently auricled (with rounded auricles projecting upward) . . . . 1. **T. angustifolia**
1. Staminate and pistillate portions of the spike not (or only slightly) separated; stigmas broad and flattened, usually very dark at the tip; pistillate flowers bractless and their hairs not enlarged or darkened at the tip; leaf blades 6–15 mm wide; sheaths usually tapered or truncate, not auricled at summit . . . . 2. **T. latifolia**

1. **T. angustifolia** L.   Fig. 20   Narrow-leaved Cat-tail

Map 14. Cat-tails occur in all kinds of wet or intermittently wet habitats, often in 1–2 feet or more of water. Spreading extensively by sturdy rhizomes with tuberous thickenings, much utilized as wildlife food, cat-tails may form extensive colonies in swales, ditches, marshes, and ponds. The records for *T. angustifolia* from the northern part of the Lower Peninsula are all recent, and there is evidence that this species is spreading northward along roadside ditches and similar habitats.

There is often a tendency for the deep green leaves to retain much of their color for a while after being dried. Although generally with the spikes and leaves narrower than in *T. latifolia*, this species cannot safely be distinguished on size alone. See also comments following the next species, including *T.* ×*glauca*.

2. **T. latifolia** L.   Plate 1-D; fig. 21   Common Cat-tail

Map 15. Common as an early successional species in bogs, and in the same wet situations as mentioned for the preceding species.

The rather pale green leaves of this species generally lose their color and turn brown in drying. A small separation of the staminate and pistillate portions of the spike occurs in f. *ambigua* (Sonder) Kronf., which is sometimes confused with *T. angustifolia*.

T. ×**glauca** Godron

The hybrid between the two cat-tails occurs in several places and may indeed be the dominant plant in some marshes, as along Lake Erie. Specimens which appear to represent the hybrid have been seen from Allegan, Cheboygan, Crawford, Grand Traverse, Huron, Kalamazoo, Leelanau, Monroe, St. Clair, Schoolcraft, and Washtenaw counties. In characteristics of the flowers, the hybrid is intermediate between the parents, or shares some of each. It may grow to a greater height than either, and the spikes may be unusually long (up to 40 cm in the pistillate portion of one collection); the foliage is blue-green.

13. Juniperus horizontalis

14. Typha angustifolia

15. Typha latifolia

SPARGANIACEAE   Bur-reed Family

This family is represented by only one genus in the world, and is sometimes included in the Typhaceae. Certain species normally have very elongate leaves which float on the surface of the water. Others normally have erect leaves but may produce rosettes of ribbon-like submersed leaves. These superficially resemble leaves of *Vallisneria* but are readily distinguished by their fine checkered venation pattern (fig. 3). Floating leaves are generally not distinctly keeled, while erect leaves are strongly keeled (± triangular in section).

1. Sparganium   Bur-reed

KEY TO THE SPECIES

1. Stigmas 2, separate, at end of beak; mature achenes broadly obpyramidal, truncate (flattish on end except for beak), 5–8 mm across at widest part . . . . . **1. S. eurycarpum**
1. Stigma 1, mostly on side of beak; mature achenes tapering about equally toward base and apex, about 1–3 mm across at widest part
  2. Beak about 0.5–1.2 mm long (including stigma); fruiting heads at most 10–12 mm in diameter (including beaks); staminate head (often gone by fruiting time) 1; anthers 0.3–0.6 mm long . . . . **2. S. minimum**
  2. Beak about 1.5–7 mm long; fruiting heads 12–30 mm in diameter; staminate heads (before falling) 2 or more; anthers various
    3. Mature fruiting heads about 15–20 mm in diameter, in a branched inflorescence; leaves flaccid, floating, (3) 5–9 mm broad; sepals short, reaching at most to middle of body of achene, attached near middle or base of stipe; achenes a dark reddish brown, with beaks ca. 2 (–3) mm long and rather strongly curved from their bases; stigmas and anthers ca. 0.3–0.7 mm long . . .
    . . . . **3. S. fluctuans**
    3. Mature heads 20–30 mm in diameter or if smaller the inflorescence unbranched (at most the lowest head peduncled); leaves erect or if floating usually not over 5 mm wide; sepals extending beyond middle of body of achene, usually attached at or near summit of stipe; achenes dark brownish (in nos. 4 & 5) or greenish to yellow-green, the beaks 2.5–7 mm long, usually rather straight or only slightly curved; stigmas and anthers various

16. Sparganium
   eurycarpum

17. Sparganium minimum

18. Sparganium fluctuans

4. Pistillate heads (or branches or stalks) all borne directly in the axils of the leaves or bracts, 20–30 mm in diameter at maturity
  5. Stigmas 0.8–1.5 (2) mm long; body of mature achene rather dull, dark greenish brown, slightly if at all ribbed at summit, 4–5 mm long, with beak 2.5–4 (5) mm long; anthers 0.6–1.1 mm long . . . . 4. S. americanum
  5. Stigmas 2–3.2 mm long; body of mature achene darkish but shiny, about 6 mm long, with numerous rather prominent ribs at summit, the beak 5–7 mm long; anthers 1.1–1.4 mm long . . . . 5. S. androcladum
4. Pistillate heads (at least some heads or stalks) supra-axillary (borne some distance above the axils), rarely all axillary in no. 6 (see text), 12–28 mm in diameter at maturity
  6. Leaves usually stiffly erect (rarely floating); beak of achene ca. 3–5 mm long, usually equalling body; fruiting heads mostly 20–28 mm in diameter (beaks shorter and heads as small as 12 mm in a form with short stem and crowded heads) . . . . 6. S. chlorocarpum
  6. Leaves floating (except in a rare semi-terrestrial form); beak of achene ca. 1.5–2.5 mm long, usually shorter than body at maturity; fruiting heads about 15–18 mm in diameter . . . . 7. S. angustifolium

1. S. eurycarpum Engelm.

Map 16. Probably our commonest species, as well as our most robust one, in favorable places a meter or more in height. In shallow water and wet ground at the edges of rivers, in marshes, swales, and bog pools, and elsewhere.

2. S. minimum (Hartman) Fries

Map 17. In contrast with the preceding, our smallest species and not an erect one. Nearly always with the leaves long and narrow, floating in shallow water; ponds, bogs, beach pools and interdunal swales, and other wet places. More "terrestrial" forms occur when water recedes.

3. S. fluctuans (Morong) Robinson   Plate 1-E

Map 18. Shallow water, usually 1–2 feet, of lakes and pools.

Other species are often misidentified as *S. fluctuans*, which is distinctive in its relatively wide *floating* leaves and *branched* inflorescence, the branches *axillary* and the mature achenes with a decided reddish cast. See also comments under *S. angustifolium*.

4. S. americanum Nutt.

Map 19. Wet places: pools, ditches, marshes, meadows, stream borders, bogs.

The inflorescence in this species and the next is often branched. Some immature specimens of *S. chlorocarpum* may be confused with *S. americanum* but the inflorescence of the former is almost never branched (at most the lower heads on short peduncles and one or more of these supra-axillary, as in fig. 22).

5. S. androcladum (Engelm.) Morong

Map 20. Apparently very local in marshes and moist ground.

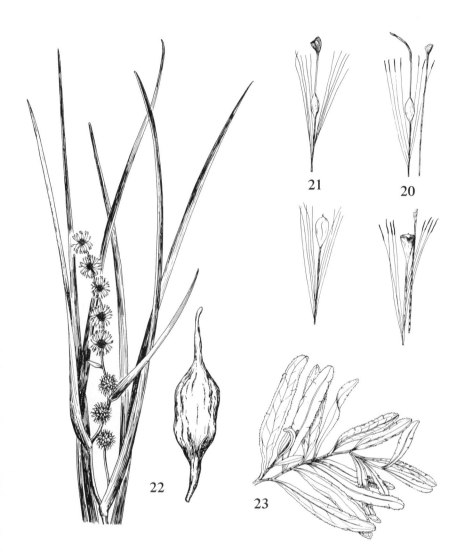

20. *Typha angustifolia,* pistillate flower (above) & flower with aborted pistil (below) ×3
21. *T. latifolia,* pistillate flower (above) & flower with aborted pistil (below) ×3
22. *Sparganium chlorocarpum* ×½ ; achene ×4
23. *Potamogeton crispus* ×½

6. S. chlorocarpum Rydb. Fig. 22

Map 21. All kinds of moist and wet places: shores, bogs, lakes (in up to 2 feet of water), beach pools, stream borders.

Next to no. 1, our most common, and certainly our most variable species, although only *very* rarely is the inflorescence branched. The stem is short (much surpassed by leaves) and the heads crowded (usually fewer and smaller) in f. *acaule* (Beeby) E. Voss. The heads are rarely all axillary in some populations, but the specimens may be placed here by the shiny, rather light green achenes with long ribbed beaks, the ribs extending onto the apical half of the achene (which is often dull and speckled with reddish on the lower half, below the constriction). The leaves of *S. chlorocarpum* often are strongly two-ranked, a helpful field character. The largest stigmas are usually 1–2 mm long.

7. S. angustifolium Michaux

Map 22. In shallow quiet water at depths up to 5 feet, rarely stranded on adjacent shores.

Normally easily recognized by the long, very narrow floating leaves (usually 2–3 mm wide, occasionally as wide as 5 mm or sheathing portions at base of plant even wider), and small supra-axillary heads of shiny, usually greenish achenes. The sepals are attached near the base of the stipe, especially in a low, semi-terrestrial form with rather curved beaks, which will run in most keys to *S. fluctuans* because of these characters. It differs from the latter species in its very narrow leaves, shiny, light brown to green, slightly smaller achenes, long sepals, supra-axillary heads, and unbranched inflorescence. The stigmas run a little shorter than in *S. chlorocarpum*, 0.8–1.5 mm long, often under 1 mm; and *S. chlorocarpum* generally has the sepals attached well to the summit of the stipe.

This species is closely related to the Old World *S. simplex* Hudson. However, specimens labeled as *S. multipedunculatum* (Morong) Rydb. (or, as long known, *S. simplex*, albeit incorrectly) have been referred to *S. chlorocarpum*. *S. multipedunculatum* — if a good species — apparently does not occur in Michigan nor do manuals include Michigan in its range.

19. Sparganium
    americanum

20. Sparganium
    androcladum

21. Sparganium
    chlorocarpum

POTAMOGETONACEAE   Pondweed Family

There has been little agreement among botanists as to the circumscription of the aquatic families usually placed in the order Najadales. Many authors would combine in one family two or more of the four genera here treated in separate families (Potamogetonaceae, Ruppiaceae, Zannichelliaceae, Najadaceae) — perhaps including one or more of them with the marine eelgrass, *Zostera*, in the Zosteraceae. Our four genera can be distinguished by the following key, which supplements the route by which these may be reached in the General Keys, Key A:

1. Leaves alternate (sometimes floating leaves present and opposite); flowers in a terminal spike or cluster (sometimes axillary spikes present), perfect; stamens 2–4
    2. Flowers several to many in a peduncled spike; perianth of 4 tepals; fruit ± sessile; leaves various; stamens 4 . . . . **Potamogeton**
    2. Flowers 2, at first enclosed in sheathing leaf base, the peduncle elongating and often spiraled or coiled at its base; perianth none; fruit long-stalked; leaves filiform (but stiff); stamens 2 . . . . **Ruppia**
1. Leaves all opposite or whorled and submersed; flowers all axillary, unisexual; stamen 1
    3. Pistils bilaterally symmetrical, ± toothed, expanded into a funnel-shaped stigma; leaves filiform, strictly entire . . . . **Zannichellia**
    3. Pistils radially symmetrical, smooth (or reticulate), with filiform style and stigma; leaves filiform or linear (dilated at base), minutely spinulose to conspicuously toothed . . . . **Najas**

1. Potamogeton   Pondweed

All species are truly aquatic, some found in deeper water than others but none terrestrial, although some species with broad leaves may survive at least temporarily on wet shores after a lowering of water levels. Species with broad leaves (whether floating or submersed) have spikes held above the surface of the water, and hence the pollen is readily wind-dispersed. Linear-leaved species mostly have spikes at or below the surface of the water, which then becomes the important dispersal medium for the pollen.

Fruiting specimens of *Potamogeton* generally give no trouble in identification. However, vegetative material is so often collected that it has become customary to expect to be able to name such specimens — even though we would scarcely expect to do so with sterile specimens of, say, *Panicum* or *Habenaria*! The present key should work fairly well with sterile material, though the effort should always be made to collect fruit.

The problem of identifying vegetative plants is made even more difficult in *Potamogeton* by the great variability associated with conditions of the water in which the plants are growing, whether deep or shallow, quiet or flowing, etc. Extreme growth forms may thus give trouble. A number of specimens of doubtful appearance, or questioned hybrids, have been checked by E. C. Ogden. His identifications based on studies of stem anatomy have been freely accepted in preparing the maps.

Several linear-leaved species (beginning with couplet 26 in the key) are characterized by having the stipule margins connate. The only sure way to determine whether these margins are connate is to moisten (if the specimen is dry) a young branch tip with a drop or two of a wetting agent (such as detergent solution), cut one or more sections about 1–1.5 mm long cleanly across it with a sharp razor blade, and examine these in water under 10x to 20x magnification. If the sections have been made where the internodes are short and the stipules overlap, at least the inner ones in those species with connate stipules (*P. foliosus, P. friesii, P. strictifolius,* and *P. pusillus*) will form cylinders or tubes which will be evident when the sections are carefully sorted out with dissecting needles. Older stipules tend to split, so this character must be determined with caution from the youngest stipules possible. The stipule margins may overlap strongly in species in which they are not connate.

In some species winter-buds (turions) are produced and, in addition to serving as a means of overwintering and vegetative propagation, they offer the botanist an added source of diagnostic characters. These structures are firm and much-shortened shoots with crowded stipules and reduced leaves.

In relatively quiet calcareous waters, there is often a considerable deposition of marl (chiefly insoluble calcium carbonate) on submersed parts of *Potamogeton* and other aquatic plants. This phenomenon results from some disturbance of the complex equilibrium between carbon dioxide, water, carbonic acid, and soluble bicarbonates. Major factors involved include utilization of bicarbonate ions and carbon dioxide by plants in photosynthesis (just as driving off of carbon dioxide in boiling hard water upsets an equilibrium and results in deposition of carbonates in a teakettle). The activities of microscopic bacteria and algae epiphytic on larger aquatic plants constitute a major source of the marl deposited. Marl encrustations often interfere with the good appearance of a herbarium specimen, but can generally be brushed off after a plant is dried. In nature these encrustations may hasten the deterioration of aquatic plants as winter approaches; in lakes with relatively little marl deposition, Potamogetons and similar plants remain in good condition later in the season. (See Robert G. Wetzel, Marl Encrustation on Hydrophytes in Several Michigan Lakes. Oikos 11: 223–236. 1960.)

REFERENCES

Fassett, Norman C. 1957. A Manual of Aquatic Plants, with Revision Appendix by Eugene C. Ogden. University of Wisconsin Press, Madison. 405 pp. [Fassett's treatment of *Potamogeton* on pp. 54–75; supplementary treatment by Ogden on pp. 364–370.]

Fernald, M. L. 1932. The Linear-leaved North American Species of Potamogeton, Section Axillares. Mem. Am. Acad. 17: 1–183 (also Mem. Gray Herb. 3).

Ogden, E. C. 1943. The Broad-leaved Species of Potamogeton of North America North of Mexico. Rhodora 45: 57–105; 119–163; 171–214 (also Contr. Gray Herb. 147).

Ogden, Eugene C. 1953. Key to the North American Species of Potamogeton. N. Y. St. Mus.

Circ. 31. 11 pp. [A key of highly unorthodox arrangement, successfully designed to be useful with fragmentary specimens.]

St. John, Harold. 1916. A Revision of the North American Species of Potamogeton of the Section Coleophylli. Rhodora 18: 121–138.

Yeo, R. R. 1965. Life History of Sago Pondweed. Weeds 13: 314–321.

KEY TO THE SPECIES

1. Leaf blades (all submersed) ± auricled at the base, (2.5) 3–8 (9) mm wide, with many (usually 20–40) nerves, strongly and rather stiffly 2-ranked, the margins ± thickened and usually minutely spinulose-serrulate (at least toward the apex – use strong lens!); sheathing portion of leaf adnate to the prominent whitish, ± fibrous stipules; inflorescence (very rarely fruiting) usually branched; nodes of lower part of stem often with slender adventitious roots . . . . 1. **P. robbinsii**

1. Leaf blades (submersed or in some species partly floating) not auricled (if clasping, then broad and without sheathing base), the width, veins, and stipules various but not combined as above; margin usually perfectly entire; inflorescence unbranched; nodes of stem very seldom with a few adventitious roots

  2. Stipules adnate to the sheathing portion of the leaves for 10–30 mm or more (occasionally some for as little as 5–6 mm); leaves all submersed, filiform to narrowly linear (up to 2 (2.5) mm wide and many timer longer than wide)

    3. Fruit (2.5) 3–4.5 mm long, not including a tiny beak (the short style persistent as a beak toward one side of end of fruit); leaves mostly sharp-pointed (the younger ones tapering slenderly to a point; older ones often more abruptly pointed) . . . . 2. **P. pectinatus**

    3. Fruit 2–3 mm long, beakless (with merely a wart-like apex, the broad stigma sessile); leaves rounded, ± blunt, or obtuse, often with a tiny notch, at the tip [use strong lens!]

      4. Sheaths (formed by leaf bases and adnate stipules) ± close around stem, less than 2 mm wide at the widest; largest leaves up to 0.5 mm wide (or to 1.2 mm in the rare var. *macounii*) . . . . 3. **P. filiformis**

      4. Sheaths ± loose, those at the base of the stem (2) 2.5–5 mm wide; largest leaves up to 1.2 (2.5) mm wide . . . . 4. **P. vaginatus**

  2. Stipules free from the leaves (or adnate for up to 4 mm in nos. 19 & 20); leaves all submersed or both floating and submersed, filiform or linear to ovate, oblong, or elliptical

    5. Margins of leaves undulating and finely but definitely and sharply toothed, especially toward the apex; fruit (very rare) with curved beak about 2 mm long; leaves all submersed, oblong, ± rounded at apex . . . . 5. **P. crispus**

    5. Margins of leaves undulating or flat but not toothed; beak of fruit shorter or absent; leaves various

      6. Submersed leaves (at least fully developed ones) with blades more than 5 mm wide

        7. Blades of principal (not the youngest) submersed leaves sessile and subcordate to clasping at base; floating leaves never developed

          8. Stipules conspicuous, persistent (sometimes fibrous), (2) 2.5–6.5 (11) cm long when fully developed; body of fruit 3.5–5 mm long, sharply keeled; tips of leaves usually boat-shaped (when pressed and flattened therefore split); rhizome white with reddish spots; leaves up to 24 cm long, often over 10 cm; peduncles (9) 18–55 (100) cm long . . . . 6. **P. praelongus**

          8. Stipules inconspicuous or soon disintegrating into fibers, less than 2 (2.3) cm long (if present); body of fruit 2.5–3 (3.5) mm long, with keel weak or absent; tips of leaves ± flat (not boat-shaped and hence not splitting when

pressed); rhizome unspotted; leaves usually not over 10 (very rarely up to 19) cm long; peduncles occasionally as much as 17 cm long but usually less than 10 cm

  9. Stipules early disintegrating into ± coarse whitish fibers (the youngest entire); leaves ovate to lanceolate, mostly 3–10 cm long and with more than 12 nerves . . . . 7. **P. richardsonii**

  9. Stipules very delicate, soon disappearing; leaves orbicular to ovate, rarely as long as 5 cm, with not over 17 nerves . . . . 8. **P. perfoliatus**

7. Blades of principal submersed leaves tapering or even petioled at base; floating leaves often developed (absent on young material or that from deep water)

  10. Submersed leaves long and ribbon-like with essentially parallel sides for most of their length; midrib of some if not all submersed leaves with a conspicuous cellular-reticulate band (i.e., appearing lighter and different in texture) on each side, totalling at least 1 mm wide and usually about a fourth or more the total width of the leaf . . . . 9. **P. epihydrus**

  10. Submersed leaves with sides ± curved most of their length; midrib with cellular-reticulate band less than 1 mm wide or if wider, still much less than a fourth the total width of leaf

    11. Submersed leaves (not transitional or floating ones) with 3–9 (11) nerves and with stipules not over 3.5 cm long

      12. Blades of floating leaves merging gradually into petioles; fully developed submersed leaves 6–13 cm long (often with conspicuous band along midrib); mature flowers usually on pedicels 0.5–1 mm long; plants usually strongly tinged with red (more evident on drying), especially in upper portions and in rachis of inflorescence . . . . 10. **P. alpinus**

      12. Blades of floating (not transitional) leaves sharply distinguished from petioles; submersed leaves often less than 6 cm long (if longer, distinguished from no. 10 by green or brownish leaves, usually with obsolete cellular-reticulate band along midrib); flowers sessile or on pedicels up to 0.5 mm long; plants green or brown, with red suffusion nearly always lacking, at least in peduncle and rachis . . . . [go to couplet 16]

    11. Submersed leaves with (7) 9–40 (52) nerves and with stipules (2) 2.5–9 (11) cm long

      13. Fully developed submersed leaves ± strongly arched (and often folded), with 24–40 (52) nerves mostly (2.5) 3.5–7.2 cm wide; fruits 4–5.5 mm long (including short beak); floating leaves with nerves about 30 or more (28–50); stipules of submersed leaves (except for the lowermost, nearest rhizome) 4–9 (11) cm long . . . . 11. **P. amplifolius**

      13. Fully developed submersed leaves flattish or at most slightly arched (undulating or crisped in no. 12), with not more than 19 nerves, with blade up to 3 (rarely 4, or very rarely 5.5) cm wide; fruits 2–4 (4.2) mm long; floating leaves with up to 27 nerves; stipules various

        14. Stem and petioles rather conspicuously black-spotted or mottled; blades of floating leaves slightly cordate at base . . . . 12. **P. pulcher**

        14. Stem and petioles not black-spotted; blades of floating leaves ± rounded to broadly tapered at base, not cordate

          15. Larger submersed leaves with blades 1–2.5 (3) cm wide, on petioles (2.5) 5–11 cm long, acute at apex but without sharp awl-like tip; fruits 3.5–4 mm long, with ± knobby or warty keel and lateral ribs; stipules of submersed leaves 4–7.5 (10.5) cm long; petioles of floating leaves up to 27 cm long, often equalling or exceeding the blades, which have 13–25 (usually 21) nerves . . . . 13. **P. nodosus**

15. Larger submersed leaves with blades various, but sessile or on petioles not over 2 cm long, acute at apex or (especially on petioled leaves) with mucronate or awl-like tip; fruits with keels rounded or sharp, but not warty; stipules and floating leaves various

   16. Submersed leaves 8–32 (50) mm wide, with (7) 9–19 nerves, the apex mucronate or with awl-like tip up to 4 (5) mm long; fruits 3–4 mm long, ± sharp-keeled; largest stipules of submersed leaves (2) 2.5–6 (7) cm long; floating leaves with petioles shorter than the blades (up to 3.5 cm long), the blades 1.7–3 (3.5) cm wide (reported as even wider) . . . . 14. **P. illinoensis**

   16. Submersed leaves up to 9 (in some vars. 14) mm wide, with (3) 5–9 (11) nerves, the apex acute or with awl-like tip up to 1.5 mm long; fruits 2–2.5 (2.8) mm long, the keel nearly always rounded or obsolete; stipules of submersed leaves 0.7–3 (4) cm long; floating leaves with petioles (up to 9 cm long) slightly shorter than to much exceeding the blades, the latter 0.8–3 cm wide . . . . . . 15. **P. gramineus**

6. Submersed leaves all less than 5 mm wide (often decayed and absent at fruiting time in nos. 16 & 17)

17. Floating leaves with well developed blades 1.2–4.7 cm wide (if rarely smaller, then narrowly elliptic and acute at both ends); submersed leaves appearing as if petioles only – rather coarse, not distinctly flattened, up to 2 (2.5) mm in diameter, often decayed and absent at flowering time; fruit 3–4.5 mm long

   18. Blades of floating leaves definitely subcordate to cordate at base, 2–4.7 cm wide and (3.2) 3.7–9 (10) cm long, at least the larger ones (18) 21–35-nerved; mature fruit (3.5) 3.7–4.5 mm long (including beak), obscurely keeled, the surface usually brownish (sometimes green) and ± irregular or puckered . . . . 16. **P. natans**

   18. Blades of floating leaves rounded to acute at base, (0.5) 1.2–2.2 (2.8) cm wide and (2.5) 3–4.7 cm long, 11–19-nerved; mature fruit 3–3.4 mm long, rather prominently keeled, the surface usually greenish and not puckered . . . . 17. **P. oakesianus**

17. Floating leaves in many species never developed – if present, their blades less than 1.2 cm broad and/or the submersed leaves with definitely flat (though narrow) blades; fruit less than 3 mm long, or larger in a few species with definitely flattened submersed leaves

   19. Leaves (all submersed) over 2 mm wide, with many (15–30 or more) fine nerves; stem strongly flattened (appearing winged) and ca. half or more the width of the leaves; fruit 3.5–5.5 mm long . . . . 18. **P. zosteriformis**

   19. Leaves with fewer nerves, of various width; stem obscurely or not at all flattened; fruit various

      20. Stipules of some or all submersed leaves adnate to the base of the leaf for up to 4 mm, the tip of the stipule projecting as a ligule; fruit suborbicular, flattened, nearly or quite beakless, the strongly coiled embryo showing through the thin wall (which appears distended by the embryo)

      21. Submersed leaves filiform, sharp-tipped, not over 0.5 mm wide (usually 0.3 mm or less), their stipules free or adnate for less than the length of the free ligule; fruit ca. 1–1.5 mm in diameter . . . . 19. **P. capillaceus**

      21. Submersed leaves broader, with blunt or rounded tip, the largest usually 0.6–1.6 mm wide, their stipules adnate for ca. 1.5–4 mm, the projecting free ligule shorter than adnate portion; fruit ca. 1.5–2 mm in diameter (including keel) . . . . 20. **P. spirillus**

20. Stipules of submersed leaves all free; fruit various but the embryo not clearly seen through thin walls of a flattened fruit, in most species not coiled more than 1 revolution; leaves various

  22. Plants of two kinds in colonies, some with small floating leaves and others with only submersed leaves; plants very delicate with submersed leaves less than 1 mm wide and 1–3-nerved

    23. Fruit maturing only on entirely submersed plants (though plants with floating leaves may produce flowers); blades of floating leaves 2–3 mm broad, ca. 2½ to 3 or more times as long, on narrowly wing-margined (flattened) petioles . . . . 21. **P. lateralis**

    23. Fruit maturing only on plants with floating leaves; blades of larger floating leaves 4.5–7.5 mm wide, ca. twice as long or even shorter, on slender petioles . . . . 22. **P. vaseyi**

  22. Plants of one kind, with only submersed leaves, these often over 1 mm wide; or floating leaves produced in the larger and coarser nos. 9 & 15, which always have submersed leaves at least 1 mm wide

    24. Floating leaves usually developed; submersed leaves at least (1) 2 mm broad, many if not all of them ± strongly 3–several-nerved; plants either with a broad cellular-reticulate band along the midrib of submersed leaves or with (usually) many bushy branches in which the internodes are mostly only 2–6 mm long [See also further descriptions in couplet 25. Narrow-leaved and often misidentified plants of species typically keying under the first half of couplet 6.]

      25. Submersed leaves ribbon-like, about 20 or more times as long as wide, distinctly parallel-sided, acute to blunt at tip, with a conspicuous cellular-reticulate band bordering the midrib and totalling a third or more the width of the leaf; plants not bushy-branched, the internodes mostly 10 mm or more in length . . . . 9. **P. epihydrus**

      25. Submersed leaves linear-lanceolate to lance-elliptic, mostly less than 15 times as long as wide, acute or often with short awl-like tip, without cellular-reticulate band; plants usually with bushy branches in which the internodes are frequently less than 6 mm long . 15. **P. gramineus**

    24. Floating leaves never developed; submersed leaves in many species less than 2 mm broad, 1–7(9)-nerved; plants with cellular-reticulate band along midrib generally narrowed or absent, without bushy branches (though winter-buds often present) unless the leaves are under 2 mm wide and/or obtuse at apex

      26. Stipules ± coarsely fibrous, whitish, the oldest tending to disintegrate into shreds; bases of winter-buds usually ± indurated and strongly ribbed

        27. Peduncles less than 1.3 (1.5) cm long, even in fruit; inflorescence few-flowered, subglobose; fruit (including short beak) ca. 3–4 mm long; stipules 6–12 (14) mm long, their margins not connate; leaves all 3-nerved, tapering to a bristle-like tip *and* the midrib bordered by a narrow cellular-reticulate band . . . . 23. **P. hillii**

        27. Peduncles 1.5–4 cm or more in length (except when very young); inflorescence definitely cylindric, the whorls of flowers becoming well separated at maturity; fruit (2) 2.2–2.8 (3) mm long; stipules (8) 10–21 mm long, their margins connate at least basally when young; leaves mostly 3–7-nerved, tapering to bristle-tip *or* the midrib bordered by cellular band (or with neither condition)

          28. Apex of leaves obtuse or rounded (usually slightly mucronate); leaves mostly 5(–7)-nerved (a few, especially the smaller, often

3-nerved), the broadest usually 1.5–2.7(3) mm wide, frequently with a very narrow cellular-reticulate band along the midrib; glands at base of leaves (on stem) usually conspicuous; peduncles ± flattened upwardly . . . . 24. **P. friesii**

28. Apex of leaves gradually tapered to a sharp, often bristle-like tip; leaves mostly 3-nerved (a few with 5 nerves often present), the broadest usually 0.8–1.3 mm wide, without any cellular-reticulate band; glands at base of leaves usually inconspicuous or apparently absent; peduncles ± terete (or flattened for only about 2 mm immediately below the inflorescence) . . . . 25. **P. strictifolius**

26. Stipules ± delicate and membranous, whitish to green or brownish, without persistent fibers on disintegration (tending to break across the delicate fibers); winter-buds, if present, not indurated or strongly ribbed at base

29. Leaves weakly 1-nerved, less than 0.3 (0.4) mm wide, very slenderly tapering to tip; peduncle solitary, 3–14 cm long (occasionally with one branch) . . . . 26. **P. confervoides**

29. Leaves wider, 1–5-nerved; peduncles up to 6 (8) cm long, often shorter, usually in axils of several leaves

30. Principal leaves 2–3.5 mm wide, usually somewhat flushed with maroon (especially when dried), very obtuse to rounded (and sometimes minutely mucronate) at apex (rarely some acute), with prominent (about 0.3–0.5 (1) mm wide) lighter-colored cellular-reticulate band bordering midrib; leaves 3-nerved, the lateral pair of nerves rather weak; glands at base of leaves mostly 0.5 mm or more across; fruit 3–4 mm long (including beak), with low but sharp keel and often rather knobby toward base . . . . 27. **P. obtusifolius**

30. Principal leaves mostly narrower, seldom flushed with maroon, obtuse or acute, 1–3(5)-nerved, with band bordering midrib generally smaller or absent; glands at base of leaves smaller or absent; fruit keeled or not, 1.8–3 mm long (except in no. 23 with very sharply acute leaves)

31. Peduncles less than 1.3 (1.5) cm long, even in fruit; fruit 3–4 mm long *or* with thin sharp keel; spike few-flowered, subglobose; glands usually absent at base of leaves; leaves acute at tip, the midrib often bordered by narrow reticulate band

32. Fruit 3–4 mm long (including short beak), without thin sharp keel; glands present or absent at base of leaves; stipules somewhat fibrous, not connate . . . . 23. **P. hillii**

32. Fruit less than 3 mm long, with a sharp often thin keel, and often knobby toward base; glands absent at base of leaves; stipules delicate, the margins connate . . . . 28. **P. foliosus**

31. Peduncles (0.8) 1.1–6 (8) cm long; fruits less than 3 mm long *and* with keel rounded or obscure, in subglobose to cylindric spikes; glands present at base of some or all leaves (usually weak, obscure, or absent in no. 29); leaves various

33. Margins of stipules connate below the middle, at least when young; largest leaves 0.8–1.2 (2) mm wide, acute; midrib lacking cellular-reticulate border . . . . 29. **P. pusillus**

33. Margins of stipules separate, often overlapping but not connate; largest leaves 0.5–2 (2.7) mm wide, obtuse or rounded to acute; midrib often with narrow cellular-reticulate border . . 30. **P. berchtoldii**

1. **P. robbinsii** Oakes   Fig. 24

Map 23. In both shallow and deep water of lakes, ponds, and rivers.

One of our most distinctive species in its many-nerved leaves, stipules adnate to the leaf bases, and usually strongly two-ranked foliage. The two-ranked appearance is less evident on plants from very deep water (15–20 feet). Occasionally flowers in Michigan; fruit is very rare throughout the range of the species. The form with perfectly entire leaves, f. *cultellatus* Fassett, occurs rarely, being known chiefly from Black Lake, Cheboygan County; Duck Lake, Calhoun County; and cited from Iron County in the original description.

2. **P. pectinatus** L.   Fig. 25   Sago Pondweed

Map 24. Widespread in lakes, ponds, creeks, and rivers, at depths to 15 feet.

The usual form of this species is rather narrow-leaved and bushy – not unlike the next species. However, especially in the waters of rivers and streams, it may become quite long-stemmed and robust, with leaves 1 mm (or even 1.7 mm) wide. Such forms superficially resemble *P. vaginatus*, from which they differ, in addition to the beaked fruit and sharp-pointed leaves, in having the largest stipular sheaths rarely over 2–2.5 mm wide (though sometimes as much as 3.5 mm).

This species is the most important pondweed for wildlife food, thanks to its abundance and copious production of both fruit and tubers. One investigator (Yeo, 1965) has found that a single plant in cultivation may in six months produce over 63,000 fruits or over 36,000 tubers.

3. **P. filiformis** Pers.

Map 25. Lakes (including the Great Lakes) and rivers, at depths up to at least 4 feet, but apparently fruiting best in shallow water over a sandy bottom.

The mature fruit of *P. filiformis* is 2.2–3 mm long, usually about 2.5 mm. The margins of the sheaths in this species are connate when young, while in the preceding and next they are not. Our plants mostly have very narrow leaves, not over 0.5 mm broad, and are var. *borealis* (Raf.) St. John. A form with wider

22. Sparganium
    angustifolium

23. Potamogeton robbinsii

24. Potamogeton
    pectinatus

leaves (0.7–1.2 mm), var. *macounii* Morong, has been found very rarely in Michigan; it is known from Roscommon and Grand Traverse counties.

Sterile specimens of *Ruppia* might be mistaken for this or the preceding species, but differ in that the open sheathing leaf bases are at most slightly auriculate at the summit (fig. 10), whereas in these species of *Potamogeton* there is a definite portion of free stipule at the summit of the sheath (fig. 9).

4. **P. vaginatus** Turcz.

Map 26. In rivers and streams, especially cold, spring-fed waters.

Mature fruiting material of this species has not been collected in the state, and there is some question about its proper status. The fruit is about 3 mm long where found. The prominent sheaths at the base of the plant may be as long as 35 mm.

5. **P. crispus** L.   Fig. 23

Map 27. Shallow to deep water of lakes, rivers, ditches, etc., including the Great Lakes. Seems to thrive in polluted waters. Generally assumed to be an introduction from Europe.

Although this species fruits very rarely, it is frequently found with prominent winter-buds (branchlets with variously modified leaves and shortened, hardened internodes), which apparently serve in propagation. Fruiting collections in Michigan have come from Little Traverse Bay of Lake Michigan, Emmet County, in 1946; Lake Erie marshes, in 1958 (see Hunt & Lutz in Jour. Wildlife Management 23: 405–408. 1959); Spring Lake, Ottawa County, in 1957; and the Au Sable River, Otsego County, in 1964.

6. **P. praelongus** Wulfen

Map 28. Usually in lakes (including the Great Lakes), often in deep water (to 22 feet), less frequently in rivers.

The leaves of this species are often merely subcordate and sessile in contrast to those of the next two, which are strongly clasping. See also comments under *P. amplifolius*.

25. Potamogeton filiformis

26. Potamogeton vaginatus

27. Potamogeton crispus

7. **P. richardsonii** (Benn.) Rydb.   Figs. 26, 27

Map 29. Known from Michigan waters of Lakes Michigan, Superior, Huron, St. Clair, and Erie, the St. Mary's, St. Clair, and Detroit rivers, and numerous inland lakes, rivers, and creeks, at depths up to 15 feet.

Even when the stipules are almost entirely gone, remnants of the fibers, giving a bearded aspect to the nodes, may usually be found (fig. 27). This species, abundant and widespread, is intermediate between the preceding and the next in several ways, and a hybrid origin has been suggested. Normally the three are well marked, but some specimens are difficult to place, especially if they lack fruit. The leaves of *P. richardsonii* are generally more coarsely textured and nerved than in the next, but occasional specimens rather closely approach *P. perfoliatus* – of which this is sometimes treated as var. *richardsonii* Bennett.

8. **P. perfoliatus** L.

Map 30. Evidently very local, in lakes, creeks, and rivers.

In longitudinal sections of the fruit (cut parallel to the broad sides), a conspicuous projection inward from the wall may be seen. This is solid in *P. perfoliatus*, but contains a cavity in the preceding two species. (However, no Michigan specimens of *P. perfoliatus* have been collected in fruit, and the map is based on sterile collections which are apparently this species, mostly determined tentatively by Ogden.) Our specimens are var. *bupleuroides* (Fern.) Farw.

9. **P. epihydrus** Raf.   Plate 1-F

Map 31. Lakes, boggy ponds, creeks, rivers, in shallow to moderately deep water (seldom over 6 feet).

Plants with narrower submersed leaves, fewer (3–7) nerves, and smaller fruits are not sharply distinguished from the typical variety, but may be called var. *ramosus* (Peck) House [var. *nuttallii* (C. & S.) Fern.]. The broad cellular-reticulate stripe, characteristic of this species, may be slightly under 1 mm wide in the narrowest leaves, but is still more than 1/3 the total width of these leaves.

28. Potamogeton
    praelongus

29. Potamogeton
    richardsonii

30. Potamogeton
    perfoliatus

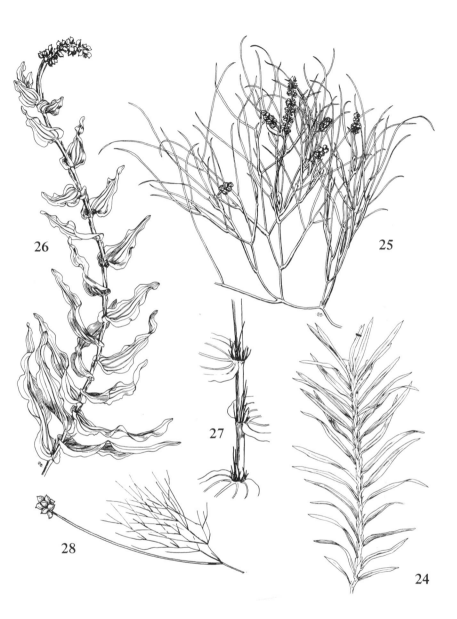

24. *Potamogeton robbinsii* ×⅖
25. *P. pectinatus* ×⅔
26. *P. richardsonii* ×⅖
27. *P. richardsonii*, leaf bases and stipular fibers ×1⅕
28. *P. confervoides* ×¾

10. **P. alpinus** Balbis

Map 32. Shallow to deep, but usually cold, water, such as spring-fed lakes and streams.

In the absence of floating leaves or fruit, this species might be confused with *P. illinoensis*, but it can be distinguished by blunter stipules and leaves and by the reticulate zone along the midrib (giving the appearance of a broad midrib), as well as by the strong reddish color which develops on drying.

11. **P. amplifolius** Tuckerman

Map 33. In lakes and larger rivers, usually at depths under 9 feet but reported in water twice that deep.

Some specimens of *P. praelongus*, especially from deep water (5–10 feet or more), may have only rounded or even somewhat tapered leaf bases, and hence could run here in the key. They may, however, generally be distinguished from *P. amplifolius* by their very strongly *whitish* stipules and at most slightly tapered subsessile leaf bases. In *P. amplifolius* the stipules, while also prominent, are strongly tinged with brownish or greenish and, particularly in deep water, the leaves are strongly narrowed at the base and/or definitely petiolate. The leaves of *P. praelongus* tend to have fewer nerves than those of *P. amplifolius* and are, of course, never floating. If fruit is present on doubtful specimens, that of *P. amplifolius* lacks the cavity mentioned above under *P. perfoliatus* and which occurs in *P. praelongus* and *P. richardsonii*.

12. **P. pulcher** Tuckerman

Map 34. Known in Michigan only from 9 feet of water over a fibrous peat bottom in Copley Lake, Cass County (*Inst. Fish. Res.* in 1938, MICH); and from ca. 2 feet of water in East Lake [one of "Twin Lakes"] north of Carp Lake, Cheboygan County (*S. G. Smith 71–121* in 1971, UMBS).

13. **P. nodosus** Poiret

Map 35. Especially in rivers, but also lakes, at depths to 6 feet.

31. Potamogeton epihydrus 32. Potamogeton alpinus 33. Potamogeton amplifolius

14. **P. illinoensis** Morong

Map 36. Lakes and rivers, in water up to 15 feet deep.

Apparent hybrids have been collected as follows: *P. illinoensis* × *P. nodosus* in Genesee, Jackson, Livingston, and Washtenaw counties; *P. illinoensis* × *P. richardsonii* in Benzie and Washtenaw counties. See also discussion under the next species.

15. **P. gramineus** L.

Map 37. Usually in ponds and lakes, including the Great Lakes, in deep to shallow water; sometimes plants with "floating" leaves stranded on a wet shore.

An extremely variable species. The distinction between *P. illinoensis* and larger-leaved plants of *P. gramineus* sometimes seems quite arbitrary, especially in the absence of fruit and floating leaves. Ordinarily, the smaller stipules and more slender submersed leaves of *P. gramineus* will distinguish it.

Several intergrading varieties have been named. Hybrids between this species and the preceding have been identified by Ogden from Lake St. Clair and the following counties (and doubtless occur elsewhere): Barry, Grand Traverse, Hillsdale, Houghton, Iron, Kalamazoo, Kent, Keweenaw, Lake, Mecosta, Newaygo, Oakland, St. Clair, Washtenaw, and Wayne. Apparent hybrids with *P. nodosus* are known from Cass County, and with *P. richardsonii* from Benzie, Chippewa, and Wayne counties.

16. **P. natans** L.

Map 38. Lakes and rivers; also ditches, ponds, bogs; usually in fairly shallow water (to 5 feet) but recorded at depths to 12 feet.

This species almost never fails to produce floating leaves, and can hardly be confused with any other (but see comments under the next species). Upper portions (lacking submersed leaves) of such species as *P. nodosus* and *P. amplifolius* with large floating leaf blades may nearly always be distinguished by their rounded or tapered bases. Furthermore, in *P. natans* the petiole generally has a rather sharp bend and a band of different (lighter) color at the junction with the cordate blade. *P. pulcher* also has cordate floating leaves, but the fruit is

34. Potamogeton pulcher   35. Potamogeton nodosus   36. Potamogeton illinoensis

prominently keeled and the broad submersed leaves are generally evident. Very rarely in flowing water a form of *P. natans* will occur with the floating leaf blades tapered at both ends, resembling those of *P. nodosus*; however, the petiole-like (phyllodial) submersed leaves and light band at summit of the petiole of floating leaves will distinguish *P. natans* (see also Thieret in Mich. Bot. 10: 121–124. 1971).

17. P. oakesianus Robbins

Map 39. Local in bogs and interdunal boggy pools, lakes, and streams.

A more delicate species in every way than the preceding. The stems of *P. oakesianus* (at least when dry) are 0.4–0.9 mm thick, while those of *P. natans* are usually 1 mm or more (0.7–2.5 mm). The submersed leaves and petioles of the floating leaves of *P. oakesianus* are likewise more delicate on the average than those of *P. natans*. The petioles in *P. oakesianus* seem rarely if ever to develop the distinct whitish band at the summit usually characteristic of *P. natans*. A form is occasionally encountered in shallow boggy pools in the Upper Peninsula and northern Lower Peninsula with floating leaves at the smallest limits of those given in the key, the blades acute at both ends. This form tends to resemble certain other small species, but the very slender and long submersed leaves (80–400 times as long as broad) with absolutely no differentiation into blade and petiole tend to be distinctive.

18. P. zosteriformis Fern.

Map 40. Shallow to deep (22 feet) water of lakes, rivers, creeks, and wet swales, including the Great Lakes system.

One of our most common and distinctive species, with its flat stem and many-nerved parallel-sided leaves mostly (2) 2.5–4.5 mm wide and acute to mucronate at the apex; unlike *P. robbinsii*, the stipules are not at all adnate to the leaves. See also discussion of *P. × longiligulatus*, following *P. strictifolius*.

When Fernald decided that the American plant was a different species from its European relative, *P. zosterifolius* Schumach., he designated as the type of *P. zosteriformis* a specimen from Alma [Gratiot County], Michigan.

37. Potamogeton
gramineus

38. Potamogeton natans

39. Potamogeton
oakesianus

19. **P. capillaceus** Poiret

Map 41. Shallow water of several small lakes in Allegan and Newaygo counties. A Coastal Plain species, very local inland.

In our plants, the peduncled dense cylindric spikes arising from axils of floating leaves are most conspicuous, though small ± globose spikes on slightly longer recurved peduncles also occur in the axils of submersed leaves of one collection. The floating leaves are very conspicuously opposite. After study of plants from the Southeast, Klekowski and Beal (Brittonia 17: 175–181. 1965) considered *P. capillaceus* not to be distinct from *P. diversifolius* Raf.; our plants seem different enough that they are here maintained under the traditional name, but the type of *P. capillaceus* has not been located to determine its correct application.

20. **P. spirillus** Tuckerman

Map 42. Usually in shallow water: lakes, ponds, wet swales, and rarely quiet river borders.

The submersed leaves are often curved, giving the whole bushy plant the aspect of a broad-leaved *Najas*. The cellular-reticulate border of the midrib may occupy most of the area of the leaf. In our plants, the few-flowered subglobose spikes are sessile or nearly so in the axils of submersed leaves, with the inflorescences in axils of floating leaves poorly or not at all developed.

21. **P. lateralis** Morong

Map 43. Known in Michigan only from "Bear Lake," Van Buren County (*E. J. Hill 54* in 1872, ILL). (See Mich. Bot. 6: 13–14. 1967.)

22. **P. vaseyi** Robbins

Map 44. Michigan collections seen have all come from fairly small lakes at depths up to 15 feet, except for specimens collected in the early 1900's in "big ditches" near Algonac, St. Clair County.

40. Potamogeton
    zosteriformis

41. Potamogeton
    capillaceus

42. Potamogeton spirillus

Sterile plants without floating leaves might resemble *P. berchtoldii*, but are generally more delicate, the winter-buds slender, elongate, long-attenuate and pointed.

23. **P. hillii** Morong

Map 45. Shallow water of small lakes, ponds, ditches, and streams.

A very local but distinctive species, of uncertain relationships. The stipules are somewhat less fibrous than is often the case in *P. friesii* and *P. strictifolius*, and the winter-buds are quite different, being similar in texture to large ones of *P. berchtoldii*, for example, and are borne at the ends of branches. The peduncle is short and the spike subglobose, as in *P. foliosus*, but there is no sharp keel on the much larger fruit. The leaves are very sharp, even bristle-like, at the apex.

The type specimens were collected by E. J. Hill at Manistee in 1880, but the pools from which they came were drained a few years later (see Mich. Bot. 4: 13–15. 1965).

24. **P. friesii** Rupr.

Map 46. Both quiet and flowing water of rivers, streams, ponds, and lakes, at depths recorded to 20 feet.

See comments under the next species.

43. Potamogeton lateralis

44. Potamogeton vaseyi

45. Potamogeton hillii

46. Potamogeton friesii

47. Potamogeton strictifolius

48. Potamogeton confervoides

25. P. strictifolius Bennett

Map 47. Lakes and streams, at depths up to 16 feet or more.

This species as here treated is largely what has been called var. *rutiloides* Fern. Type material of typical *strictifolius* (from Indiana) has the apices of the leaves as acute as much of what has supposedly been the var. *rutiloides*. Most of what has recently been called "typical" *P. strictifolius* (of which there have been very few Michigan specimens) with a rounded or obtuse apex on the leaves is apparently merely *P. friesii*. *P. strictifolius* has the broadest leaves rather more often 5-nerved than most keys and manuals imply. It tends to have more rigid, revolute, deeper (almost olive) green leaves than does *P. friesii*, but no clear distinction can be made on these characters. Leaves of plants otherwise clearly *P. friesii* may have revolute margins, although they are typically flat. The two species as here treated are usually easily separable although there is some tendency to intergrade.

P. × longiligulatus Fern. appears, from field observations in Michigan, to result from hybridization of *P. strictifolius* (probably) with *P. zosteriformis*. In the Black River and Black Lake, Cheboygan County, it grows with these putative parents, and a few plants intergrade with them. The firmly sharp-tipped leaves, very fine nerves in addition to the main nerves, somewhat flattened stem, and coarse winter-buds from which the plants arise, all suggest that *P. longiligulatus* is related to *P. zosteriformis* (see Mich. Bot. 4: 11–13. 1965). *P. longiligulatus* was originally described as having the young stipules not connate, and the leaves with 5–9 veins and a sharp-pointed apex. We have a few specimens with 5–7 veins and often slightly wider leaves than in *P. strictifolius*, which could be *P. longiligulatus* except that the young stipules are connate. These plants may be coarse *P. strictifolius* with more veins than usual (e. g., specimens from Beaver Island and from Barry, Cheboygan, Roscommon, and Schoolcraft counties).

Specimens of *P. × longiligulatus*, with non-connate stipules, have been seen from Sarnia Bay, Ontario, and from several counties in Michigan: Beaver Island (Charlevoix County), Cheboygan, St. Clair, Schoolcraft, Van Buren, and Washtenaw. Apparently it is to be expected in marly or hard-water lakes and rivers with the putative parents. Mature fruit is unknown in Michigan, as elsewhere.

26. P. confervoides Reichenb.   Fig. 28

Map 48. Local in acid or soft-water bogs and lakes.

A very distinctive species with fan-shaped branches of extremely delicate leaves and a long solitary peduncle.

27. P. obtusifolius Mert. & Koch

Map 49. Quite local, in lakes, ponds, and streams — frequently in bog waters.

Some sterile collections from southern Michigan (Livingston, Van Buren, and Washtenaw counties) resemble this species and have been referred to it in the past; they differ in such characters as their fewer-nerved or browner stipules and probably represent coarse forms of *P. berchtoldii* (R. R. Haynes, pers. comm.).

28. **P. foliosus** Raf.   Fig. 29

Map 50. Shallow to deep water (recorded to 12 feet) of lakes, ponds, rivers, and streams; sometimes rooted in organic mud of bog lakes.

Bushy plants with narrow leaves (up to 1.5 mm wide) and no (or almost no) cellular-reticulate border along the midrib are commonest, and have been called var. *macellus* Fern. Typical var. *foliosus* has somewhat coarser leaves and stipules. It is more frequent in flowing water.

Sterile specimens of var. *macellus* can often not be distinguished from *P. pusillus*. Both have delicate, connate stipules, but *P. pusillus* can ordinarily be distinguished easily by the longer peduncles and absence of a distinct sharp, narrow keel on the fruit. *P. pusillus* can supposedly be identified vegetatively by the presence of glands on the stem at the base of the leaves, but in our specimens (including good fruiting ones), such glands are usually absent, weak, or obscure.

29. **P. pusillus** L.

Map 51. Usually in lakes and ponds, at depths to 7 feet; occasionally in rivers.

The connate stipule margins are the only certain means of distinguishing this species from some individuals of the next, and sterile specimens may be indistinguishable from *P. foliosus* (see above).

This species was formerly widely known as *P. panormitanus* Biv. Opinion has differed as to the correct type of *P. pusillus*, some botanists holding that this name should be restored to what is here called *P. berchtoldii*. The confusion that would result from shifting the application of "*P. pusillus*" once again might suggest the advisability of rejecting the name and returning to *P. panormitanus* for the present species, leaving the next as *P. berchtoldii*. Some authorities believe that the two species should not be maintained, in which case *P. pusillus* is considered as having either connate or non-connate stipules.

49. Potamogeton obtusifolius

50. Potamogeton foliosus

51. Potamogeton pusillus

30. **P. berchtoldii** Fieber

Map 52. In lakes, bogs, streams, and rivers, including bays of the Great Lakes; often abundant, usually in shallow water (up to 4 feet, sometimes as much as 15 feet deep).

The plants of this species fall into two series: those with the leaf tips rounded or obtuse (though often mucronate) and those with acute to sharp-pointed leaves. The former include var. *polyphyllus* (Morong) Fern. (mature leaves 8–25 mm long) and the typical var. *berchtoldii* (with longer leaves). The acute-leaved series has been divided into three varieties, based largely on width of the leaf and of its cellular-reticulate band; these are distinguished only with great difficulty, except for the extreme [var. *lacunatus* (Hagström) Fern.] with the broadest band along the midrib.

In the wider leaves, in which three veins are evident the full length, the lateral veins tend to join the midrib at right angles about 0.3–1 (2) mm below the tip of the leaf; in the wider leaves of *P. pusillus*, the lateral veins tend to join the midrib at a sharper angle. In our specimens, as fundamentally distinguished by their stipules, *P. berchtoldii* almost invariably has glands at the base of the leaves (see comments under *P. foliosus* above). The wider, obtuse-leaved plants of *P. berchtoldii*, with conspicuous band along the midrib, are easily distinguished from *P. pusillus*. The smaller plants, with acute, narrow, 1–3-veined leaves and with very little or no border along the midrib must be identified by the nature of the stipules.

RUPPIACEAE   Ditch-grass Family

1. Ruppia

1. **R. maritima** L.   Figs. 10, 30   Ditch-grass

Map 53. Known in Michigan only from Manistique Lake, where collected in up to 8 feet of water and often abundant washed on shore after a storm (see Mich. Bot. 4: 15–16. 1965).

52. Potamogeton berchtoldii

53. Ruppia maritima

54. Zannichellia palustris

Our plants have the sheaths at the bases of the leaves about 14–52 mm long, and therefore are var. *occidentalis* (S. Watson) Graebner (*R. occidentalis* Watson – doubtfully distinct). Because of the spiraled or coiled peduncle, some authors would separate these plants as *R. spiralis* Dum. Normally, four separate fruits, each on a long stalk, develop from each of the two flowers; flowering material may be abundant in Manistique Lake, but fruiting specimens have not yet been seen.

ZANNICHELLIACEAE   Horned Pondweed Family

1. Zannichellia

1. Z. palustris L.   Plate 2-A; figs. 31, 32   Horned Pondweed
Map 54. A local and easily overlooked species in a diversity of ponds and streams, muddy lake bottoms, the Detroit River, and even Lake Erie and Lake Michigan (Green, Little Traverse, and Grand Traverse bays, and St. James harbor of Beaver Island); often largely buried in mud or silt.

Pollination occurs entirely under water. The peculiar peltate, funnel-shaped stigma is well shown in fig. 31. The fruits have bodies about 2.5 mm long, slightly flattened and curved, ± toothed on the convex margin (fig. 32). The plant grown from a single seed planted May 1 in Montana yielded a total of over 2 million seeds in five months (Yeo, Weeds 14: 113. 1966), giving some idea of the great productive capacity of this rather insignificant-looking plant.

NAJADACEAE   Naiad Family

1. Najas   Naiad
Unlike most aquatics, the species of this genus are all annual, and spread by seed, which is often produced abundantly. Pollination occurs entirely under water.

55. Najas marina

56. Najas minor

57. Najas gracillima

REFERENCE

Clausen, R. T. 1936. Studies in the Genus Najas in the Northern United States. Rhodora 38: 333–345.

KEY TO THE SPECIES

1. Leaves appearing conspicuously toothed to the naked eye, the broad-based teeth as much as 0.5–1 mm long; fruits very plump, 4–5 mm long at maturity . . . 1. **N. marina**
1. Leaves entire or very minutely toothed or spinulose when seen under a lens; fruits more slender, 2–3.5 (4) mm long
  2. Leaves with auriculate or broadly truncate basal lobes (these toothed at summit), the setaceous blade 0.2–0.5 mm wide
    3. Teeth of leaf margin visible with lens – sometimes to naked eye; fruits about 2–2.5 mm long, with longitudinal striations more prominent than the transverse ones, giving a ladder-like appearance to the reticulations; lobes at base of leaf ± truncate and herbaceous . . . . 2. **N. minor**
    3. Teeth of leaf margin merely microscopic spinules; fruits 2.5–3.5 mm long, evenly reticulate (the longitudinal striations not more prominent); lobes at base of leaf often definitely auriculate and scarious . . . . 3. **N. gracillima**
  2. Leaves expanded at the base, but expanded portion tapering to the slender blade (not auriculate or broadly truncate), minutely spinulose; blades 0.3–2.2 mm wide
    4. Styles (including stigmas) 1–2.5 mm long; fruit (2) 2.5–3.5 (4) mm long, smooth and glossy, with very fine and rather obscure reticulate pattern; leaves, at least most of the well developed ones, very slender (0.3–1 mm wide at the middle, 20 or more times as long – sometimes up to ca. 3 cm long), tapering to apex . . . . 4. **N. flexilis**
    4. Styles (including stigmas) ca. 0.3–0.6 mm long; fruit ca. 2 mm long with a rather conspicuously reticulate and shallowly pitted surface; leaves 0.6–2.2 mm wide at middle, nearly always 7–15 (20) times as long (up to 1.8 cm long), thus appearing more abruptly acute at tip . . . . 5. **N. guadalupensis**

1. N. marina L.

Map 55. First collected in Michigan in 1938 in Ogemaw County, and appears to be somewhat aggressive. Local in lakes, especially in Newaygo County, at depths to 4 feet, and in the marshes of Lake Erie.

In one collection from 2–3 feet of water in Peach Lake, Ogemaw County, the leaves are as long as 12 cm, and in some cases almost entire, with normal shorter toothed leaves also occurring.

2. N. minor All.

Map 56. A native of the Old World, becoming aggressive in some parts of the United States (Wentz & Stuckey, Ohio Jour. Sci. 71: 298–300. 1971), but thus far collected in Michigan only in marshes of the Pte. Mouillée State Game Area, Monroe County (*M. McDonald 5318* & *5783* in 1949 & 1950, MICH, MSC).

3. N. gracillima (A. Br.) Magnus

Map 57. Very local, generally in shallow water of mucky-bottomed lakes. First collected in Michigan in 1936.

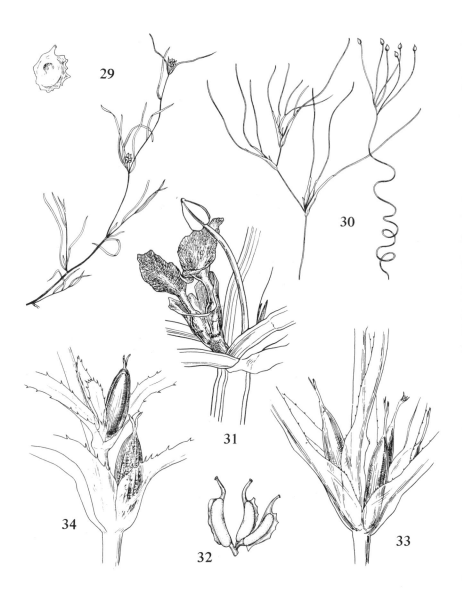

29. *Potamogeton foliosus* ×½ ; fruit ×5
30. *Ruppia maritima* ×½
31. *Zannichellia palustris*, staminate flower and spathe with 2 pistillate flowers in leaf axil ×16
32. *Z. palustris*, mature fruit ×5
33. *Najas flexilis*, pistillate flowers in leaf axils ×8
34. *N. guadalupensis*, pistillate flowers and young fruit in leaf axils ×8

4. **N. flexilis** (Willd.) Rostk. & Schmidt   Plate 2-B; fig. 33

Map 58. By far our commonest species, probably occurring in almost every pond and lake in the state, as well as many rivers and streams. It occurs in the Great Lakes system as well as inland and in bogs (on false bottoms). Collections are reported at depths to 16 feet, but the species has been found at 27 feet (Bromley, ms.). A tender annual that fruits abundantly, the plant is an important duck food.

This is an exceedingly variable species, the differences between individuals probably due in large part to ecological factors. The short, much-branched, condensed and bushy extreme may be called var. *congesta* Farw. (TL: Detroit River at Belle Isle [Wayne County]). At the other extreme, plants with stout stems, long internodes, and wider leaves have been called var. *robusta* Morong; however, most of our plants previously identified as this variety appear more properly to be the next species.

5. **N. guadalupensis** (Sprengel) Magnus   Fig. 34

Map 59. Usually in lakes at depths to at least 12 feet, less commonly in rivers.

Specimens from Barry, Livingston, Monroe, St. Joseph, Schoolcraft, and Van Buren counties have young or mature fruit and are clearly this species. The others on which the map is based are sterile, but are probably also *N. guadalupensis*, although some of them have been called *N. olivacea* Rosend. & Butt. These include several specimens annotated as the latter by Clausen shortly after publishing his paper on *Najas*. However, in Kalamazoo County Mr. and Mrs. Hanes collected a *Najas* which in many years of searching they were never able to find in fruit; this was considered by Clausen to be *N. olivacea*, but specimens were sent by the Haneses to Rosendahl who claimed that it was not his species because the leaves were wider, longer, thinner, and much less firm and stiff. Rosendahl referred these specimens to *N. guadalupensis*. The fruiting collection of *N. guadalupensis* from Fine Lake, Barry County, is an excellent match vegetatively for most of what has passed as *N. olivacea* in Michigan. A specimen with immature fruit, referred to *N. guadalupensis* by the short style, from the north end of Whitmore Lake, Livingston County, helps to suggest that similar

58. Najas flexilis

59. Najas guadalupensis

60. Scheuchzeria palustris

sterile specimens from this lake previously labeled as *N. olivacea* are not that species. Until fruiting specimens of *N. olivacea* (which is usually sterile) are found in the state, it seems best to consider our wide-leaved specimens of *Najas* as *N. guadalupensis. N. olivacea* has a long style, like *N. flexilis*, but the fruit is dull, intermediate in size between *N. flexilis* and *N. guadalupensis*.

JUNCAGINACEAE   Arrow-grass Family

KEY TO THE GENERA

1. Pedicels bracted; carpels 3, widely divergent in fruit; leaves mostly cauline, each with a terminal pore . . . . 1. **Scheuchzeria**
1. Pedicels bractless; carpels 3 or 6, erect and ± adherent to a central axis at maturity; leaves all basal or nearly so, without a terminal pore . . . . 2. **Triglochin**

1. Scheuchzeria

Sometimes segregated into a separate family, Scheuchzeriaceae.

1. S. palustris L.   Fig. 35

Map 60. Almost entirely restricted to bogs, where it tends to thrive in wetter sphagnum areas, although sporadic in abundance from year to year.

The distinctive pore at the tip of each leaf makes this an easy bog plant to recognize in vegetative condition. American plants have been segregated from the Eurasian ones as var. *americana* Fern.

2. Triglochin   Arrow-grass

REFERENCE

Löve, Askell, & Doris Löve. 1958. Biosystematics of Triglochin maritimum agg. Nat. Canad. 85: 156–165.

KEY TO THE SPECIES

1. Carpels and stigmas 6; groups of mature fruits broadly rounded at base, about 2 (−3) times as long as broad (or even shorter), the central axis not winged . . . . .
. . . . 1. **T. maritimum**
1. Carpels and stigmas 3; groups of mature fruits narrowly tapered to base (the individual follicles very sharp-pointed basally), about 5–7 times as long as broad, the central axis winged . . . . 2. **T. palustre**

1. T. maritimum L.   Plate 2-C; fig. 36

Map 61. Gravelly and marly shores, wet sandy beaches and interdunal swales, marshes and bogs, stream borders.

Löve and Löve have urged recognition of the 24-ploid eastern American plants of this circumpolar complex as a species distinct from the octoploid Eurasian *T. maritimum*; it would then be called *T. elatum* Nutt.

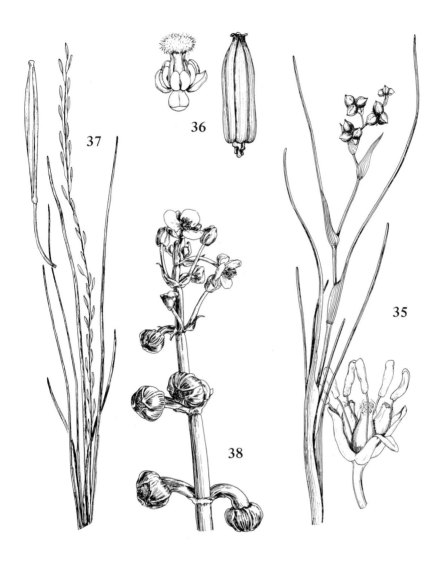

35. *Scheuchzeria palustris* ×½ ; flower ×4
36. *Triglochin maritimum,* flower ×4; fruits ×5
37. *T. palustre* ×½ ; fruits ×5
38. *Sagittaria montevidensis* ×⅔

2. T. palustre L.   Fig. 37

Map 62. In similar places to the preceding, although seldom in bogs (except very marly sites) and usually on wet sandy shores and marshy flats; also at salt springs.

In all aspects, usually a smaller and more slender plant than *T. maritimum*.

ALISMATACEAE   Water-plantain Family

REFERENCE

Beal, Ernest O. 1960. The Alismataceae of the Carolinas. Jour. Elisha Mitchell Sci. Soc. 76: 68–79.

KEY TO THE GENERA

1. Pistils in a dense globose head; uppermost flowers mostly staminate, with the stamens numerous; leaves unlobed or sagittate . . . . 1. **Sagittaria**
1. Pistils in a single ring on a flattish receptacle or few in a tiny loose head; flowers all perfect, with 6 (–9) stamens; leaves never lobed
   2. Plants less than 10 cm tall; achenes ca. 1 mm long, plump . . . . 2. **Echinodorus**
   2. Plants nearly always larger; achenes larger, flattened . . . . 3. **Alisma**

61. Triglochin maritimum

62. Triglochin palustre

63. Sagittaria montevidensis

64. Sagittaria rigida

65. Sagittaria graminea

66. Sagittaria latifolia

1. Sagittaria   Arrowhead

REFERENCE

Bogin, Clifford. 1955. Revision of the Genus Sagittaria (Alismataceae). Mem. N. Y. Bot. Gard. 9: 179–233.

KEY TO THE SPECIES

1. Sepals closely appressed; pedicels thick (the lower ca. 3–5 mm in diameter); leaves very broadly sagittate, much broader than long; lowermost flowers mostly perfect
. . . . 1. **S. montevidensis**
1. Sepals of mature pistillate flowers reflexed; pedicels more slender; leaves various; lowermost flowers ordinarily pistillate
  2. Filaments pubescent with ± flattened or scale-like hairs; leaves not sagittate (or small basal lobes occasionally present in *S. rigida*)
    3. Flowers sessile or nearly so; axis of the inflorescence strongly bent above the lowest flowers; achenes with beak ca. 1–1.5 mm long . . . . 2. **S. rigida**
    3. Flowers pedicelled; axis of inflorescence straight; achenes with beak scarcely over 0.5 mm long . . . . 3. **S. graminea**
  2. Filaments glabrous; leaves usually sagittate when emersed
    4. Mature achenes with a ± horizontal beak projecting to one side and attaining 0.5–1.5 (2.3) mm in length; faces of achene not winged, at most with a visible resin duct; bracts 4–9 mm long (or up to 11 mm in some robust plants) . . . .
. . . . 4. **S. latifolia**
    4. Mature achenes with an erect or curved apical beak up to ca. 0.7 mm long; faces of achene sometimes with a low wing or ridge; bracts usually (8) 10–16 mm long
      5. Achenes with a very tiny erect beak scarcely 0.2–0.3 mm long, terminating a swollen or convex margin of the achene and thus appearing to project from summit of the achene, definitely set in from the margin . . . . 5. **S. cuneata**
      5. Achenes with beak ca. 0.4–0.7 mm long, appearing as a continuation of the ± straight margin of the achene, straight or curves outward or over apex of achene . . . . 6. **S. brevirostra**

1. **S. montevidensis** Cham. & Schlect.   Fig. 38
    Map 63. This species is known in Michigan only from mud flats and banks in Monroe County. Earlier reports from "Grand Rapids" have been erroneously attributed to the state (see Mich. Bot. 6: 14–15. 1967).
    The North American representatives of this typically South American species have long been segregated as *Lophotocarpus calycinus* (Engelm.) J. G. Sm. or *Sagittaria calycina* Engelm. The filaments are pubescent.

2. **S. rigida** Michaux
    Map 64. Shallow water, edges of rivers and lakes, in marshes and swales, etc.; rather local.

3. **S. graminea** Michaux   Fig. 39
    Map 65. Shallow water and sandy-mucky exposed shores.

Most Michigan specimens are var. *cristata* (Engelm.) Bogin, with the anthers shorter than the filaments. The anthers about equal the filaments or are longer in var. *graminea*. Frequently produces rosettes of stiff underwater (juvenile?) leaves suggestive of knife blades in shape (not the elongate limp ribbon-like ones of *S. cuneata*).

4. **S. latifolia** Willd.   Figs. 40, 41   Wapato; Duck-potato
    Map 66. Wet and moist places generally: shallow water and shores of lakes, ponds, ditches, streams, rivers, swamps, marshes, and bogs.
    Our most common and variable species. Although mature or almost-mature achenes are needed for positive identification, some immature specimens can be placed by a combination of short bracts and very narrow leaf lobes. The terminal lobe of the leaf may exceed 10 cm with a width often over 2 cm (see comments under next species). Apparently does not produce floating leaves and usually does not have distinctive juvenile submersed leaves as in our other common species, *S. graminea* and *S. cuneata*.

5. **S. cuneata** Sheldon   Fig. 42
    Map 67. Creeks, rivers, ditches, lakes, and streams; usually in shallow water or on wet shores. In deep water only the ribbon-like phyllodial leaves may be produced.
    A variable species, but evidently not exhibiting the very narrowly linear leaf lobes often found in *S. latifolia*. The terminal lobe is usually under 10 cm in length; but if longer (not necessarily if shorter), the width does not exceed 2 cm. Plants bearing leaves having the three lobes about equal have been called f. *equiloba* Fern. (TL: Isle Royale). Plants of this species (and those referred to the next) often produce floating leaves on the surface of the water and, when young and submersed, long ribbon-like phyllodial leaves. These somewhat resemble leaves of *Vallisneria* or submersed forms of *Sparganium*, but differ in venation (cf. figs. 2, 3, 4) and in not being so definitely parallel-sided. The floating leaves are arrow-shaped (or some only elliptical); the other species in the genus apparently do not produce floating leaves.

67. Sagittaria cuneata

68. Sagittaria brevirostra

69. Echinodorus tenellus

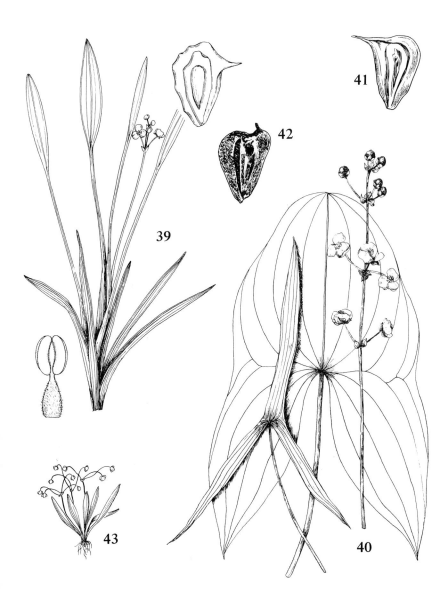

39. *Sagittaria graminea* ×½ ; stamen ×10; achene ×10
40. *S. latifolia* ×½
41. *S. latifolia,* achene ×6
42. *S. cuneata,* achene ×6
43. *Echinodorus tenellus* ×½

6. **S. brevirostra** Mack. & Bush

Map 68. Muddy shores and shallow water along the Black River at Port Huron (*S. A. Cain* in 1948, BLH) and along the Rainy River near its mouth at Black Lake, Presque Isle County (*Ehlers 5190* in 1932, MICH, UMBS).

The specimens referred here differ from *S. cuneata* as described in the key, but may be only variants of that species. This taxon may also be called *S. engelmanniana* J. G. Sm. ssp. *brevirostra* (Mack. & Bush) Bogin (typical *S. engelmanniana* is found along the coast from Massachusetts to South Carolina).

2. Echinodorus   Burhead

1. **E. tenellus** (Mart.) Buch   Fig. 43   Dwarf Burhead

Map 69. Collected at White Pigeon, St. Joseph County, August 11, 1837, during the first Geological Survey of Michigan, presumably by Abram Sager, who was the botanical and zoological assistant to Douglass Houghton, the first State Geologist (see Mich. Bot. 6: 16—17. 1967; 9: 241. 1970).

This species is restricted to Central and South America except for var. *parvulus* (Engelm.) Fassett (Rhodora 57: 185. 1955), which occurs along the Gulf and Atlantic coasts, with very few inland stations. The type locality for this variety (originally described as a distinct species, *E. parvulus*) is "Michigan," although it is doubtful whether any Michigan material was seen by Small when he selected this state from the wider range given by Engelmann in the original description. The First Survey specimens (MICH, GH) are apparently the only ones ever found in Michigan.

3. Alisma

1. **Alisma plantago-aquatica** L.   Plate 2-D   Water-plantain

Map 70. Shallow water and low wet ground generally: marshes, borders of streams and ponds, ditches, muddy banks, floodplains, wet thickets, older parts of bogs, etc.

The name is used here in an inclusive sense. Opinions of authorities differ as to the status of American plants, most recognizing one or two species, subspecies, or varieties distinct from the European. An extremely variable species in size, sometimes plants only a few cm tall bearing fruit. The largest plants I have seen (ca. 2 m tall) were in a marsh dominated by *Scirpus fluviatilis* along the Maple River in southern Gratiot County.

Plants with larger parts (achenes 2.2—3 mm long, fruiting heads 4—7 mm in diameter, sepals in anthesis 3—4 mm long, flowers 7—13 mm broad) have been called *A. triviale* Pursh or *A. plantago-aquatica* var. *americanum* Schultes & Schultes. Smaller plants (achenes 1.5—2 mm long, fruiting heads 3—4 mm in diameter, sepals 2—2.5 mm long, and flowers 3—3.5 mm broad) have been known as *A. subcordatum* Raf. or *A. plantago-aquatica* var. *parviflorum* (Pursh) Torrey (see Fernald, The North American Representatives of Alisma plantago-aquatica. Rhodora 48: 86–88. 1946). While the extremes are quite distinct, too many specimens fail to fit the combinations of characters which supposedly distinguish the two entities. Plants corresponding to *A. triviale* are found throughout the state; those referable to *A. subcordatum* apparently do not occur north of the Straits of Mackinac.

BUTOMACEAE   Flowering-rush Family

1. Butomus

A Eurasian genus with a single species, locally naturalized in North America.

REFERENCE

Stuckey, Ronald L. 1968. Distributional History of Butomus umbellatus (Flowering-rush) in the Western Lake Erie and Lake St. Clair Region. Mich. Bot. 7: 134–142.

1. **B. umbellatus** L.   Plate 2-E   Flowering-rush
Map 71. Locally established in wet ground, common only in the marshes and river borders near Lake Erie and Lake St. Clair. First collected in Michigan in 1930, although reported to have been here long before that date.

HYDROCHARITACEAE   Frog's-bit Family

A rather small family of aquatic plants, some of them marine and most of them native in warm or tropical regions. Our two genera are readily distinguished vegetatively from each other – less readily from certain other aquatic plants. All of our species are dioecious and are pollinated at the surface of the water. *Elodea* is sometimes segregated into a separate family, Elodeaceae (see St. John in Rhodora 67: 155–156. 1965).

KEY TO THE GENERA

1. Leaves very long and ribbon-like (mostly 3–11 mm wide and many times as long),
   in a basal rosette; anthers usually 2 . . . . 1. **Vallisneria**
1. Leaves up to 2 cm long, whorled (occasionally some opposite); anthers usually 9 . .
   . . . . 2. **Elodea**

1. Vallisneria

The solitary pistillate flower is carried to the surface of the water on a long peduncle (which becomes ± spiraled in age). Staminate plants produce a large number of flowers, borne densely in a spathe on a short peduncle (sometimes as long as 33 cm). These staminate flowers are released under water, the perianth tightly closed and hence containing a gaseous "bubble"; at the surface, the perianth opens, acting as a sail. The staminate flowers (fig. 45) drift upon the surface of the water, and when one reaches the dimple formed by a pistillate flower at the surface, it slides quickly to it and pollination occurs – barely above the surface of the water.

The characteristic three-zoned appearance of the leaves of *Vallisneria* will distinguish it from other plants which may produce rosettes of ribbon-like submersed leaves (cf. figs. 2, 3, 4, but also Mich. Bot. 10: 124. 1971). The leaves of *Potamogeton epihydrus* are similarly three-zoned, but are shorter and alternate.

1. **V. americana** Michaux   Figs. 2, 44, 45   Tape-grass; Wild-celery

Map 72. Submersed in lakes, ponds, and rivers, often in deep water (recorded to 21 feet), spreading by rhizomes with tuberous tips which, like the fruit and other parts, are relished by waterfowl.

By many older authors, not distingushed from *V. spiralis* L.

2. Elodea   Waterweed; Elodea

Although there is now general agreement among authors that we have but two native species of this genus (often called *Anacharis*), the precise characters used to distingush them differ, especially in regard to measurements, in several manuals.

Elodea densa (Planchon) Caspary [*Egeria densa* Planchon] is the common cultivated *Elodea* of aquaria and fish ponds. It is a native of South America, reported as naturalized in several places in the United States but not yet from Michigan. It has longer leaves (in whorls usually of 4–6) and larger flowers than our native species.

Plants of *Callitriche hermaphroditica* have sometimes been mistaken for sterile *Elodea*. The leaves in *C. hermaphroditica* are strictly opposite and have a very characteristic bidentate apex.

The pistillate flowers of both our species are solitary and are carried to the surface (if water conditions are right) by elongation of the floral tube, which appears as a thread-like stalk. Staminate flowers differ, as described in the key, but in both species the anthers dehisce after the perianth opens, spraying pollen onto the surface of the water, where it drifts and may reach a pistillate flower in

70. Alisma
    plantago-aquatica

71. Butomus umbellatus

72. Vallisneria americana

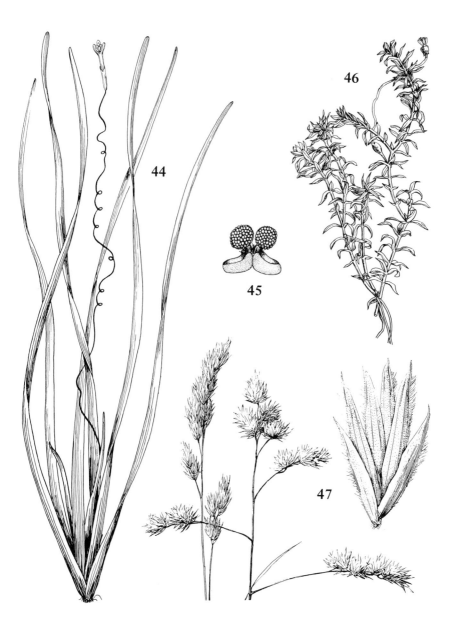

44. *Vallisneria americana* (pistillate) ×½
45. *V. americana*, staminate flower ×15
46. *Elodea canadensis* (staminate) ×⅔
47. *Dactylis glomerata* ×½ ; spikelet ×5

a manner similar to the drifting entire flower of *Vallisneria*. Most reproduction is certainly vegetative, however, by fragmentation of the stem. Staminate plants of *E. canadensis* are relatively uncommon; however, pistillate plants of *E. nuttallii* appear to be less common than staminate ones in the state.

REFERENCE

St. John, Harold. 1965. Monograph of the Genus Elodea: Part 4. Rhodora 67: 1–35.

KEY TO THE SPECIES

1. Well developed leaves generally 0.8–1.5 mm wide (mostly about 1 mm, very rarely to 2 mm, but averaging less) and (4) 5–10 (13) times as long as wide; staminate flower in a ± globose spathe with body about 2–3 mm long, sessile, breaking free from spathe at maturity; sepals of pistillate flowers ca. 1–1.5 mm long . . . **1. E. nuttallii**
1. Well developed leaves generally 1.5–4 (5) mm wide (mostly about 2 mm) and 2–5 times as long as wide (or sometimes a little longer, especially in deep-water plants); staminate flower in an elongated spathe 4–8.5 (14) mm long, sessile when immature, at length remaining attached by a long extremely delicate filiform stalk (floral tube); sepals of pistillate flowers ca. 2–4.5 mm long. . . . . **2. E. canadensis**

1. E. nuttallii (Planchon) St. John

Map 73. In lakes, pools, and rivers, much more local than the next species; recorded at depths to 24 feet.

Usually a distinctive small- (or at least narrow-) leaved plant more delicate than the much more common *E. canadensis*. Occasionally the two may grow in the same body of water, where they are readily compared. In some forms the leaves are rather delicate, narrow, 5–10 times as long as wide; in others the leaves (especially on flowering branches) are short (scarcely 5 mm long), stiffish and often ± recurved, the length/width ratio as in *E. canadensis* but the leaves smaller and less imbricate than in that species. Plants of both forms have been collected with the characteristic small globose sessile staminate spathes of *E. nuttallii*. Sterile material matching these forms vegetatively could presumably be depauperate *E. canadensis*, and a few of the most ambiguous sterile specimens are not mapped. The possible correlation of variable growth forms with sex of

73. Elodea nuttallii   74. Elodea canadensis   75. Phragmites australis

plant, depth of water, and other factors should be investigated. Staminate material is apparently more common than in *E. canadensis*, and careful examination will usually reveal spathes, making identification possible.

2. **E. canadensis** Michaux.   Plate 2-F; fig. 46

Map 74. In waters of almost every kind and depth, including the Great Lakes, but cannot survive exposure; recorded at depths to 27 feet.

Staminate plants are much less common than pistillate, and may resemble the preceding species vegetatively. The staminate spathe is generally said to have a stipitate base and a gaping 2-cleft apex, although our staminate specimens (as determined by size and shape of spathe and of leaves) mostly have ± immature spathes ranging from sessile to pediceled; the spathe elongates considerably as it matures — on fresh flowering material from Emmet County it has been seen as long as 14 mm when the flower is exserted on its long stalk. On pistillate plants there is a tendency for the leaves to be broad, crowded, ascending, and strongly imbricate on short branches. Forms with internodes longer than the length of the leaves may occur in deep water, but the large leaves and stout stem will readily distinguish such plants from the delicate *E. nuttallii*.

GRAMINEAE (POACEAE)   Grass Family

In order to facilitate reference to other floras and manuals, the sequence of tribes here used is the same as in most standard floristic works. However, for convenience in the use of the present keys, the genera within each tribe, and the species within each genus, are in the same sequence as they appear in the keys. This pattern, as in other families in the Flora, makes possible ready comparison of taxa whenever notes are given in the text to supplement the diagnoses in the keys.

The circumscriptions of the traditional tribes are also here maintained for convenience, although much recent research has led to the alteration of several tribes, the recognition of some additional tribes and subfamilies, and the shifting of various genera — all with the goal of a more natural classification. A complete and well documented modern treatment, assigning all United States genera to tribes on the basis of an up-to-date synthesis of agrostological research and including a key to the genera, appeared soon after the manuscript for this family was completed (Gould, 1968). Differences in Gould's assignment of genera to tribes are mentioned at the beginning of the treatment of each tribe below, as indicative of the directions in which grass classification is moving. For example, it will be noted that all of the Phalarideae and most of the genera formerly in the Agrostideae including *Agrostis* itself are placed in the Aveneae; and the Tripsaceae are included in the Andropogoneae.

In addition to Gould's text, a good introduction to the interpretation of grass structure for the beginner is *First Book of Grasses* by Agnes Chase. The third edition was published in 1959 by the Smithsonian Institution in honor of the

90th birthday of its author, who continued her research on grasses at the Smithsonian until shortly before her death in 1963.

The chief problem with grasses for the novice is knowing what he is looking at. Once one becomes familiar with the specialized parts of the grass inflorescence, these plants are no more difficult than others to identify — and easier than many.

The overall inflorescence of a grass is described in the usual terms, such as "panicle" or "spike," although each unit whether sessile (in a spike) or pedicelled is not a single flower alone. The basic unit of the grass inflorescence is a *spikelet* containing one or more flowers, each called a *floret*. Except in the rare instances when one or both are absent, there is a pair of scales or bracts called *glumes* at the base of each spikelet. In addition, at the base of each floret are two often similar scales or bracts enclosing the reproductive parts. The lower (outer) scale is the *lemma* and the upper (inner) one the *palea*. The palea is generally two-nerved or two-keeled and is often smaller and less firm in texture than the lemma — and sometimes difficult to distinguish without a careful dissection. Most grass flowers — within the lemma and palea — consist of 3 stamens and a single pistil. Some flowers are sterile (lacking a pistil or all reproductive parts), some have 1, 2, or 6 stamens, and some are unisexual. The axis of a spikelet, to which the florets are attached, is the *rachilla*. The hard often enlarged area at the base of a floret, sometimes bearing hairs, is the *callus*. Tiny swollen scale-like structures at the base of the stamens are *lodicules* and may represent a very much reduced perianth; they are not mentioned in any of the keys here. A *pulvinus* is a swelling at the base of a branch of the inflorescence. The fruit produced by a grass flower is almost always a seedlike *grain* or *caryopsis*, a dry indehiscent one-seeded fruit with the pericarp adnate to the seed (in an achene the pericarp is free).

The leaf of a grass consists of a *blade*, generally narrow and elongate, and a *sheath* which surrounds the stem like a sleeve. The sheath is usually *open*, i. e., with the margins separate, but in some genera (including the common *Bromus* and *Glyceria*) it is *closed*, i. e., with the margins connate. At the inside summit of the sheath is almost always a small appendage, the *ligule*, which may be either a fringe of hairs or a membranous collar (or, rarely, a combination). At the junction of the blade and sheath in some species is a pair of prolongations called *auricles*; these are lateral and often tend to clasp the stem (fig. 81).

The following notes may help in interpreting statements in the keys and other comments. Remember that the florets in a spikelet are basically alternate in arrangement.

The first (or lowest) glume is below the lemma in 1-flowered spikelets; below the lowest lemma in several-flowered spikelets. The second glume is below the palea in 1-flowered spikelets; below the second lowest lemma in several-flowered spikelets.

In order to avoid the necessity of specifying "first glume" or "second glume," the keys often refer merely to the "larger" or "smaller" glume, the relative size being more readily seen for quick identification; such references to size carry no implication as to whether the larger glume is the first or second one.

When it is necessary to determine the place of articulation (whether above or below the glumes at the summit of the pedicel) or whether there is articulation at all, it can often be determined in dried specimens, even those not yet mature enough to be disarticulating naturally, by attempting a break with a dissecting needle or fine forceps. The break will usually occur at the same place as it normally would.

Ligule measurements are based on ligules from the upper and middle portion of a plant, not necessarily the lower; and lengths are of the free portion.

REFERENCES

Chase, Agnes, 1959. First Book of Grasses. Ed. 3. Smithsonian Inst., Washington. 127 pp.

Deam, Charles C. 1929. Grasses of Indiana. Dep. Conservation Publ. 82, Indianapolis. 356 pp.

Dore, William G. 1959. Grasses of the Ottawa District. Canada Dep. Agr. Publ. 1049, Ottawa. [101] pp.

Fassett, Norman C. 1951. Grasses of Wisconsin. Univ. Wis. Press, Madison. 173 pp.

Gould, Frank W. 1968. Grass Systematics. McGraw-Hill, New York. 382 pp.

Hitchcock, A. S. 1951. Manual of the Grasses of the United States. Ed. 2, rev. by Agnes Chase. U. S. Dep. Agr. Misc. Publ. 200. 1051 pp.

Pohl, Richard W. 1947. A Taxonomic Study on the Grasses of Pennsylvania. Am. Midl. Nat. 38: 513–600.

KEY TO THE TRIBES OF GRAMINEAE

Note: Tribes 9, 10, and 11 have regularly been placed in the subfamily Panicoideae, the others in the subfamily Pooideae (usually given the nomenclaturally incorrect name Festucoideae).

1. Summit of plant with a large "tassel" or spike-like raceme bearing staminate florets in pairs, the pistillate florets either sunken in indurate joints of rachis below the staminate portion (fig. 167) or in separate "ears" lower on the plant (fig. 163). . . . . 11. TRIPSACEAE (p. 242)
1. Summit of plant not as above: upper portion of inflorescence bearing pistillate or perfect florets
  2. Spikelets unisexual, segregated into different parts of the inflorescence (upper panicle branches bearing awned pistillate spikelets, lower branches bearing staminate spikelets); stamens 6; glumes absent (fig. 136) . . . . 8. ZIZANIEAE (p. 212)
  2. Spikelets perfect, or if unisexual then scattered among perfect ones; stamens 1–3; glumes present or absent
    3. Spikelets forming a simple spike or spikes, directly sessile or subsessile on main axis of inflorescence or at most on secondary branches
      4. Spike solitary, terminal (its rachis a continuation of the culm), the spikelets on opposite sides of rachis (one-sided only in *Nardus,* which lacks glumes). . . . . 2. TRITICEAE (p. 149)
      4. Spikes several, one-sided (spikelets in two rows on one side of rachis)
        5. Glumes keeled and ± equal (or the smaller half or more as long as the larger) . . . . 5. CHLORIDEAE (p. 204)
        5. Glumes rounded on the back (not keeled), very unequal . . 9. PANICEAE.(p. 214)

3. Spikelets not forming simple spikes as above: pedicelled and/or on tertiary or further branches of the inflorescence; in some species congested and hence spike-like, but not directly sessile or subsessile (reduced panicle branches visible on close examination and removal of some spikelets)
  6. Spikelets containing only 1–2 florets
    7. Glumes or lemmas (or both) ± laterally compressed or keeled (lateral nerves, if present, less prominent than midnerve) . . . . [go to couplet 13]
    7. Glumes and lemmas rounded on back, not keeled (nerves, if present, about equally prominent)
      8. Spikelets enclosed in a spiny bur . . . . 9. PANICEAE (p. 214)
      8. Spikelets without spiny bur
        9. Glumes very unequal in length, one of them minute, or absent, or at most about half as long as the spikelet*
          10. Spikelets disarticulating below the glumes (except in *Setaria italica*), ± elliptic (less than 3 times as long as wide); a sterile lemma resembling the larger glume present . . . . 9. PANICEAE (p. 214)
          10. Spikelets disarticulating above the glumes, ± lanceolate (3–10 times as long as wide); no sterile lemma present . . . . . . 4. AGROSTIDEAE (p. 172)
        9. Glumes ± equal in length, neither of them much reduced or absent
          11. Spikelets paniculate, all (or mostly) 1-flowered, perfect, the florets all alike (no sterile lemmas nor separate sterile pedicels); spikelets less than 4 mm long except in *Stipa* with awns over 5 cm long . . . .
          . . . . 4. AGROSTIDEAE (p. 172)
          11. Spikelets paniculate or racemose, basically 2-flowered (the lower floret staminate or sterile with often suppressed palea); spikelets 3 mm or more long
            12. Spikelets all alike, not paired with a pedicel bearing a rudimentary, staminate, or no floret (fig. 143) . . . . . 9. PANICEAE (*Panicum virgatum*)
            12. Spikelets of two kinds: one sessile and with a perfect floret, the other a hairy pedicel with or without a staminate or rudimentary floret (rarely 2 stalked florets with 1 sessile one) (cf. figs. 161, 164, 165, 166). . . .
            . . . . 10. ANDROPOGONEAE (p. 238)
  6. Spikelets containing 3 or more florets, including any sterile ones [Note: couplet 13 is also reached from couplet 7 if the spikelets are 1- or 2-flowered]
    13. Spikelets all or mostly containing 1 perfect floret and no sterile or vestigial ones below it
      14. Glumes both completely absent; spikelets strongly flattened, appressed and ± overlapping, the lemmas scabrous or hispid-ciliate (fig. 134). . . . .
      . . . . 7. ORYZEAE (p. 211)
      14. Glumes (one or both) usually present; spikelets various but not as above. .
      . . . . 4. AGROSTIDEAE (p. 172)
    13. Spikelets all or mostly containing 2–several florets, the lower ones sometimes staminate or rudimentary (scale-like or reduced to tiny hairy appendages)
      15. Glumes shorter than the lowest floret (excluding awns if present); awn of lemma none, terminal, arising from between terminal teeth and not twisted, or at most subterminal . . . . 1. POEAE (p. 113)

---

*In the Paniceae, a sterile lemma is present which closely resembles the large second glume opposite it and might easily be misinterpreted as the first glume. Look for the true first glume as a small, sometimes minute and membranous, even deciduous, scale at the very base of the spikelet; a reduced palea, usually associated with the sterile lemma, will also help to identify the latter as part of a sterile floret and not a glume.

15. Glumes (at least one of them, not necessarily the first glume) longer than lowest floret; awn of lemma none or arising from between terminal teeth and strongly twisted below, or (the usual condition) inserted on the middle or lower part of the lemma
16. Spikelets containing one perfect awnless floret (the lemma sometimes membranaceous) with two additional, often dissimilar (sometimes awned) staminate, sterile, or vestigial lemmas below it (cf. figs. 131, 133) . . . . .
. . . . 6. PHALARIDEAE (p. 208)
16. Spikelets usually containing 2 or more perfect florets (staminate or sterile florets, if present, above the fertile one and/or fertile lemma awned) . . .
. . . . 3. AVENEAE (p. 164)*

1. POEAE

Since this tribe includes the genus which is the type of the family, *Poa*, it cannot be called Festuceae as is often done. Similarly, the subfamily which includes *Poa* must be called Pooideae.

Gould (1968) places several of our genera elsewhere: *Phragmites* is our sole representative of the tribe Arundineae; *Diarrhena* is the only American genus in the Diarrheneae; *Schizachne*, *Melica*, and *Glyceria* comprise our genera of the tribe Meliceae; *Eragrostis*, *Triplasis*, and *Tridens*, along with four genera from the Agrostideae in our flora, are placed in Eragrosteae, in a separate subfamily (Eragrostoideae); *Uniola latifolia* is placed in *Chasmanthium*, a genus of the Centotheceae in the subfamily Arundinoideae.

KEY TO THE GENERA OF POEAE

1. Rachilla (above the lowest floret) with silky beard about equalling or exceeding the lemmas; plants tall and stout (usually over 1.5 m tall) with larger leaf blades 1–3.5 cm wide (plate 2-G); spikelets ca. 11–17 mm long; ligule a densely ciliate brown band . . . . 1. **Phragmites**
1. Rachilla with beard shorter or absent; plants generally shorter with mostly narrower leaves; spikelets and ligule various
  2. Spikelets sessile or at most very short-pedicelled, crowded into dense clusters, these either at the ends of elongate panicle branches or in a single congested, rather spike-like inflorescence
    3. Lemma with a prominent, somewhat twisted or spreading dorsal awn; rachilla villous . . . . AVENEAE (*Trisetum spicatum*)
    3. Lemma with awn absent or short and strictly terminal; rachilla not villous
      4. Clusters of spikelets at the ends of elongate naked branches of the panicle (fig. 47); sheaths closed much of their length; ligule ca. 2–8 mm long . . . . . 2. **Dactylis**
      4. Clusters of spikelets all crowded into a congested, rather spike-like inflorescence; sheaths open their entire length; ligule ca. 1 mm or less in length
        5. Spikelets of two kinds in a cluster: normal fertile and special sterile fan-like ones; fertile lemmas mostly short- or long-awned; foliage glabrous . . 3. **Cynosurus**
        5. Spikelets all similar, fertile; lemmas not awned; foliage (at least the lowermost sheaths) pubescent or puberulent . . . . AVENEAE (couplet 3, p. 164)

*Some Aveneae will run in this key to the Poeae and are also included in the key to genera of that tribe.

2. Spikelets short- to long-pedicelled in a ± open panicle
  6. Callus at base of floret with dense beard of straight hairs 0.5 mm or more in length
    7. Awn of lemma arising near base . . . . AVENEAE (*Deschampsia*)
    7. Awn of lemma absent, terminal, subterminal, or arising between terminal teeth
      8. Lemmas awnless, weakly 5-nerved; sheaths open . . . .
    . . . . AVENEAE (*Trisetum melicoides*)
      8. Lemmas with short or long awn, either 3-nerved or plants with closed sheaths
        9. Sheaths open; lemmas 3-nerved, ± truncate (ragged or lobed) at apex, the nerves hairy
          10. Panicles terminal and axillary, small, the former often partly and the latter entirely included in the swollen sheaths (fig. 49); palea villous on apical half; nodes bearded; plants annual, with the leaves, sheaths, and culms toward base upwardly scabrous . . . . **4. Triplasis**
          10. Panicles terminal, large, exserted; palea not villous on apical half (fig. 50); nodes glabrous; plant a stout perennial smooth toward base . . . . . . **5. Tridens**
        9. Sheaths closed; lemmas 5–7-nerved, tapering to an apparently 2-lobed or sharply bifid apex, glabrous or hairy
          11. Callus with distinct beard, the lemma glabrous or nearly so (fig. 51); grain glabrous . . . . **6. Schizachne**
          11. Callus pubescent like the lemma; grain pubescent at the summit. . . **13. Bromus**
  6. Callus glabrous, minutely puberulent, or cobwebby (not bearded with straight hairs)
    12. Glumes (at least one of them) and usually also lemmas strongly keeled (lemmas in a few species rounded on the back); awns absent or not over 2 mm long
      13. Larger glumes ca. 4.5–7 mm long
        14. Spikelets nearly sessile, numerous, ascending in a ± crowded panicle; ligules ca. 2–8 mm long; sheaths closed much of their length . . . . **2. Dactylis**
        14. Spikelets on long pedicels, ascending to drooping in a much expanded raceme or panicle; ligules less than 1.5 mm long; sheaths open [if confronted at this point with a plant bearing an open panicle and closed sheaths, try *Bromus*]
          15. Lemmas smooth except on keel; spikelets very strongly flattened and keeled; lowest lemma sterile; leaf blades mostly more than 1 cm wide, flat
          . . . . **7. Uniola**
          15. Lemmas scabrous over the back; spikelets not strongly flattened, the glumes and lemmas ± obscurely keeled; lowest lemma fertile; leaf blades less than 1 cm wide, the lower ones ± involute . . . . . . **14. Festuca (scabrella)**
      13. Larger glumes not over 4.4 mm long
        16. Larger glumes ± obovate, broadest above the middle . . . .
        . . . . AVENEAE (couplet 3, p. 164)
        16. Larger glumes broadest at or below the middle
          17. Ligule a fringe of hairs; lemmas with 3 prominent nerves, glabrous; spikelets 2–30-flowered. . . . . **8. Eragrostis**
          17. Ligule a membranous scale, the cilia, if any, shorter than the scale; lemmas with 3–5 nerves, glabrous, hairy, and/or cobwebby at base; spikelets various [if specimens will not fit here, try couplet 19]
            18. Glumes very unequal, the smaller (first) usually only slightly more than half the length of the larger; lemmas obscurely 2-toothed, with minute awn between teeth, the nerves silky-pubescent basally; callus glabrous
            . . . . CHLORIDEAE (*Leptochloa fascicularis*)

18. Glumes slightly unequal; lemmas rounded or pointed, but neither toothed nor awned, the nerves pubescent or glabrous; callus often with a tuft of cobwebby hairs (figs. 57–59) . . . . **9. Poa**

12. Glumes *and* lemmas ± rounded on the back, not keeled (or obscurely so toward apex); awns absent or present

  19. Lemmas distinctly 3-nerved, thick and coriaceous (fig. 56) . . . . . . **10. Diarrhena**

  19. Lemmas 5- (or many-) nerved (the nerves sometimes very indistinct), in most species thin

    20. Lemmas at least as broad as long; mature glumes and florets spreading nearly at right angles to rachilla, the spikelets nearly or quite as broad as long (fig. 52) . . . . 11. **Briza**

    20. Lemmas longer than broad; glumes and florets not so widely spreading, the spikelets in most species much longer than broad

      21. Lemmas usually 2-toothed or minutely 2-lobed at the apex and usually with at least a short awn arising from just below or between the teeth (if teeth apparently united, as in some species of *Bromus,* the awn thus subterminal); sheaths closed nearly to their summit

        22. Spikelets narrowly linear-lanceolate on much shorter, densely hispid pedicels; lemmas awned, minutely strigose or scabrous, at least on the nerves; sheaths retrorsely scabrous; ligules 3–6 (7) mm long; grain glabrous . . . . 12. **Melica**

        22. Spikelets broadly linear to oblong, usually on ± elongate pedicels; lemmas various; sheaths glabrous or pubescent but not scabrous; ligules less than 2.5 (4) mm long; grain pubescent at the summit . . . . . . 13. **Bromus**

      21. Lemmas not 2-lobed or -toothed at apex, awnless or with strictly terminal awn; sheaths open (or closed in *Glyceria,* with prominently nerved and awnless lemmas, and in the youngest shoots of *Festuca rubra*)

        23. Lemmas acute at the tip, awned . . . . 14. **Festuca**

        23. Lemmas acutish or obtuse, awnless

          24. Nerves of lemma prominent, straight and becoming parallel at the tip (figs. 71, 72); sheaths closed or open

            25. Sheaths closed much of their length (but easily splitting); second (larger) glume with 1 distinct nerve; plants rhizomatous . . . . . 15. **Glyceria**

            25. Sheaths completely open; second glume with 3 (−5) nerves distinct at its base; plants without rhizomes (though culms may be decumbent or prostrate) . . . . 16. **Puccinellia**

          24. Nerves of lemma very weak (or if visible, then converging, not parallel, at the tip); sheaths open

            26. Lemmas ca. 2 mm long . . . . 16. **Puccinellia**

            26. Lemmas 2.5–8 mm long . . . . 14. **Festuca**

76. Dactylis glomerata   77. Cynosurus cristatus   78. Cynosurus echinatus

1. Phragmites

1. P. australis (Cav.) Steudel   Plate 2-G   Reed
Map 75. Marshes, wet shores, ditches and swales, tamarack bogs; often in water (sometimes as deep as 6 feet).

A cosmopolitan species, found around the world, and long known as *P. communis* Trin. until the earlier name was found to apply. Fertile seed is often not developed, the plant customarily reproducing vegetatively — often forming large colonies by rhizomes or stolons, the latter noted as long as 43 feet by L. H. Harvey on the flats at Cecil Bay, Emmet County. The spikelets are normally several-flowered, the lowest floret staminate or sterile. At maturity (late summer) the rather dense panicle, at least on fertile plants, has a feathery aspect because of the elongate beard; the rachilla disarticulates above the glumes and between the florets. Occasionally a plant will have such reduced spikelets that only one glume is developed and a single sterile floret is apparent, the characteristic long-bearded rachilla thus being absent; such plants might be mistaken for some unknown member of the Agrostideae by one not familiar with the distinctive vegetative appearance of *Phragmites* with its large pennant-like leaves.

2. Dactylis

1. D. glomerata L.   Fig. 47   Orchard Grass
Map 76. Thoroughly established as an escape from cultivation (native of Eurasia): roadsides, fields, clearings, etc., sometimes spreading into natural habitats such as woods and shores.

Pollen of orchard grass is one of the most important causes of early season "hay fever."

3. Cynosurus   Dogtail
Both species are introduced from Europe, and are rarely established locally in waste places.

79. Triplasis purpurea

80. Tridens flavus

81. Schizachne
purpurascens

1. Inflorescence elongate, more than 4 times as long as broad; awns less than 2 mm long . . . . 1. **C. cristatus**
1. Inflorescence ± ovoid, less than 4 times as long as broad (excluding awns); awns (at least the longest) 1 cm or more . . . . 2. **C. echinatus**

1. **C. cristatus** L.   Fig. 48

Map 77. Grassy waste places, quite local in Michigan; half of our collections were made before 1900.

2. **C. echinatus** L.

Map 78. Collected in 1966 on a dry hillside in the Kellogg Bird Sanctuary, Kalamazoo County (*S. N. Stephenson 1180*, MSC, MICH).

4. Triplasis

1. **T. purpurea** (Walter) Chapman   Fig. 49   Sand Grass

Map 79. Sandy shores and low dunes.

Old culms tend to disarticulate at the nodes into short segments (the internodes), each with a reduced fruiting inflorescence in the swollen sheath.

5. Tridens

1. **T. flavus** (L.) Hitchc.   Fig. 50   Purpletop

Map 80. Oak woodland, sandy clearings, roadsides and railroads; perhaps adventive in Michigan from farther south, although collected in Berrien County by the First Survey as early as 1838.

A large, handsome grass, often included in an Australian genus as *Triodia flava* (L.) Smyth. The 3 nerves of the lemma are slightly excurrent as short awns or bristles. The axis and branches of the inflorescence are covered with sticky dots.

6. Schizachne

1. **S. purpurascens** (Torrey) Swallen   Fig. 51   False Melic

Map 81. In woods of all kinds, especially in openings and on sandy or rocky soil (mixed woods, jack pines, oak, aspen), but also in rich deciduous woods and in groves of pine and cedar.

A distinctive and attractive grass, once widely known as *Melica striata* (Michaux) Hitchc. The spikelets, especially the glumes, are usually strongly flushed with purplish, and the awns tend to be spreading.

7. Uniola

1. **U. latifolia** Michaux   Fig. 53   Wild-oats
Map 82. Floodplain of the Galien River at Warren Woods, Berrien County
(*Billington* in 1920, BLH, WUD, MICH, MSC).

8. Eragrostis   Love Grass

Many of our specimens have been checked by L. H. Harvey during his
preparation of an unpublished monograph of the genus. Several of the species
are quite difficult to distinguish, and the characteristics given in manuals do not
all hold up under examination. Those employed here seem to be the most
consistent. Anthers in all of our species are very small, about 0.2–0.3 (0.4) mm
long.

REFERENCE

Koch, Stephen Douglas. 1969. The Eragrostis pectinacea-pilosa Complex in North and
Central America. Ph.D. thesis, Univ. Mich. 191 pp.

KEY TO THE SPECIES

1. Plants prostrate basally, rooting at lower nodes; nodes of culm bearded (very rarely
glabrate) . . . . 1. **E. hypnoides**
1. Plants ± erect or spreading from the base, not rooting at the nodes; nodes of culm
glabrous
  2. Margins (often inrolled) of leaves and also (usually at least sparsely) pedicels and
keels of lemmas and glumes ± glandular-warty
    3. Well developed spikelets 1.5–2 (2.2) mm wide; larger glume 1–1.5 (1.8) mm
long; sheaths sparsely pilose . . . . 2. **E. poaeoides**
    3. Well developed spikelets 2.5–3.5 (4) mm wide; larger glume 1.7–2.5 mm long;
sheaths essentially glabrous except at summit . . . . 3. **E. cilianensis**
  2. Margins of leaves, pedicels, and keels of lemmas and glumes not glandular-warty
    4. Primary branches of panicle with a prominent long-pilose yellowish (to red)
pulvinus in the axil; spikelets reddish to purplish; plants perennial, with hard
knotty base . . . . 4. **E. spectabilis**
    4. Primary branches of panicle with pulvinus glabrous to sparsely pilose (or
obscure or lacking); spikelets greenish gray to dark lead-colored (occasionally
with purplish flush besides); plants annual, with relatively soft base
      5. Larger spikelets mostly 6–11(15)-flowered, usually on ± appressed pedicels
(though panicle branches may be widely spreading); lowest lemma ca.
1.4–2 mm long; lateral nerves of lemma distinct (in nos. 6 & 7)
        6. Lateral nerves of lemma usually obscure; larger mature spikelets ca.
1–1.4 mm wide; axils of lower primary branches of panicle sparsely pilose;
panicle branches whorled or fascicled at one of the two lowest nodes of
inflorescence . . . . 5. **E. pilosa**
        6. Lateral nerves of lemma distinct, at least on lower half; larger mature
spikelets (1.3) 1.5–2 mm wide; axils of panicle glabrous or rarely the
lowermost with a few hairs; panicle branches usually alternate or subopposite
at lowest two nodes of inflorescence
          7. Pedicels appressed (at least their lower portions nearly or quite parallel to
the panicle branches from which they arise) . . . . 6. **E. pectinacea**

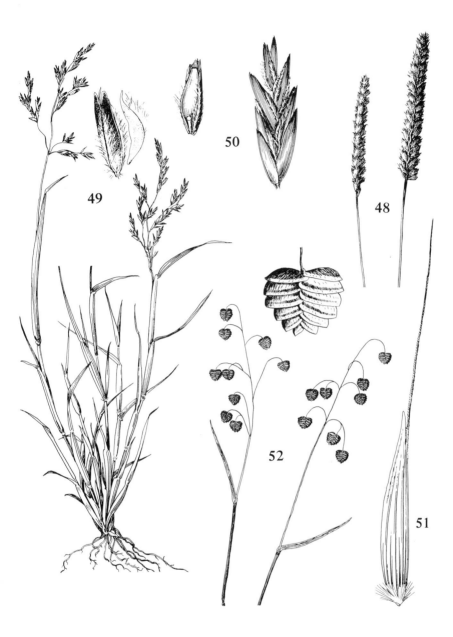

48. *Cynosurus cristatus* ×½
49. *Triplasis purpurea* ×½ ; lemma and palea ×5
50. *Tridens flavus,* floret and spikelet ×5
51. *Schizachne purpurascens,* floret ×5
52. *Briza media* ×½ ; spikelet ×3

7. Pedicels mostly spreading . . . . 7. **E. tephrosanthos**
5. Larger spikelets mostly 2–4(6)-flowered, on spreading pedicels; lowest lemma
    ca. 1.2–1.4 (1.6 or very rarely 1.9) mm long; lateral nerves of lemma obscure
    8. Sheaths pilose; grain with a groove the length of one edge; length of culm
    below lowest branch of terminal panicle less than the height of the panicle. .
    . . . . 8. **E. capillaris**
    8. Sheaths essentially glabrous except at summit; grain not grooved; length of
    culm below lowest branch of terminal panicle usually more than the panicle
        9. Axils of at least the lower primary branches of panicle sparsely long-pilose;
        tip of second glume usually much shorter than the lowest lemma (across
        from it) . . . . 5. **E. pilosa**
        9. Axils glabrous; tip of second glume ± opposite tip of lowest lemma . 9. **E. frankii**

1. **E. hypnoides** (Lam.) BSP.   Fig. 54

Map 83. Muddy or sandy shores, river banks, ditches, etc., often locally abundant, especially on exposed ground after a lowering of water level.

2. **E. poaeoides** R. & S.

Map 84. Chiefly along railroads, but also in cindery places along roadsides, dumps, etc. Naturalized from Europe.

3. **E. cilianensis** (All.) Mosher   Fig. 55   Stink Grass

Map 85. Fields, roadsides, and dry waste places; a garden weed, sometimes in low ground. Naturalized from Europe, but collected in Michigan as early as 1838, at St. Joseph, in Berrien County.

Also widely known as *E. megastachya* (Koeler) Link (see Shinners in Rhodora 56: 26–27. 1954; and Pohl in Iowa St. Jour. Sci. 40: 476. 1966). Occasional culms have unusually small and few-flowered spikelets. This species tends to have prominent pilose pulvini in the panicle, as in *E. spectabilis,* but the aspect of the plant is otherwise very different.

4. **E. spectabilis** (Pursh) Steudel   Tumble Grass

Map 86. Dry fields, sand barrens, roadsides, and railroads; often a very conspicuous plant of roadsides and dry open ground, forming large colonies.

82. Uniola latifolia   83. Eragrostis hypnoides   84. Eragrostis poaeoides

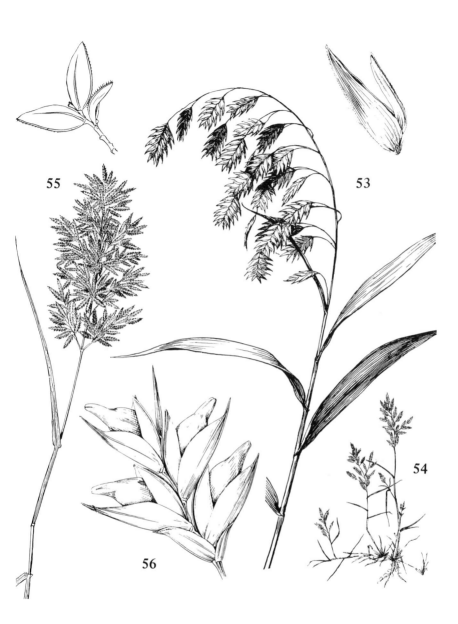

53. *Uniola latifolia* ×½; floret ×3
54. *Eragrostis hypnoides* ×½
55. *E. cilianensis* ×½; portion of spikelet (2 florets and a persistent palea) ×10
56. *Diarrhena americana,* spikelet ×5

Apparently native in the southern part of the state, spreading northward. The large purplish inflorescences become detached and act as tumbleweeds.

Most if not all of our plants have at least the lower sheaths sparsely to densely pilose and hence are var. *sparsihirsuta* Farw. (TL: Royal Oak [Oakland County]).

5. **E. pilosa** (L.) Beauv.

Map 87. A plant of weedy habits like others in the genus, but much less common.

Our very few collections have the spikelets few-flowered, so they do not resemble the next species as much as this one normally does. *E. pilosa* tends to have a more open, spreading panicle than *E. pectinacea*, with less persistent paleas. This species is tetraploid ($n = 20$), as is *E. frankii*, while *E. pectinacea* and *E. tephrosanthos* are hexaploid ($n = 30$) and *E. capillaris* is decaploid ($n = 50$).

6. **E. pectinacea** (Michaux) Nees

Map 88. Sandy ground, roadsides, railroads, fields, and waste places, either dry or moist; riverbanks. Presumably native, at least in the southern part of the state.

A species often confused with the next three and with the preceding one, distinctive in its larger spikelets on appressed pedicels, the lemmas strongly nerved. The paleas are generally even more persistent than in other species, and are conspicuous on old panicles from which the lemmas and grains have fallen (cf. persistent lemma on *E. cilianensis* in fig. 55). A robust form with more branched panicle has been segregated as *E. diffusa* Buckley, and is reported from Kalamazoo County.

Widespread older usage of the name *E. pectinacea* applied it to *E. spectabilis*, while the present species was included in *E. pilosa* or known as *E. purshii*.

7. **E. tephrosanthos** Schultes

Map 89. Four collections previously referred to *E. pectinacea* have been placed in this more southern species by Koch, who indicates that the only truly consistent difference (holding even under uniform greenhouse conditions) is the

85. Eragrostis cilianensis   86. Eragrostis spectabilis   87. Eragrostis pilosa

spreading of mature pedicels in *E. tephrosanthos*. Our specimens are from the vicinity of Lake Linden (in 1936), Monroe (in 1923), and Belle Isle (1896–1899), all collected by Farwell (BLH, MICH), without habitat data but presumably from waste ground.

8. **E. capillaris** (L.) Nees

Map 90. A species of dry ground, collected by Farwell (BLH, MICH, WUD) at Detroit in 1905 and 1922, without habitat data.

A rather distinctive plant, with small spikelets on elongate capillary pedicels.

9. **E. frankii** Steudel

Map 91. Roadsides, fields, and muddy places.

A collection from Kalamazoo County (mud flat on border of Kalamazoo River in Cooper Tp., *C. & F. Hanes 446* in 1944, WMU) must be this species, but has unusually long (1.7–1.9 mm), sharp- and scabrous-keeled lemmas.

This species differs from somewhat similar ones with which it might be confused (*E. pectinacea, E. pilosa, E. tephrosanthos*) in the tendencies for the tip of the larger (second) glume to be nearly or quite opposite the tip of the first lemma, for the grain to be more plumply globose, and for the sheaths to be longer than the internodes.

88. Eragrostis pectinacea

89. Eragrostis tephrosanthos

90. Eragrostis capillaris

91. Eragrostis frankii

92. Poa annua

93. Poa autumnalis

9. Poa   Bluegrass

This is an easily recognized genus, most of the species having a tuft of crinkled, cobwebby hairs on the callus below each lemma, and the leaves ending in a boat-shaped tip (as they do, however, in several other genera). The species, on the other hand, often cause considerable difficulty in identification. The lemma must be very carefully examined, for hairs along the margin and keel may be so closely appressed as to be nearly invisible until disturbed with a needle. Specimens of some species of *Festuca* might run here in the key, but have firm glabrous lemmas nearly or quite nerveless. In some species, at least, of *Poa* the young sheaths may be closed, splitting later into the characteristic open sheaths.

KEY TO THE SPECIES

1. Callus at base of lemma without a cobwebby tuft of crinkled hairs (although the keel may be villous)
   2. Lemmas mostly with 5 distinct nerves (sometimes 3-nerved in the annual and small-anthered *P. annua*); lower glume 1-nerved
      3. Anthers not over 1 mm long; lemma glabrous between the villous nerves; plant usually a low annual, less than 30 cm tall, the panicle 2–8 (11) cm high . . 1. **P. annua**
      3. Anthers mostly slightly over 1 mm long; lower part of lemma sparsely and finely pubescent between the nerves; plant a tall perennial, 30 cm or more in height, the panicle ca. 6–15 cm high . . . . 2. **P. autumnalis**
   2. Lemmas with only 3 distinct nerves (the intermediate nerves very obscure or obsolete); lower glume 3-nerved
      4. Lower part of lemma villous between the keel and the margins; glumes broadly ovate to oblong or obovate (more than half as wide as long); panicle about twice as high as broad, or broader . . . . 3. **P. alpina**
      4. Lower part of lemma glabrous, scabrous, or at most short-pubescent between the keel and the margins; glumes ± narrowly ovate-lanceolate (about half as wide as long or narrower); panicle usually ± contracted, 3 or more times as high as broad (open in some forms)
         5. Clums strongly flattened (nodes normally compressed); exposed nodes (those exserted beyond the sheaths) mostly 2–3 (4) per culm; plants rhizomatous; larger glumes ca. 2–2.8 (3) mm long; rachilla smooth and glabrous . 10. **P. compressa**
         5. Culms slightly if at all flattened (nodes terete or nearly so, even on pressed specimens); exposed nodes mostly 0–1 per culm; plants not rhizomatous; larger glumes (2.7) 3–4.4 mm long; rachilla usually ± pubescent
            6. Lemma scabrous or minutely pubescent at base between and on the keel and margins; spikelets roundish on back, the glumes and lemmas rather obscurely keeled . . . . 4. **P. canbyi**
            6. Lemma glabrous except for the usually strongly villous keel and margins (and sometimes a few long hairs at the base); spikelets ± flattened, the glumes and lemmas distinctly keeled . . . . [go to couplet 16]
1. Callus at base of lemma with a small or large cobwebby tuft of crinkled hairs
   7. Margins of lemma completely glabrous
      8. Keel of lemma glabrous and smooth (lemmas therefore completely glabrous except for web on callus); lower branches of panicle 1–2 (very rarely 3) per node
         9. Ligules 0.6–1.5 (3) mm long; lemmas acute (as seen from the side, the margin meeting the tip of the keel at an angle of about 45° or less); anthers ca. 0.9–1.5 mm long . . . . 5. **P. saltuensis**

9. Ligules (2.1) 2.4–4 mm long; lemmas obtuse (margin on most or all of them meeting the tip of the keel at an angle greater than 45°); anthers ca. 0.7–1 mm long . . . . 6. **P. languida**
8. Keel of lemma scabrous to minutely or strongly pubescent below; lower branches of panicle 4–8 per node
  10. Ligules 0.7–2.2 (3) mm long; lemmas mostly 3-nerved (intermediate nerves obscure); anthers 0.4–0.7 mm long . . . . 7. **P. alsodes**
  10. Ligules (3.2) 4–11 mm long; lemmas distinctly 5-nerved (intermediate nerves prominent); anthers ca. 0.8–1.8 mm long . . . . 8. **P. trivialis**
7. Margins of lemma ± hairy, at least toward base
  11. Lemmas 5-nerved (the nerves between the keel and margin prominent); ligules 0.6–2 (2.6) mm long; anthers 1.2–1.8 (2.1) mm long; rachilla glabrous
    12. Keel of lemma hairy nearly to the end of the green portion; surface and intermediate nerves of lemma usually ± pubescent; plants without rhizomes . . . . 9. **P. sylvestris**
    12. Keel of lemma hairy on about the basal 2/3, only scabrous or smoothish on distal third; lemmas glabrous between keel and margins (or marginal nerves); plants rhizomatous . . . . 11. **P. pratensis**
  11. Lemmas mostly 3-nerved (intermediate nerves obscure or obsolete); ligules, anthers, and rachilla various, but not combined as above (except in *P. compressa*, with flattened nodes and culms)
    13. Upper ligules (2.3) 2.6–5 mm long; anthers 0.8–1.2 mm long; rachilla glabrous . . . . 12. **P. palustris**
    13. Upper ligules 0.2–2 (2.5) mm long; anthers mostly less than 0.8 or more than 1.2 mm long; rachilla glabrous or pubescent
      14. Rachilla smooth and glabrous; branches of panicle 1–4, usually 2, per node; larger glumes ca. 2–2.8 (3) mm long
        15. Panicle ± lax and open, the elongate capillary branches bearing branchlets or spikelets mostly above the middle; culms not flattened; plants not rhizomatous; anthers ca. 0.5–0.7 mm long; upper ligules ca. 0.5–1 (1.5) mm long . . . . 13. **P. paludigena**
        15. Panicle ± stiff and contracted, the branches mostly bearing spikelets nearly to the base; culms strongly flattened (note especially the nodes); plants rhizomatous; anthers 1.1–1.7 mm long; upper ligules (0.5) 0.8–1.8 (2.5) mm long . . . . 10. **P. compressa**
      14. Rachilla ± puberulent, short-hairy, or strongly scabrous [a black background will help reveal this]; branches of panicle mostly 2–5 per node; larger glumes (2.5) 3–4.4 mm long
        16. Ligules 1.2–1.7 (2.5) mm long; exposed nodes of culm 0–1; leaf blades ± strongly ascending, the uppermost at or below the middle of the plant . . . . 14. **P. glauca**
        16. Ligules 0.2–0.7 (1) mm long; exposed nodes usually 1–4; leaf blades often divergent from culm, the uppermost often above the middle of the plant. . . . . 15. **P. nemoralis**

1. **P. annua** L. Annual Bluegrass

Map 92. Roadsides, lawns, and waste ground; trails in woods and clearings; shores and stream banks.

Occasionally rooting at the nodes and hence ± perennial by stolons. The nerves of the lemma are normally villous basally, especially the keel, but occasionally they are nearly or quite glabrous.

2. P. autumnalis Ell.

Map 93. A single depauperate collection from Michigan has been seen, from woods at Rochester, Oakland County (*Farwell 1537* in 1896, BLH).

The spikelets occur at the ends of the long panicle branches in this species, whereas they are crowded on the whole outer half of the branches in the preceding.

3. P. alpina L.

Map 94. Rock crevices near Lake Superior, and thence northward to the Arctic.

Distinctive in its broad open panicle of large spikelets and its comparatively broad leaves.

4. P. canbyi (Scribner) Piper

Map 95. Rock crevices on Isle Royale (*J. B. McFarlin 2175* in 1930, MICH).

5. P. saltuensis Fern. & Wieg.   Fig. 57

Map 96. Deciduous or mixed woods, pine groves, wooded dunes, rock openings.

6. P. languida Hitchc.   Fig. 58

Map 97. Woods, shores, and thickets.

94. Poa alpina

95. Poa canbyi

96. Poa saltuensis

97. Poa languida

98. Poa alsodes

99. Poa trivialis

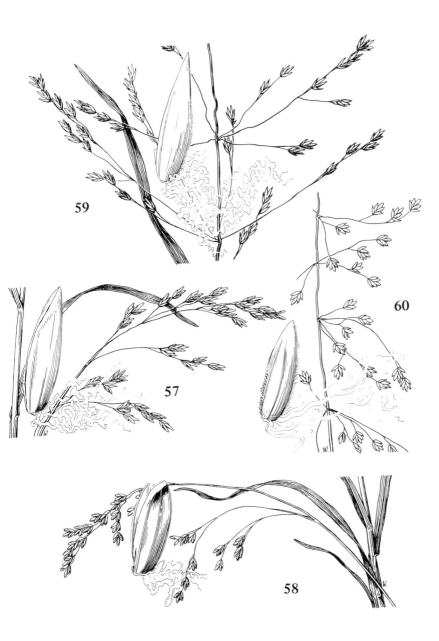

57. *Poa saltuensis* ×1; floret ×10
58. *P. languida* ×1; floret ×10
59. *P. alsodes* ×1; floret ×10
60. *P. sylvestris* ×1; floret ×10

7. **P. alsodes** Gray   Fig. 59

Map 98. Rich deciduous or mixed woods and thickets.

8. **P. trivialis** L.

Map 99. Moist places both wooded and open; quite local. Generally considered to be introduced from Europe.

9. **P. sylvestris** Gray   Fig. 60

Map 100. Rich deciduous woods.

The lemmas are somewhat more obtuse than in *P. pratensis*.

10. **P. compressa** L.   Canada Bluegrass

Map 101. Widespread in old fields, roadsides, waste ground, rocky or sandy woods and openings (oak, aspen, jack pine), usually in dry places but sometimes on shores and in damp woods. Introduced from Europe.

This species and the next are two of our commonest grasses. They are generally segregated from our other species of *Poa* by the presence of creeping rhizomes, the plants thus forming large colonies. While a useful field character, the rhizomes may not be evident on specimens merely picked or pulled hastily from the ground. Hence, the key has de-emphasized the rhizome. *P. compressa* differs from *P. pratensis* in the flattened, less tufted culms, lemmas with obscure intermediate nerves and sparse or even absent web at the base, and tendency toward a more slender panicle with fewer branches at each node.

11. **P. pratensis** L.   Kentucky Bluegrass

Map 102. Almost ubiquitous in all but the wettest places — waste ground, woods, dunes, fields. A complex species mostly introduced from Eurasia and a very important cultivated pasture grass; plants of northern shores, rocks, and open woods are presumably native.

Included here are plants with very narrow or involute basal leaves (narrower than the culm) sometimes segregated as var. *angustifolia* (L.) Gaudin (or even as a distinct species, *P. angustifolia* L.). Standard descriptions of the species as glabrous notwithstanding, the sheaths and leaf blades toward the base of the

100. Poa sylvestris   101. Poa compressa   102. Poa pratensis

plant are occasionally pubescent. Some specimens of *P. palustris* might run here if the lemmas are ± 5-nerved, but can be distinguished by their longer ligules, shorter anthers, and less copious callus web. See also comments under *P. compressa* and *P. glauca*.

12. **P. palustris** L. Fowl Meadow Grass
Map 103. Shores, meadows, and low damp ground generally; pond and stream banks; openings in deciduous or coniferous woods and swamps; bogs and marshes; sometimes on drier sandy or rocky ground, as in aspen woods or openings on rocks near Lake Superior.

A common and variable species, often superficially recognized by the golden tips of the narrow florets. Rarely appears to be rhizomatous. Distinctive in the long ligule (rarely some culms with ligules a little shorter than in the key), short anthers, and ± open panicle at maturity. *P. trivialis* is somewhat similar, but the lemmas are very distinctly 5-nerved and the margins are glabrous; in *P. palustris*, the lemmas are at most obscurely 5-nerved and the margins hairy basally. A few collections, especially from the northern shores of Lake Huron, have the rachilla minutely pubescent as in *P. glauca*, but the long ligule (ca. 4 mm) and aspect of the plant are as in *P. palustris*.

13. **P. paludigena** Fern. & Wieg.
Map 104. Bogs, swamps, and wet woods, usually in sphagnum or other moss.

14. **P. glauca** Vahl
Map 105. Rock crevices and rocky shores, including the limestone areas of northern Lakes Huron and Michigan.

A complex group of forms, no effort being made here to recognize as segregate species or varieties the entities which have been proposed. Typically, the callus at the base of the lemma is glabrous, but in many of our specimens (as noted by Hulten also in Alaska) there is a small crinkled beard. Such specimens may be confused with *P. pratensis*, but differ in the generally longer glumes, pubescent rachilla, and long peduncle. (The rachilla may be nearly or quite glabrous in some specimens which also lack the beard on the callus; the beard is

103. Poa palustris   104. Poa paludigena   105. Poa glauca

well developed in *P. pratensis*.) If a node of the culm is exposed in *P. glauca*, it is at a point only a fourth to a third the height of the plant. If only one node is exposed in *P. pratensis*, it is only slightly below the middle of the plant. Ordinarily *P. glauca* has a compact, purple-tinged panicle (occasionally an open greenish one) on a very long peduncle. Plants rarely approach *P. nemoralis* in having more leafy culms (uppermost blade on upper half of plant). See also comments under the next species.

15. **P. nemoralis** L.

Map 106. Usually in dry, often shallow or rocky soil, such as woodland borders and clearings; rock crevices and shores along Lake Superior.

This species and the preceding form a group distinctive (at least in our area) in the minutely pubescent rachilla. The major difference between the two, ligule length, is quite consistent but not thoroughly correlated with other characters. Typical *P. nemoralis* is often considered to be an introduction from Europe. Some of our plants, placed as *P. nemoralis* by their very short ligule, seem surely to be native on the rocks of the Lake Superior region; they occasionally approach *P. glauca* in having long-exserted panicles and ligules up to 1 mm long. Some of these may be the controversial *P. interior* Rydb. [*P. nemoralis* var. *interior* (Rydb.) Butters & Abbe], but are not a good match for *P. interior* from farther west in the United States. (See F. K. Butters & E. C. Abbe, The Genus Poa in Cook County, Minnesota. Rhodora 49: 1–21. 1947.)

10. **Diarrhena**

1. **D. americana** Beauv.   Fig. 56

Map 107. Very local: floodplain swamp forests, riverbanks, and creek bottoms.

The other three species of this small genus are all natives of northeastern Asia. The uppermost florets in each spikelet are sterile. The firm coriaceous lemma and palea are widely spread at maturity by the conspicuous, bilaterally symmetrical swollen fruit — which is not a true grain or caryopsis as the ovary wall is not adnate to the seed (Pohl in Iowa St. Jour. Sci. 40: 396. 1966).

106. Poa nemoralis

107. Diarrhena americana

108. Briza media

11. Briza

1. **B. media** L.   Fig. 52   Quaking Grass
Map 108. Collected at Bay City by Bradford 1894—1895 and by Farwell at
Grosse Pointe in 1906; none of the specimens bear habitat data, but such plants
are to be expected in waste places as, presumably, escapes from cultivation. An
attractive ornamental grass, not native in North America.
B. maxima L. has been reported from Michigan but specimens have not been
found. It has spikelets at least 10 mm long and ligules over 1.5 mm long, whereas
in *B. media* the spikelets are less than 6 mm long and the ligules are less than 1.5
mm.

12. Melica
The genus is normally distinguished in having the uppermost lemmas of a
spikelet empty and convolute into a club-shaped mass. However, this characteris-
tic is not true of the section (or subgenus, or sometimes segregated as a genus),
Bromelica, to which our species belongs. (See O. A. Farwell, Bromelica
(Thurber): A New Genus of Grasses, Rhodora 21: 76—78. 1919.)

1. **M. smithii** (Gray) Vasey   Fig. 61
Map 109. Deciduous woods, especially rich beech—maple—hemlock stands,
wooded dunes, rarely under white-cedar. Reported from Isle Royale in the
original description but no specimens from there have been located. The species
is wide-ranging in the western United States and adjacent Canada, re-occurring in
the northern Great Lakes area and the Ottawa, Ontario, district. First recognized
in Michigan (TL: Sault Ste. Marie [Chippewa County]).
The panicle branches are deflexed at maturity. The base of the stem is ±
enlarged and bulbous; this character, together with the densely antrorse-hispid
pedicels and retrorsely scabrous sheaths, makes even over-ripe specimens readily
identifiable.

13. Bromus   Brome Grass
Poa canbyi might run to this genus because of the weakness of the keel on its
glumes and lemmas; it may be distinguished from our species of *Bromus* by its
short (less than 5 mm) awnless lemmas in small few-flowered spikelets, as well as
by its open sheath.
Bromus scoparius was recorded from Michigan in the *Manual of Grasses*, but
no specimen can now be located and the basis of the report (if there once was a
supporting specimen) is unknown, being presumably re-identified.

REFERENCE

Wagnon, H. Keith. 1952. A Revision of the Genus Bromus, Section Bromopsis, of North
America. Brittonia 7: 415—480.

KEY TO THE SPECIES

1. First glume with one distinct nerve; second glume with 3 (−5) nerves
   2. Awns (8) 10−30 mm long, as long as or longer than their lemmas; apex of lemma beyond insertion of awn 1.5−2.7 mm long; plants annual weeds
      3. Lemmas 16−19 mm long; awns mostly ca. 20−30 mm long; second glume 13−17 mm long . . . . 1. **B. sterilis**
      3. Lemmas (8) 10−12 (14) mm long; awns ca. (8) 10−17 (19) mm long; second glume 7−12 mm long . . . . 2. **B. tectorum**
   2. Awns absent or up to 7 (9) mm long, shorter than their lemmas; apex of lemma less than 1.5 mm long; plants perennial, mostly native (nos. 3 & 5 introduced)
      4. Plants with elongate rhizomes; lemmas (at least when fresh) usually ± flushed with purplish, especially toward the margins, the awns absent or less than 4 (5.5) mm long; anthers 3.3−4.7 (6) mm long
         5. Lemmas glabrous or scabrous to short-hispid; culms glabrous or finely pubescent at the nodes; leaf blades glabrous (rarely ± pilose on both surfaces or at least on lower surface); awns absent or up to 2.5 (3.1) mm long; common introduced species . . . . 3. **B. inermis**
         5. Lemmas pubescent with distinct long hairs (0.5 mm or more) at least toward the margins; culms usually pubescent with long hairs at or immediately adjacent to the nodes (occasionally glabrous); leaf blades pubescent on upper surface, glabrous or sparsely pubescent below; awns mostly (1) 1.5−4 (5.5) mm long; native species localized on beaches or dunes . . . . 4. **B. pumpellianus**
      4. Plants without rhizomes; lemmas (when fresh) green (very rarely flushed with purple), the larger awns 3−7 (9) mm long; anthers various
         6. Branches of nearly simple panicle erect or strongly ascending; anthers ca. 4.5−5.5 mm long; leaf blades involute; plant a rare adventive of waste ground . . . . 5. **B. erectus**
         6. Branches of the compound panicle loosely ascending, spreading, or nodding at maturity; anthers ca. 0.9−5 mm long; leaf blades flat; plants native, in woodlands and thickets
            7. Nodes (= also number of leaves) usually 8−15; leaf sheaths longer than the internodes, thus overlapping and covering all the nodes, the summit of the sheath with a band of dense pubescence and (when intact) with a pair of prominent tooth-like auricles (fig. 63); anthers 1.5−2.2 mm long . . 6. **B. latiglumis**
            7. Nodes usually not more than 6 (−8 in *B. ciliatus*); leaf sheaths shorter than at least the upper internodes, exposing one or more of them, the summit of the sheath glabrous or pubescent but lacking auricles (fig. 64); anthers various
               8. Lemmas ± uniformly hairy (very rarely glabrous); anthers ca. 2.5−5 mm long; glumes pubescent at least on keel (sometimes only scabrous) . . . . 7. **B. pubescens**
               8. Lemmas with long hairs along the margin, especially toward the base, glabrous or only minutely pubescent on the back; anthers 0.9−1.7 mm long; glumes glabrous or at most scabrous to minutely hispid . . . . 8. **B. ciliatus**
1. First glume with 3 (−5) distinct nerves; second glume with 5−7 nerves
   9. Lemmas pubescent all across the back, at least apically; glumes pubescent; awns straight (except in the rare no. 11)
      10. Larger awns 2−2.8 mm long; primary branches of inflorescence mostly longer than the spikelets; anthers (0.9) 1.5−2 mm long; ligule less than 0.7 mm long, glabrous on the back; plant a native perennial . . . . 9. **B. kalmii**
      10. Larger awns 3.5−14 mm long; branches of inflorescence mostly much shorter than the spikelets; anthers 0.5−1.2 mm long; ligule ca. 0.5−2 mm long, pubescent on the back (side next to the blade); plants introduced annuals

11. Awn straight, inserted at most 1.5 mm below tip of lemma . . . . .10. **B. mollis**
11. Awn, at least on upper lemmas, becoming divaricate, inserted 1.5–3 mm
below tip of lemma . . . . 11. **B. molliformis**
9. Lemmas glabrous or scabrous on the back; glumes likewise glabrous; awns usually
± divaricate or undulate (or absent)
  12. Lemma equalling or slightly shorter than the tip of the mature palea; sheaths
  glabrous (or occasionally the lowermost with some short hairs); margins of ripe
  lemmas strongly inrolled, exposing the rachilla; lemmas ca. 7–8.2 mm long,
  the awns ± undulate, sometimes as long as lemma but usually much shorter,
  rudimentary, or occasionally absent . . . . 12. **B. secalinus**
  12. Lemma at least slightly exceeding tip of palea; sheaths of at least middle and
  lower leaves ± densely (though sometimes finely) hairy; lemmas and awns
  various
    13. Sides of larger lemmas (from middle of back to margin) (2.5) 3–3.5 mm
    wide, including a broad hyaline border; tips of at least the upper lemmas in a
    spikelet exceeding their paleas by more than 2 mm; awns either absent (or
    less than 1 mm long) or very strongly divaricate (even recurved) at maturity
      14. Awns absent or less than 1 mm long; spikelets broad (the larger 8–12 mm
      wide) and flat . . . . 13. **B. briziformis**
      14. Awns elongate and very strongly divaricate at maturity (fig. 67); spikelets
      mostly not so broad and flat . . . . 14. **B. squarrosus**
    13. Sides of larger lemmas not over 2.5 mm wide; tips of upper lemmas exceeding
    paleas by less than 2 mm (occasionally 2.5 mm)
      15. Longest awns in a spikelet longer than their lemmas and more than twice as
      long as awn on lowest lemma of the spikelet; branches of inflorescence lax
      and flexuous; hairs of sheath very fine and delicate, and tending to be ±
      crooked or tangled toward their tips (though basically ± retrorse); tip of
      palea 1–2.5 mm shorter than tip of lemma; anthers 0.5–1.5 mm long . . .
      . . . . .15. **B. japonicus**
      15. Longest awns in a spikelet about as long as their lemmas or shorter.
      and usually less than twice as long as awn on lowest lemma of the spikelet;
      branches of inflorescence rather stiff (whether spreading or ascending); hairs
      of sheath usually fine but stiffish and straight (spreading to retrorse); tip of
      palea less than 1.5 mm shorter than tip of its lemma (rarely 2.5 mm in *B.
      racemosus*); anthers (1) 1.2–2 (2.5) mm long
        16. Branches of panicle, at least the lower ones, ± widely spreading or even
        drooping at maturity, forming a broad open inflorescence usually nodding
        at the summit; larger lemmas ca. (8) 8.5–10 mm long, attached ca.
        1.5–2 mm apart on the rachilla . . . . 16. **B. commutatus**
        16. Branches of panicle erect or very strongly ascending, the entire
        inflorescence narrow and compact; larger lemmas ca. 6.5–7.5 mm long,
        attached ca. 1–1.5 mm apart on the rachilla . . . . 17. **B. racemosus**

1. **B. sterilis** L.

Map 110. Waste places. A European species sparingly introduced in North
America, and rarely collected in Michigan.

2. **B. tectorum** L.   Fig. 62   Downy Chess

Map 111. Roadsides, fields, and waste places, especially common along
railroad tracks. First collected in Michigan around Grand Rapids in 1894 and
noted as "recently introduced"; now our most widespread weedy *Bromus*. A

native of Europe, becoming well naturalized in the United States and southern Canada.

The foliage and inflorescence of this species tend to be more pubescent than in the preceding.

3. **B. inermis** Leysser   Smooth Brome

Map 112. Roadsides, fields, and waste ground, sometimes spreading to woods, shores, and other habitats; the rhizomes make this an excellent grass to hold the soil of roadside banks. Naturalized from Europe.

4. **B. pumpellianus** Scribner

Map 113. Sandy shores and dunes in the northern part of Lake Michigan. The main range of this species is in Alaska and northwestern Canada, southward into the Rocky Mountains and Black Hills; stations in Michigan and Ontario are strikingly disjunct.

By Wagnon treated as a native subspecies of the preceding species, with which it apparently hybridizes. Usually easily recognized by the conspicuously villous lemmas, pubescent leaves and nodes, and also often well developed auricles at the summit of the leaf sheath; in *B. inermis* auricles are absent or rudimentary. Occasional specimens have glabrous nodes (possibly a result of hybridization with *B. inermis*?) but these may have the lemmas very villous and even the glumes ± hairy (e. g., some specimens from Emmet and Leelanau counties).

5. **B. erectus** Hudson

Map 114. Despite earlier reports from Michigan, the only specimen seen was collected in 1970 by Louis Ludwig (MICH) in sandy soil on the Botanical Gardens grounds, University of Michigan. Another Old World species locally established in North America.

Even if basal parts are lacking, so the absence of rhizomes cannot be noted, plants of this species look quite different from *B. inermis* because of the erect inflorescence and the very narrow, mostly involute leaves.

109. Melica smithii

110. Bromus sterilis

111. Bromus tectorum

6. **B. latiglumis** (Shear) Hitchc.   Fig. 63

Map 115. Riverbanks, floodplain woods, thickets, and wooded ravines.

This is the species to which Wagnon has applied the name *B. purgans*, but see comments under *B. kalmii* below. *B. latiglumis* ripens its fruit generally toward the latter part of August, while *B. pubescens* matures a month or so earlier.

7. **B. pubescens** Willd.   Fig. 64   Canada Brome

Map 116. Chiefly in either low or upland woods and wooded riverbanks, including beech—maple, oak, and oak—hickory. The Keweenaw County record is based on an old Farwell collection (*563* in 1887, BLH) which may well be mislabelled.

This is the species known in most recent manuals as *B. purgans*. The pubescence (or its absence) on blades and sheaths in both this species and the preceding is extremely variable. Very rarely the lemmas and glumes are all completely glabrous [f. *glabriflorus* (Wieg.) E. Voss]. An occasional specimen of *B. kalmii* in which the 3-nerved condition of the first glume may be obscure might run here, but can be distinguished by the shorter awn (2–2.8 mm long) and shorter anthers.

B. nottowayanus Fern. was reported from southern Michigan by Wagnon, who recognized it as somewhat doubtfully distinct from *B. pubescens* (characterized by 3-nerved second glumes and glabrous to pilose sheaths) on the basis of 5-nerved second glumes and sheaths densely pubescent at the summit (as in *B.*

112. Bromus inermis

113. Bromus pumpellianus

114. Bromus erectus

115. Bromus latiglumis

116. Bromus pubescens

117. Bromus ciliatus

latiglumis, but lacking auricles). However, our specimens of *B. pubescens* clearly show that the second glume is distinctly 5-nerved on the majority of them and weakly so on most of the remainder; many of these also have the sheaths more densely pubescent at the summit than below. Separate recognition of *B. nottowayanus* does not seem warranted in our region.

Very rarely specimens of *B. pubescens* or *B. latiglumis* have the first glume 3-nerved and the second glume either 3- or 5-nerved. They will not run well to any species in the other half of the key, but will rather obviously belong here on the basis of all other characteristics.

8. **B. ciliatus** L.   Fig. 65   Fringed Brome
Map 117. Stream banks, thickets, moist shores, openings in cedar bogs, ditches, and wet places generally, including marshes, bogs, and low woods.

Extremely variable in pubescence. Very rarely the upper nodes are barely included in the sheaths.

9. **B. kalmii** Gray
Map 118. Chiefly associated with jack pine, including swampy areas, but also with oaks, on dolomite pavement of Drummond Island, in bogs, and on banks of streams and lakes.

A rare specimen of *B. pubescens* with the first glume weakly 3-nerved might run here but could be readily distinguished by its longer awns and anthers.

Baum (Canad. Jour. Bot. 45: 1848. 1967) has pointed out that the specimen which should serve as the type of *B. purgans* L. is the same as the type of the later-named *B. kalmii*. However, Wagnon applied the name *B. purgans* to what others have called *B.latiglumis*, and most previous authors have used *B. purgans* in the sense of what now is called *B. pubescens*. Rather than apply the name *B. purgans* in a third sense, replacing what has long been known as *B. kalmii*, I reject it as a *nomen confusum* under Art. 69 of the International Code of Botanical Nomenclature.

10. **B. mollis** L.   Soft Chess
Map 119. Clearings, fields, lawns, roadsides, and waste places. Introduced from Europe.

Normally hairy on sheaths, blades, and culms, as well as spikelets. An occasional specimen with glabrate spikelets can be distinguished from *B. racemosus* by the short anthers. A taller plant than the next species.

11. **B. molliformis** Lloyd
Map 120. Distinctive-looking small plants with divaricate awns collected in waste ground at the former site of the University of Michigan Botanical Gardens in Ann Arbor (*Hermann 6014* in 1934, MICH) seem to be this European species, which is locally adventive in North America. A specimen from the same place (*Hermann 6225*, US) is the basis of the report of *B. alopecuros* Poiret from

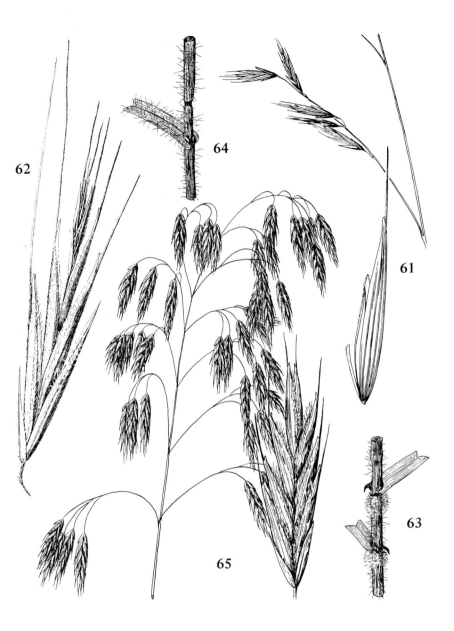

61. *Melica smithii,* panicle branch ×1; floret ×5
62. *Bromus tectorum,* spikelet ×5
63. *B. latiglumis,* stem showing leaf sheath ×2
64. *B. pubescens,* stem showing leaf sheath ×2
65. *B. ciliatus* ×1; spikelet ×4

Michigan (*Manual of Grasses*). These specimens differ from *B. mollis* as indicated in the key, but resemble the latter in having the lemmas strongly 7-nerved and anthers ca. 1 mm long. (The European *B. alopecuros* is said to have the lemmas 9-nerved and anthers ca. 1.5 mm long.) The group of *Bromus* including species 10–17 is in a state of taxonomic and nomenclatural turmoil and the treatment here is by no means definitive.

12. **B. secalinus** L.   Fig. 66   Cheat; Chess
    Map 121. Roadsides, fields, banks, and waste ground. Native of Europe.
    A distinctive plant when ripe, with strongly inrolled lemmas and tendency for the tip of the grain and palea to be slightly exserted. Occasional plants have the lowermost sheaths with short hairs, but can usually be identified by the long palea and (if ripe) strongly inrolled lemmas, combined with the relatively short and strongly undulate awns.

13. **B. briziformis** Fischer & Meyer
    Map 122. Roadsides and other waste places. Introduced from Europe.
    Superficially resembling *Briza*, and similarly sometimes cultivated for orna-ment.

14. **B. squarrosus** L.   Fig. 67
    Map 123. First collected in Michigan in 1932 at Niles, Berrien County (*Hebert*, ND), and spreading along railroad tracks. Another European species.
    This species resembles *B. japonicus* in the short anthers, the fine, soft, ± crooked or matted hairs on the sheaths, and the shortness of the awn of the lowest lemma in a spikelet (less than half as long as the longest awn in the same spikelet). However, the lemmas of *B. japonicus* are not over 2.2 mm wide on a side, whereas in *B. squarrosus* at least the larger ones are 2.5–3.5 mm wide, including a hyaline border which may be as broad as 0.7–1 mm at the prominent angled "shoulder" near the middle of the margin of the lemma. Well developed mature awns in *B. squarrosus* spread ± at right angles to the spikelet or are somewhat recurved; in *B. japonicus* they are seldom so widely spreading.

118. Bromus kalmii

119. Bromus mollis

120. Bromus molliformis

15. **B. japonicus** Murray   Japanese Brome

Map 124. Roadsides, railroads, fields, and other waste places. Native of Europe and Asia.

See remarks under *B. squarrosus* above.

16. **B. commutatus** Schrader   Hairy Chess

Map 125. Roadsides, fields, railroads, and other waste places. Another introduction from Europe.

Occasional specimens closely approach the next species or the preceding one.

17. **B. racemosus** L.

Map 126. Waste places, as with the other weedy species in the genus, and, like others, a native of Europe.

The panicle is much smaller, narrower, and with fewer spikelets than the preceding, and the whole plant is smaller (ca. 30–40 cm tall, compared with 50–90 or even 120 cm in *B. commutatus* although depauperate individuals are sometimes smaller).

This and the last few species have been subject to varying interpretations, and are not as clear-cut as might be desired. The treatment here follows basically the concepts of the *Manual of Grasses*.

121. Bromus secalinus

122. Bromus briziformis

123. Bromus squarrosus

124. Bromus japonicus

125. Bromus commutatus

126. Bromus racemosus

14. Festuca

Fescue Grass

In many specimens the tip of the palea is narrow, ± inrolled, and projecting slightly beyond the tip of the lemma. A superficial glance at a pressed specimen may thus suggest a bifid tip to the lemma, when this is in fact not the case.

KEY TO THE SPECIES

1. Blades of leaves flat (or merely once-folded), at least the larger ones (2.5) 3–8 mm broad; lemmas awnless or rarely with awn less than 0.8 mm long
    2. Larger lemmas 2.5–4.5 mm long; anthers 0.8–1.4 mm long; spikelets mostly containing 2–4 (5) florets and borne beyond the middle of the primary panicle branches . . . . 1. **F. obtusa**
    2. Larger lemmas 5.5–8 mm long; anthers 2.2–3.5 (3.8) mm long; spikelets often containing 5 or more florets, borne (except in *F. scabrella*) below as well as above the middle of the primary panicle branches
        3. Lemmas scabrous; larger glumes ca. 5.5–7.5 mm long . . . . 4. **F. scabrella**
        3. Lemmas smooth; larger glumes less than 5.5 (6) mm long
            4. Auricles at summit of leaf sheath ciliate; larger lemmas 7–8.5 mm long; branches of panicle mostly 2 at each node, both with several spikelets . . . . .
            . . . . 2. **F. arundinacea**
            4. Auricles glabrous; larger lemmas 5.5–7 mm long (very rarely some to 8 mm); branches of panicle mostly 1 at each node, or the second if present bearing usually only one spikelet . . . . 3. **F. pratensis**
1. Blades of leaves ± strongly involute, less (usually much less) than 3 mm broad; lemmas awned or awnless
    5. Second (larger) glume ca. 5.5–7.5 mm long; larger lemmas ca. 6–7.5 mm long, somewhat keeled, ± strongly scabrous over the back, merely acute or at most very short-awned; plants tall (ca. 5–8 dm) and stout, the blades of the lower leaves disarticulating from their stiff persistent sheaths . . . . 4. **F. scabrella**
    5. Second glume less than 5 mm long; larger lemmas ca. 2.5–6.5 mm long, not at all keeled, scabrous to smooth or somewhat hairy, distinctly awned (except in *F. tenuifolia*); plants rather more slender and usually shorter than the preceding, the lower sheaths persistent or not
        6. Lemmas awnless, ca. 2.5 (–3) mm long; leaves delicately capillary, at most 0.2 mm thick, mostly more than half as high as the culm . . . . 5. **F. tenuifolia**
        6. Lemmas awned, the body ca. 3–6.5 mm long; leaves often stiff, up to 0.8 mm thick, and in some species mostly less than half as high as the culm
            7. Plants annual, usually in small tufts or solitary; florets cleistogamous, with usually one included anther (rarely 3 anthers) . . . . 6. **F. octoflora**
            7. Plants perennial, usually in dense tufts including numerous dry sheaths of previous years; florets open at anthesis, with 3 anthers
                8. Margins of lemmas conspicuously thin and membranous; summit of ovary bristly-pubescent; awns (at least the longer ones) more than 3 mm long, nearly equalling or longer than the bodies of their lemmas; mature panicle open and lax . . . . 7. **F. occidentalis**
                8. Margins of lemmas at most very narrowly membranous-bordered, the lemmas ± firm and thick throughout; summit of ovary glabrous; awns all less than 3 (4) mm long, shorter than the bodies of their lemmas; mature panicle rather narrow, crowded, and compact, the branches strongly ascending or, if spreading, very short

9. Sheaths closed in young leaves, the old ones ± dark reddish brown basally, becoming fibrous by splitting between the prominent pale veins; basal shoots usually arising laterally, the culms thus tending to be strongly curved or bent at the base; anthers (1.7) 2.1–3.5 (3.7) mm long . . . . 8. **F. rubra**
9. Sheaths open most of their length even in young leaves (margins ± overlapping), the old ones mostly pale or drab brown, not becoming fibrous; basal shoots erect, the culms thus nearly or quite straight from the base upwards; anthers various
    10. Anthers (1.8) 2–3 mm long; lower panicle branches often spreading . . .
    . . . . 9. **F. ovina**
    10. Anthers 1–1.6 (1.8) mm long; lower panicle branches strongly ascending
    . . . . 10. **F. saximontana**

1. **F. obtusa** Biehler   Fig. 68   Nodding Fescue

Map 127. Rich, often moist woods, usually beech–maple but also oak–hickory; occasionally in wet coniferous woods.

The panicle branches are deflexed at maturity, as in *Melica smithii*, which may be growing in the same woods.

2. **F. arundinacea** Schreber   Tall Fescue

Map 128. Probably more widespread than suggested by our few records from roadsides, riverbanks, and lawns. A recent introduction from Europe, sparingly established in North America.

The lemmas often have a very short awn formed by the excurrent midvein, whereas in the next species the lemmas are usually awnless. This species is reported to be a hexaploid, taller and coarser than the diploid *F. pratensis*.

3. **F. pratensis** Hudson   Fig. 69   Meadow Fescue

Map 129. Roadsides, shores, meadows, and waste ground, often damp. A native of Europe, long known as *F. elatior* L. (a name which really applies to the preceding species and hence is a source of confusion).

4. **F. scabrella** Torrey   Rough Fescue

Map 130. Local, sometimes common, on the jack pine plains. A northern and prairie species, found from North Dakota and Colorado northwestward to

127. Festuca obtusa

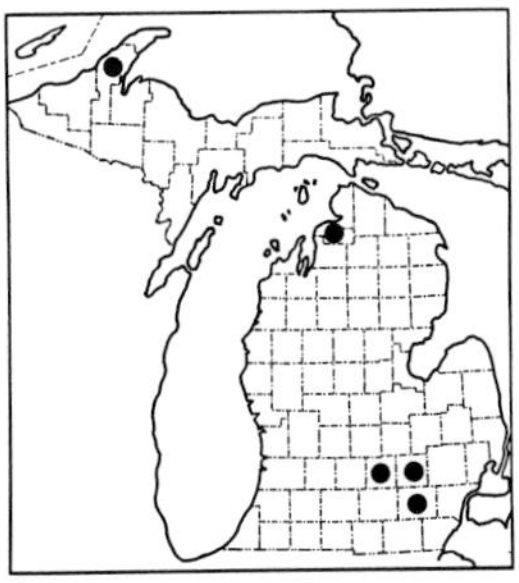

128. Festuca arundinacea

129. Festuca pratensis

Alaska; also in Newfoundland and Quebec, and isolated in the Great Lakes region.

A larger and coarser variety, var. *major* Vasey, is sometimes recognized, and includes our plants. The species is by some authors included with *F. altaica* of northeastern Asia.

5. **F. tenuifolia** Sibth.

Map 131. A weed of dry open ground, introduced from Europe; sometimes in lawn seed.

Often called *F. capillata* Lam., an illegitimate name.

6. **F. octoflora** Walter   Fig. 70   Six-weeks Fescue

Map 132. Usually in sandy, often disturbed places: dunes and shores, roadsides, oak woods.

Often placed in a segregate genus, as *Vulpia octoflora* (Walter) Rydb.

7. **F. occidentalis** Hooker   Western Fescue

Map 133. Open often rocky woods, wooded dunes (with pines), cedar–fir woods and thickets (often in calcareous sites), aspens, and hardwoods. A species of the northwestern United States and adjacent Canada, with outlying stations in the northern Great Lakes region.

The characteristic bristly pubescence on the summit of the ovary should not

130. Festuca scabrella

131. Festuca tenuifolia

132. Festuca octoflora

133. Festuca occidentalis

134. Festuca rubra

135. Festuca ovina

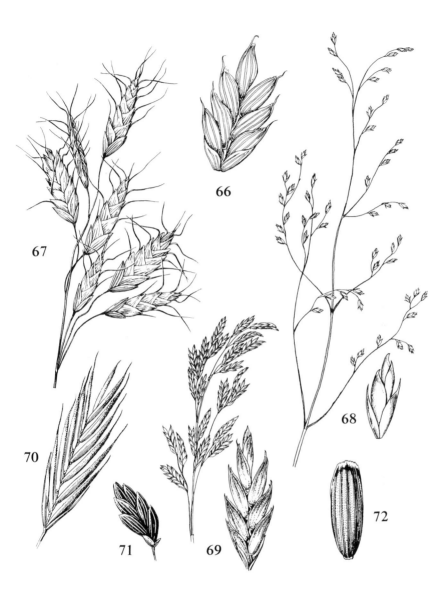

66. *Bromus secalinus,* spikelet ×2
67. *B. squarrosus* ×1¼
68. *Festuca obtusa* ×½ ; spikelet ×5
69. *F. pratensis* ×½ ; spikelet ×2
70. *F. octoflora,* spikelet ×5
71. *Glyceria striata,* spikelet ×5
72. *Puccinellia pallida,* floret (lemma) ×10

be confused with the feathery styles which are present in this as well as other species.

8. **F. rubra** L.

Red Fescue

Map 134. Roadsides, fields, shores, meadows, woodlands — almost anywhere in open places. Both native and introduced strains are presumably included in our material.

The youngest green shoots available should be examined for the distinctive closed sheaths of this species. Otherwise, *F. rubra* is difficult to distinguish from *F. ovina* which it approaches in var. *commutata* Gaudin, in which the new shoots are erect, rather than bursting laterally through the base of the old fibrous sheaths. In *F. rubra* there is a strong tendency for the lowermost panicle branches (at least one of them) to be 5 (rarely only 4) mm long before the first pedicel; whereas in the next two species, the lowermost panicle branches are usually all less than 5 (rarely 7) mm long before the first (sometimes aborted) pedicel. *F. rubra*, unlike the next two species, is sometimes conspicuously rhizomatous, especially in lawns.

9. **F. ovina** L.

Sheep Fescue

Map 135. Roadsides, fields, and waste places; dry woods (e. g., oak, pine). An introduced species, naturalized from the Old World; sometimes in lawn seed mixtures.

Many of our specimens are a coarser form sometimes segregated as var. *duriuscula* (L.) W. D. J. Koch. See also notes under the next species and the preceding.

10. **F. saximontana** Rydb.

Map 136. Dunes, dry woods (aspen, pine, oak–hickory), shores, and waste places; rock crevices and rocky summits in the Lake Superior region.

A native species very similar to the preceding, with which it seems to intergrade and from which some authors do not distinguish it. *F. brachyphylla* Schultes has been reported from Keweenaw County; it supposedly differs from

136. Festuca saximontana   137. Glyceria acutiflora   138. Glyceria borealis

F. saximontana in having even smaller anthers (not over 1 mm) and a shorter panicle (1–3 cm). It is doubtful whether our plants are anything other than depauperate specimens of *F. saximontana*. These last three species (nos. 8, 9, & 10) are superficially very similar. Spikelets in all three may be flushed with reddish or purplish, though this condition is most frequent and pronounced in *F. rubra*.

15. Glyceria   Manna Grass

KEY TO THE SPECIES

1. Spikelets (7) 9–31 (38) mm long, linear-cylindric, on erect, straightish, strongly ascending pedicels mostly shorter than the spikelets; ligules mostly 6–16 mm long (often infolded and/or strongly lacerate apically)
   2. Lemmas acute at the tip, exceeded ca. 1.5–2.5 mm by the palea . . . . 1. **G. acutiflora**
   2. Lemmas obtuse at the tip, at most exceeded 0.5 mm by the palea
      3. Back of lemma smooth or nearly so, at least between the nerves; anthers 0.5–1.1 mm long; grain 1.2–1.6 mm long; wider leaf blades 2–4.5 (7) mm broad . . . . 2. **G. borealis**
      3. Back of lemma scabrous to hispidulous; anthers 1–1.6 (1.8) mm long; grain 1.8–2.6 mm long; wider leaf blades 5–11 mm broad . . . . . . 3. **G. septentrionalis**
1. Spikelets 2–6.5 (7) mm long, ovoid to oblong or short-cylindric, mostly on spreading, lax, or strongly undulate pedicels often longer than the spikelets; ligules 2–5 (6.5) mm long
   4. Lemmas smooth, not corrugated (the nerves visible but not strongly raised), the larger ones (2.8) 3–3.5 (4) mm long; spikelets becoming 3–5 mm broad at maturity . . . . 4. **G. canadensis**
   4. Lemmas with prominent raised nerves, giving a ± corrugated appearance (fig. 71), the larger ones 1.5–2.5 mm long; spikelets becoming 1.5–2.6 mm broad at maturity
      5. Larger (second) glume 1.6–2.2 (2.5) mm long; smaller (first) glume 1.2–1.7 (1.8) mm long; anthers ca. 0.7–1 mm long. . . . . 5. **G. grandis**
      5. Larger glume less than 1.5 mm long; smaller glume less than 1 (1.2) mm long; anthers ca. 0.3–0.5 mm long . . . . 6. **G. striata**

1. G. acutiflora Torrey

Map 137. Known in Michigan only from collections (WMU, MICH, GH) made by Mr. and Mrs. Hanes in marshes and wet ground near Schoolcraft, Kalamazoo County; otherwise ranges south of Michigan. This species grows in eastern Asia and eastern North America.

2. **G. borealis** (Nash) Batch.   Fig. 73

Map 138. Shallow water (seldom over 2 feet) and wet borders of ponds, lakes, ditches, marshes.

This species and the next develop limp floating leaves with a papillose-pubescent non-wettable upper surface when growing in water.

3. G. septentrionalis Hitchc.

Map 139. Wet places, including woodland pools, swampy hollows, marshes, ditches. Not as common as most species in the genus.

4. G. canadensis (Michaux) Trin.   Fig. 74   Rattlesnake Grass

Map 140. In the same situations as *G. borealis*, but also frequently in bogs, tamarack and cedar swamps, and on peaty shores.

The lemmas appear distinctly pointed in this species, and exceed their paleas — often by as much as 0.5 mm or slightly more. In the next two species, the lemmas are about the same length as their paleas or exceed them slightly with broad scarious tips. *G. canadensis* apparently does not produce the aquatic leaf form of *G. borealis*.

5. G. grandis S. Watson

Map 141. River and stream margins, ditches and swales, marshes, shores, bogs, meadows — often in shallow water.

A much more robust plant in all aspects than the next, with generally stouter culm, broader leaf blades, and much larger panicles, the latter usually 22—40 cm high (rarely as small as 13 cm). The spikelets are about 4—6.5 mm long.

6. G. striata (Lam.) Hitchc.   Fig. 71   Fowl Manna Grass

Map 142. The commonest species of the genus throughout the state, in almost all wet situations: hollows and ravines in woods, shores, bogs, swamp forests (both coniferous and deciduous).

Smaller in stature than our other species. The panicle is usually 12—17 cm high (rarely as small as 5 cm or as large as 25 cm). The spikelets are about (2) 2.5—4 (5.5) mm long. Probably our most variable species in color of inflorescence, the spikelets ranging from completely green to deeply suffused with purple. The two-ranked leaves are often strongly displayed in the shade of wet woods and open thickets.

139. Glyceria
   septentrionalis

140. Glyceria canadensis

141. Glyceria grandis

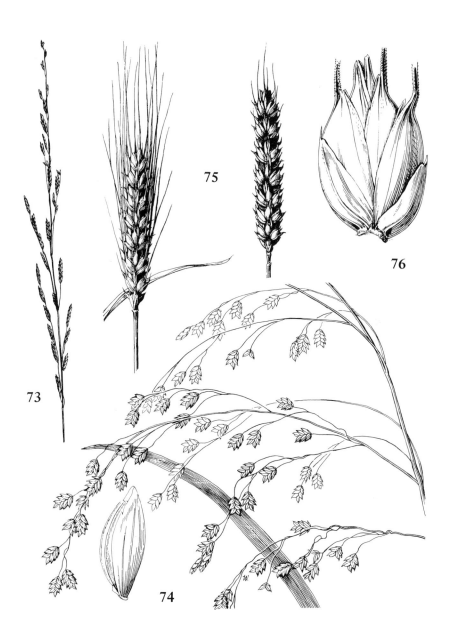

73. *Glyceria borealis* ×⅖

74. *G. canadensis* ×1; floret ×10

75. *Triticum aestivum* ×½ (awned or "bearded" at left; nearly awnless or "beardless" at right)

76. *T. aestivum,* spikelet (awns cut short) ×3

16. Puccinellia

REFERENCES

Church, George L. 1949. A Cytotaxonomic Study of Glyceria and Puccinellia. Am. Jour. Bot. 36: 155–165.

Church, George L. 1952. The Genus Torreyochloa. Rhodora 54: 197–200.

Clausen, Robert T. 1952. Suggestion for the Assignment of Torreyochloa to Puccinellia. Rhodora 54: 42–45.

Fassett, Norman C. 1946. Glyceria pallida and G. fernaldii. Bull. Torrey Bot. Club 73: 463–464.

KEY TO THE SPECIES

1. Lemmas smooth, usually somewhat hairy at the base, the nerves very indistinct; ligules less than 2 (2.5) mm long . . . . 1. **P. distans**
1. Lemmas minutely hispidulous, not hairy at the base, the nerves ± prominent; larger ligules 2.5–9 mm long
   2. Larger leaves with blade (4) 5–10 mm wide and ligule 5–9 mm long; anthers (0.8) 1.1–1.5 (1.7) mm long; larger lemmas 2.7–3.5 mm long . . . . 2. **P. pallida**
   2. Larger leaves with blades 1.5–3.5 mm wide and ligule 2.5–6.5 mm long; anthers 0.3–0.5 (0.6) mm long; larger lemmas ca. 2.1–2.5 mm long . . . . 3. **P. fernaldii**

1. **P. distans** (Jacq.) Parl.   Alkali Grass

Map 143. Local in salty waste ground, especially along roadsides and ditches. Introduced from Eurasia.

2. **P. pallida** (Torrey) Clausen   Fig. 72

Map 144. Pond borders, wet hollows in woods, bogs, cat-tail marshes, often in shallow water. A Farwell collection labeled (not by him) from Keweenaw County (in 1888, ILL) is suspicious and is not mapped, although this basically southern species in Michigan is known from one authentic Upper Peninsula collection: Au Train Tp., Alger County (*Churchill* in 1964, MSC).

This species and the next have long been placed in the genus *Glyceria* because of their strongly nerved lemmas. They differ, however, in their open sheaths and 3-nerved second glumes; in these characters, as well as in basic chromosome

142. Glyceria striata   143. Puccinellia distans   144. Puccinellia pallida

number, they belong with *Puccinellia* (as confirmed by Gould), although a separate genus, *Torreyochloa*, has also been described for them. *P. pallida* is one of the plants which is found in eastern North America and eastern Asia.

3. **P. fernaldii** (Hitchc.) E. Voss

Map 145. Wet creek borders, bogs, alder thickets, and wet coniferous thickets.

More slender and smaller than the preceding, which it closely resembles and of which it is often considered a variety. Specimens may superficially be confused with *Glyceria striata*, from which they differ not only in the open sheaths but also in the more strongly scabrous branches of the inflorescence.

2. TRITICEAE

This is the tribe long known as Hordeae, a more recent name. Containing wheat, barley, and rye, it is one of the most important groups of economic plants in the world. The spikelets and their parts in some species are oriented asymmetrically, leading to easy confusion in their interpretation: glumes may be obsolete, reduced to mere awns, or apparently beside each other on one side of the spikelet; keels may not be centered, the inward-turned portion of glumes or lemmas then narrower and sometimes much more membranous than the outward portion. The key below is based on the superficial appearance of our species and avoids as far as possible the morphological problems of the spikelets. Intergeneric hybrids have frequently been reported, and the lines between genera are not always clear.

Nardus is segregated by Gould (1968) into a separate tribe, Nardeae; and *Lolium* is included in the Poeae [Festuceae]. *Parapholis incurva* (L.) Hubbard, also classified as *Pholiurus incurvus* (L.) Schinz & Thell., is placed by Gould in the tribe Monermeae but it has traditionally been in the Triticeae. It was reported by Farwell from waste ground in Detroit (1914), but search for specimens has been fruitless. The spikelets are 1-flowered, partly embedded in the distinctively curved spikes, covered by the glumes, forming a cylindrical spike that disarticulates into segments each of which contains one spikelet.

145. Puccinellia fernaldii

146. Secale cereale

147. Triticum aestivum

REFERENCE

Bowden, Wray M. 1959. The Taxonomy and Nomenclature of the Wheats, Barleys, and Ryes and their Wild Relatives. Canad. Jour. Bot. 37: 657–684.

KEY TO THE GENERA OF TRITICEAE

1. Lemmas smooth and glabrous except for a spiny-ciliate keel and exposed margin, tapering into a long awn . . . . 1. **Secale**
1. Lemmas smooth to scabrous or pubescent, but not simply with spiny-ciliate keel and margin, awned or awnless
  2. Larger glumes 3.3–6.5 mm broad with at least 3 prominent nerves, the keel or midnerve not centered
    3. Glumes glabrous, or pubescent toward the base on nerves and margins (rarely pubescent throughout), the larger ones (3.7) 5–6.5 mm wide, less than 3 times as long (excluding awns if present); lemmas awned or awnless . . . . 2. **Triticum**
    3. Glumes softly hairy or glabrous throughout, 3.3–4.2 mm wide, ca. 6–10 times as long; lemmas awnless . . . . 3. **Elymus** (couplet 2)
  2. Larger glumes less than 2.5 mm broad, variously nerved (or glumes absent)
    4. Spikelets mostly 2–3 at each node of the rachis; if this arrangement is obscured by reduction of some spikelets and/or their asymmetric position, the basic structure is still evident by the presence of a total of 4–6 glumes subtending the entire group of spikelets (glumes awn-like or narrow, resembling an involucre by their position side by side rather than opposite each other on the two sides of each spikelet, usually obsolete in *Hystrix* with mostly 2 easily recognized narrow spikelets at each node)
      5. Spikelets 2 at each node of the rachis (or at some nodes, only 1, rarely 3, but total number of glumes (awn-like or broader) developed at a node not more than 4)
        6. Glumes present, awn-like to lanceolate, of about equal length; spikelets ascending at maturity, usually concealing much of the rachis . . . . 3. **Elymus**
        6. Glumes absent or obsolete, or if present slenderly awn-like their entire length and at least one of those at a node much shorter than others; spikelets horizontally spreading at maturity (± ascending when young), well separated, clearly revealing the entire rachis (fig. 82) . . . . 4. **Hystrix**
      5. Spikelets basically 3 at each node of the rachis (the lateral 2 in commonest species reduced to bristles), this arrangement most easily recognized by the presence of 6 awn-like or narrowly lanceolate and awn-tipped glumes at a node
        7. Body of larger lemmas ca. 3.5–6 mm long; rachis of spike readily disintegrating at maturity . . . . 5. **Hordeum**
        7. Body of larger lemmas ca. 8–12 mm long; rachis not disintegrating
          8. Awn of lemmas much stouter than awn of glumes, ± straight . 5. **Hordeum (vulgare)**
          8. Awn of lemmas as slender as awn of glumes, spreading to recurved at maturity . . . . 3. **Elymus** (couplet 5)
    4. Spikelets clearly 1 at each node (or most nodes) of the rachis; glumes variously arranged (or absent), but not more than 2
      9. Glumes absent; spikelets 1-flowered, in two rows appearing as one, all on one side of rachis . . . . 6. **Nardus**
      9. Glumes (1 or 2) present; spikelets 1–many-flowered, on opposite sides of the rachis
        10. Glumes 1 (except on terminal spikelet), the narrow edge of the spikelet against the rachis and lacking a glume . . . . 7. **Lolium**
        10. Glumes 2, the broad side of the spikelet against the rachis . . . . . . 8. **Agropyron**

1. Secale

1. **S. cereale** L.   Fig. 77   Rye
Map 146. Chiefly along roadsides, where in part presumably sown for erosion control on newly made shoulders and embankments; also on shores, dunes, along railroads, and in old fields. A cultivated annual, freely escaping but not long persisting. Doubtless more widespread than the map suggests.

2. Triticum

1. **T. aestivum** L.   Figs. 75, 76   Wheat
Map 147. Roadsides, railroads, and other waste ground, occasionally on shores. A familiar cultivated annual, escaping less often than rye.
   There are numerous cultivated strains, and no effort has been made to distinguish those which escape in Michigan; all of our wheat, because of its basic hybrid origin, may be grouped under the name *T.* × *aestivum* L. (*pro sp.*) *emend.* Bowden (see Bowden, 1959). "Bearded wheat" has long-awned lemmas; both bearded and beardless forms occur among our waifs and are shown in fig. 75.

3. Elymus   Wild-rye
   Hybridization among the species of *Elymus*, and even with other genera, has been the subject of considerable speculation and experimentation by cyto-taxonomists, some of whose work is included in the references below. *Elymus interruptus* Buckley has been reported from Michigan, but specimens from our area, if not variants of other species, are presumably of some sort of hybrid origin (see Church, 1954). George L. Church has been of distinct aid in clarifying many of our specimens.
   The total number of leaves (i. e., nodes) on a culm is often a useful character in distinguishing species; care must be taken to include the often shriveled leaf at the lowest node. Our species are quite distinct, with very few hybrids apparently having been collected in Michigan. However, the range of variation in any one "key character" often meets or slightly overlaps that of another species, so that a plant which does not run well under one half of a couplet in the key should be sought under the other half, where it may better fit the descriptive statements.

REFERENCES

Bowden, Wray M. 1957. Cytotaxonomy of Section Psammelymus of the Genus Elymus. Canad. Jour. Bot. 35: 951–993.
Church, George L. 1954. Interspecific Hybridization in Eastern Elymus. Rhodora 56: 185–197.
Church, George L. 1958. Artificial Hybrids of Elymus virginicus with E. canadensis, interruptus, riparius, and wiegandii. Am. Jour. Bot. 45: 410–417.
Church, George L. 1967. Taxonomic and Genetic Relationships of Eastern North American Species of Elymus with Setaceous Glumes. Rhodora 69: 121–162.

1. Glumes lanceolate, 3.3–4.2 mm wide, usually hairy; lemmas awnless (at most with
firm pointed tip); culms usually pubescent at the summit
  2. Culms finely and densely pubescent at the summit; glumes hairy . . . . 1. **E. mollis**
  2. Culms essentially glabrous at the summit; glumes glabrous or scabrous . 2. **E. arenarius**
1. Glumes very narrowly lanceolate, not over 2 mm wide, glabrous to very scabrous
or villous-hispid; lemmas distinctly awned; culms glabrous at summit (or at most
antrorsely hispid for ca. 2 mm)
  3. Larger paleas (lowest in each spikelet) 8.6–12.7 (14) mm long; awns of lemmas
usually widely spreading to recurved at maturity
    4. Body of glume about twice as long as its awn, or longer; awns of lemmas usually
straight at maturity; spike curved to erect . . . . [go to couplet 7]
    4. Body of glume about equalling its awn, or shorter; awns of lemmas ± curved at
maturity (straight when young); spike curved to strongly nodding
      5. Leaves 5–8 on a culm, the broadest blades rarely as much as 15 (17) mm wide,
glabrous above . . . . 3. **E. canadensis**
      5. Leaves 10–12 on a culm, the broadest blades (14) 15–19 mm wide, finely
hairy above . . . . 4. **E. wiegandii**
  3. Larger paleas 5.5–8.5 (9) mm long; awns of lemmas mostly straight
    6. Glumes, at least the broadest, 1–1.7 (2) mm wide, clearly expanded and
flattened above the base
      7. Base of glumes not conspicuously bowed out, but flattened, indurated for less
than 1 mm; glumes not thickened above the base on inner face, with very
narrow, thin, translucent margins, often slightly overlapping; culm leaves 5–6;
largest palea 7.5–10.5 (12) mm long . . . . 5. **E. glaucus**
      7. Base of glumes ± bowed out, terete, and indurated for 1 mm or more; glumes
also swollen, pale, and indurated on inner face for about the basal half or
more, with firm margins, not at all overlapping; culm leaves (6) 7–10; largest
palea 6.8–8.5 (9) mm long . . . . 6. **E. virginicus**
    6. Glumes less than 1 mm wide, scarcely if at all widened above the base
      8. Palea (of lowest floret in spikelet) (6.5) 7–8 mm long; leaves ca. (8) 9–10,
glabrous . . . . 7. **E. riparius**
      8. Palea 5.5–6.7 (7) mm long; leaves (5) 6–7, the sheaths and upper surface of
blades ± finely villous . . . . 8. **E. villosus**

1. **E. mollis** Trin.   Plate 3-A   Dune Grass
    Map 148. Rather local on sandy beaches and dunes along Lake Superior from
Chapel Beach to Whitefish Point. A very large, glaucous, and handsome grass,
standing out even among the *Ammophila* near which it is usually growing.
Closely related to the European octoploid *E. arenarius*, the tetraploid *E. mollis* is
found on shores from Korea and Japan eastward through Alaska and northern
Canada to Greenland and Iceland, ranging southward to California, eastern
Canada, Hudson Bay, and Lake Superior.

2. **E. arenarius** L.   Lyme Grass
    Map 149. An introduced species which, like the preceding, is an excellent
sand binder and may be planted for that purpose. Known from sand dunes in
Allegan County near Saugatuck (*J. G. Guerin* in 1952, MSC) and several places
in Berrien County where collected as early as 1941 (*G. N. Jones 14194*, ILL,

MO). Known from Wisconsin and Illinois shores, and to be expected elsewhere on Lake Michigan.

3. **E. canadensis** L.   Fig. 78
   Map 150. Most often seen on sandy (or even somewhat marshy) shores and on sand dunes and in associated thickets; also in woods, especially along trails, riverbanks, and streams; occasionally with weedy tendencies along roadsides and in waste ground.
   The spikelets are often 3 at a node, especially toward the base of the spike, in this and the next species. The lemmas are usually ± strongly hirsute, very rarely glabrous or merely scabrous [f. *glaucifolius* (Willd.) Fern., known from Ingham, Kalamazoo, Menominee, and Wayne counties]. The spikes range from about 6 cm long in depauperate plants to 30 cm or more. A very variable species, the spikelets and foliage sometimes strongly glaucous. Some specimens of *E. riparius* may resemble young *E. canadensis* superficially, but differ in narrower glumes with a more distinct terete portion at the indurated base, and in more numerous, less involute leaves. See also comments under the next species.

4. **E. wiegandii** Fern.
   Map 151. In woods along rivers and creeks.
   Not distinguished from *E. canadensis* by some authors, but recognized by others with the support of cytological evidence. Collections from Michigan can be easily distinguished as given in the key. *E. canadensis* supposedly tends to have shorter paleas than *E. wiegandii* (8.6–11 mm in the former); however, our specimens which on all other characters are clearly *E. canadensis* have the larger (lower) paleas usually 8.6–12 mm and rarely as much as 14 mm long. Only one or two collections of *E. canadensis* have hairs on the upper side of the leaf blades, which are narrow and few in number, so the leaves seem to offer the best separation in our material. The larger glumes in *E. canadensis* are 0.7–1.6 mm broad, while in our *E. wiegandii* they are 0.4–7 mm. The broad leaves of *E. wiegandii* are thinner and mostly flat, while in *E. canadensis* the leaves are stiff and often involute, especially toward the end.

148. Elymus mollis

149. Elymus arenarius

150. Elymus canadensis

5. E. glaucus Buckley

Map 152. Seemingly in a diversity of habitats, mostly near Lake Superior: sandy thickets and woods among dunes, deciduous woods, especially in rocky places; crevices of rock shores. Referred to this species (with confirmation by Church) are a Fernald & Pease collection from a sandy place in Chippewa County originally determined as *E. virginicus* var. *jejunus* and a Keweenaw County roadside collection (*Hermann 1587*, MICH) with divergent awns and more indurate bases than usual.

The back of the lemma is essentially glabrous or very minutely strigose in all our specimens. This is an extremely variable species in width and stiffness of inflorescence and straightness of awns; but apparently it is distinctive among our species (except for the easily identified nos. 1 & 2) in the very narrow thin hyaline borders on the glumes. The spikes of this species rather closely resemble those of some species of *Agropyron*, from which they are distinguished by the presence of 2 spikelets at all, or nearly all, nodes. (See also notes under *Agropyron trachycaulum* and *A. repens*.)

6. E. virginicus L.   Fig. 79

Map 153. Low deciduous or sometimes coniferous woods and thickets, especially along stream banks and floodplains; open marshy shores and meadows.

The back of the lemmas and glumes is rarely hirsute [f. *hirsutiglumis* (Scribner) Fern.]; an awnless form has not yet been found in Michigan. Another extremely variable species, but normally distinct in the strongly indurate glumes (basally and on inner face), conspicuously outcurved at their bases (more slender usually than in fig. 79); the spikelets disarticulate *below* the glumes. In the typical form, to which almost all of our material belongs, the base of the spike is included, or nearly so, in the summit of the ± inflated sheath of the uppermost leaf. In f. *jejunus* Ramaley, the spike is strongly exserted.

A very few collections appear intermediate with *E. riparius* and may represent hybrids.

7. E. riparius Wieg.

Map 154. Usually in moist ground along creeks, borders of woods, and riverbanks; occasionally in somewhat drier places.

The back of the lemmas is ± hispidulous in all of our specimens except one from Cass County. The glumes somewhat resemble narrow ones of *E. virginicus*, but are ± straight, not bowed, at the base.

8. E. villosus Willd.

Map 155. Woods, usually ± swampy ones, as along rivers.

The very slender glumes and the lemmas are usually conspicuously villous-hispid, although nearly smooth in f. *arkansanus* (Scribner & Ball) Fernald; in one Oakland County collection (*Farwell 8582*, MICH, BLH) which appears to be this form, the leaves are glabrous also. The spikelets are nearly always only 1-flowered in *E. villosus*.

4. Hystrix

1. **H. patula** Moench   Fig. 82   Bottlebrush Grass
Map 156. Usually in rich deciduous or mixed woods and riverbanks, especially in wet or slightly disturbed areas in beech—maple—hemlock stands; in swamp forests and sometimes oak—hickory woods; rarely in coniferous swamps.

Reported as tall as 5 feet (nearly 2 meters). The lemmas are glabrous in the typical form; in f. *bigeloviana* (Fern.) Gl., they are rather strongly pubescent. Natural hybridization between species of *Elymus* and *Hystrix* has been postulated (see Church, 1954; 1967), and this species is sometimes included in the preceding genus as *Elymus hystrix* L.

5. Hordeum

KEY TO THE SPECIES

1. Body of larger lemmas ca. 8—11 mm long; leaves glabrous, with prominent auricles at base of blade; awns of lemmas much stouter than those of glumes; rachis of spike not disintegrating . . . . 1. **H. vulgare**
1. Body of larger lemmas ca. 3.5—6 mm long; leaves (at least lower sheaths) ± pubescent, without auricles; awns of lemmas as slender as those of glumes; rachis of spike readily disintegrating as it matures
 2. Awns less than 2 cm long; glumes in part broadened (± lanceolate) . . . . 2. **H. pusillum**
 2. Awns much longer; glumes all bristle-like (reduced to awns) . . . . 3. **H. jubatum**

1. **H. vulgare** L.   Barley
Map 157. Railroads, roadsides, and fields, escaped from cultivation and not long persisting.

The taxonomy of barley is discussed by Bowden (1959), who treats all of the cultivated strains as belonging to a single species. In some plants ("2-rowed barley") the lateral spikelets of each group of 3 are sterile; in others ("6-rowed barley") the lateral spikelets are well developed and fertile. Both forms are

151. Elymus wiegandii

152. Elymus glaucus

153. Elymus virginicus

found in waste ground in Michigan. The narrowly lanceolate glumes of this species are nearly always silky.

2. **H. pusillum** Nutt.

Map 158. Collected in 1900 on Belle Isle (*Farwell 1674½*, BLH) and in 1968 in waste ground along railroad tracks in White Pigeon (*S. N. Stephenson 68-17*, MSC, MICH). Generally considered native only in the southern United States, but like the next species, aggressively weedy.

A small annual, quite unlike *H. jubatum* in overall appearance, although the spikelets are basically similar except for the expanded glumes (both glumes of the central floret and the inner glumes of the 2 lateral florets in each group of 3 florets).

3. **H. jubatum** L.   Plate 3-B   Squirrel-tail Grass

Map 159. Roadsides, railroads, and waste ground generally; shores and disturbed places. Probably throughout the state although not well collected.

The florets of the lateral spikelets of each group of 3 in this species are reduced to inconspicuous bristles; these spikelets are short-pedicelled, the awn-like glumes on the pedicel thus forming a Y-shaped structure with the bristle-like reduced floret at the fork. The awns of the narrow fertile central floret are similar to the glumes of all three spikelets, and the entire spike appears to be a large mass of upwardly scabrous awns, making most unmanageable specimens when the dry rachis separates between each group of spikelets.

154. Elymus riparius

155. Elymus villosus

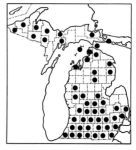
156. Hystrix patula

157. Hordeum vulgare

158. Hordeum pusillum

159. Hordeum jubatum

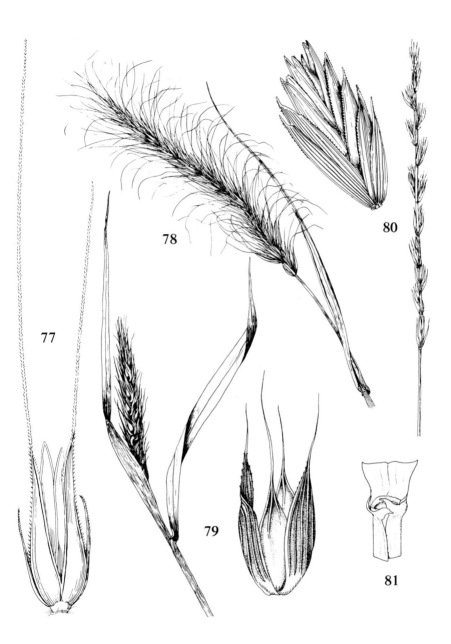

77. *Secale cereale,* spikelet ×3
78. *Elymus canadensis* ×½
79. *E. virginicus* ×½ ; spikelet ×3
80. *Agropyron repens* ×½ ; spikelet ×3
81. *A. repens,* summit of sheath with ligule and auricles ×3

6. Nardus

Placed in a separate tribe, Nardeae, by many modern authors.

1. **N. stricta** L.   Fig. 83   Mat Grass
   Map 160. Introduced from Europe. Found by Farwell in low ground at two
localities in Houghton County, 1937–1940 (MICH, BLH, MSC, WUD, WIS).
   A densely cespitose grass, with narrow involute leaves.

7. Lolium

REFERENCE

Terrell, Edward E. 1968. A Taxonomic Revision of the Genus Lolium. U. S. Dep. Agr.
   Tech. Bull. 1392. 65 pp.

KEY TO THE SPECIES

1. Glume about equalling or slightly exceeding spikelet (excluding awns). . 1. **L. temulentum**
1. Glume distinctly shorter than spikelet . . . . 2. **L. perenne**

1. **L. temulentum** L.   Fig. 84   Darnel
   Map 161. Reported several times from Michigan, but apparently the only
collections are from Hancock (*J. W. Robbins* in 1860, GH) and a dump in
Cheboygan (*L. H. Harvey 671a* in 1938, Harvey). A native of Europe.
   The longest glumes in spikes of this species are nearly always at least 15 mm,
often 2 cm, whereas in *L. perenne* they are less than 15 mm long. This is the
plant referred to as "tares" or simply "weeds" sown by the enemy in the parable
in the 13th chapter of Matthew. The species has a bad reputation because of a
fungus which infests the grains and renders them poisonous.

2. **L. perenne** L.   Ryegrass
   Map 162. Roadsides, yards, and waste places; spreading in woods and on
shores. An occasional weed or waif, not so common out of cultivation as many
of our forage grasses. This species has been cultivated longer than any other

160. Nardus stricta   161. Lolium temulentum   162. Lolium perenne

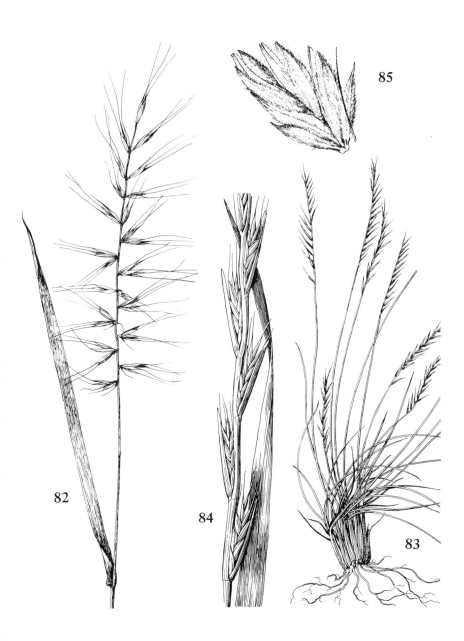

82. *Hystrix patula* ×½
83. *Nardus stricta* ×½
84. *Lolium temulentum* ×1
85. *Agropyron dasystachyum*, spikelet ×3

single forage grass (in contrast to pastures of mixed native species), having been sown in England for almost 300 years.

We have apparently both var. *perenne* and var. *aristatum* Willd., the latter often recognized as a distinct species, *L. multiflorum* Lam., and also known as var. *italicum* Parnell. It supposedly differs from typical *L. perenne* in the following characters: lemmas awned (not awnless), 10 or more in a spikelet (not 10 or fewer); rachis scabrous on convex surface as well as edges (not on edges only); auricles prominent, elongate (not obsolete or small); leaf blades in young shoots inrolled at the margins (not merely folded lengthwise); plants annual or biennial (not perennial). The two taxa are known to hybridize, and there is considerable doubt as to whether they are more than agricultural strains. Our plants include a few specimens which might be clearly placed as one or the other, but most have one or more characters not in line with the rest of the assemblage attributed to a given variety, so they might be placed in either depending on the characters chosen as most important. The inflorescence is usually simple in *Lolium*, but is occasionally branched.

8. Agropyron

The summit of the leaf sheath is often auricled in species of this genus (fig. 81).

One of the crested wheatgrasses, *A. pectiniforme* R. & S., was collected once in 1939 as a weed in the rose garden at the former site of the University of Michigan Botanical Gardens in Ann Arbor (*Ludwig 679*, Ludwig). This species has glabrous spikelets in a strongly two-ranked pectinate spike; the body of the glumes is less than 4 mm long. In all our other species, the body of the glumes exceeds 4 mm. *A. pectiniforme* has often been confused with *A. cristatum* (L.) Gaertner, a name that applies to a similar species with villous spikelets (see Bowden, 1965). Its status at the Ann Arbor locality is dubious.

REFERENCES

Bowden, Wray M. 1965. Cytotaxonomy of the Species and Interspecific Hybrids of the Genus Agropyron in Canada and Neighboring Areas. Canad. Jour. Bot. 43: 1421–1443.
Gillett, J. M., & H. A. Senn. 1961. A New Species of Agropyron from the Great Lakes. Canad. Jour. Bot. 39: 1169–1175.
Pohl, Richard W. 1962. Agropyron Hybrids and the Status of Agropyron pseudorepens. Rhodora 64: 143–147.

KEY TO THE SPECIES

1. Lemmas with awns strongly divergent or recurved at maturity
  2. Anthers ca. 4–5 mm long; glumes with body mostly shorter than the internodes of the spike; broadest leaf blades not over 3 mm wide, ± involute . . . . 1. **A. spicatum**
  2. Anthers ca. 1–2.2 mm long; glumes with body mostly longer than the internodes of the spike; broadest leaf blades ca. 4–7.5 (8.5) mm wide, ± flat . . 2. **A. trachycaulum**
1. Lemmas with awns ± straight or absent

3. Lemmas densely hairy; leaves with narrow (rarely as much as 4.5 mm wide) often involute blades, the whole plant usually strongly glaucous . . . . . 3. **A. dasystachyum**
3. Lemmas glabrous (rarely slightly pubescent), smooth or scabrous; leaves various
  4. Anthers 1−2.2 (2.4) mm long; rachilla readily disarticulating between the florets when mature (on dry specimens, the florets very easily dislodged and empty glumes often remaining on older plants); culms cespitose, rhizomes absent . . . . 2. **A. trachycaulum**
  4. Anthers (2.7) 3−5 (6.1) mm long; rachilla often not readily disarticulating (florets not easily dislodged on dry specimens except over-ripe ones, and empty glumes seldom if ever present); culms from elongate rhizomes
    5. Leaf blades ± strongly involute when dry, deeply grooved above between the prominent raised nerves (and usually strongly scabrous above); glumes mostly with margin minutely ciliate toward base; cartilaginous belt (sharply defined, usually darker, non-green zone) at upper nodes of culm usually less than half as long as its diameter . . . . 4. **A. smithii**
    5. Leaf blades mostly broad and flat, slightly or not at all involute when dry, no more deeply grooved above than below between the numerous fine nerves (not strongly scabrous, usually with scattered long hairs above); glumes completely eciliate toward the base; cartilaginous belt at upper nodes nearly or fully as long as its diameter . . . . 5. **A. repens**

1. **A. spicatum** (Pursh) Scribner & J. G. Smith

Map 163. A western species, found once in the Lake Superior region: bluffs in Keweenaw County (*Farwell 85lb* in 1895, BLH, US).

Another western species, *A. bakeri* E. Nelson, has also been reported from the Upper Peninsula. The only collection labeled as such (*Farwell 533* in 1888, US) has very short anthers and is referred to *A. trachycaulum. A. bakeri* has strongly divergent awns, as in *A. spicatum*, as well as long anthers and a short-hispid rachilla; otherwise, it resembles *A. trachycaulum*.

2. **A. trachycaulum** (Link) Malte   Wheatgrass

Map 164. Widespread, but perhaps most often in dry or rocky woods and woodlands (oak, jack pine, hickory), sand barrens, shores, and dunes; also recorded for bogs and tamarack swamps; sometimes along roadsides and in other waste places.

163. Agropyron spicatum

164. Agropyron
    trachycaulum

165. Agropyron
    dasystachyum

Fortunately, this extremely variable species can usually be easily recognized, if young by the short anthers and if old by the very readily disintegrating spikelets (as also in *A. spicatum*). In addition, the rachilla is nearly always strongly villous with fine, flexible hairs — especially in those forms most likely to be otherwise confused with other species (none of which have villous rachillas).

The varieties all run into each other, but are helpful in describing the range of variation. The awns of the lemmas are less than half as long as their bodies, or absent, in three varieties:

> var. *trachycaulum*: tips of spikelets (excluding short awns if present) barely or not at all overlapping the bases of the next ones above on the same side of rachis. Local throughout the state.

> var. *majus* (Vasey) Fern.: spikelets more strongly overlapping; body of glumes mostly more than 10 mm long. Apparently local in the Upper Peninsula.

> var. *novae-angliae* (Scribner) Fern.: spikelets overlapping as in var. *majus*; body of glumes mostly less than 10 mm long. One of our commonest varieties, throughout the state.

The awns of the lemmas are well developed in var. *glaucum* (Pease & Moore) Malte, in which the body of the glumes is not over 12 mm long. A few of our specimens approach var. *unilaterale* (Cassidy) Malte in having slightly longer glumes. Like var. *novae-angliae*, var. *glaucum* is frequent throughout the state. Plants with awned lemmas are sometimes segregated as a separate species, *A. subsecundum* (Link) Hitchc. Occasionally the awns are somewhat divergent, as in the preceding species, but these plants may be distinguished as described in couplet 2 of the key and also by the rachilla, which is ± villous rather than smooth to hispid.

Occasionally some nodes of the spike may bear 3 spikelets, but such plants have only 1 spikelet at most nodes; they may also be distinguished from *Elymus glaucus*, which they tend to resemble, by the villous rachilla and the generally narrower leaf blades.

Certain plants from the Huron Mountain area, Marquette County, and elsewhere on the Lake Superior shore appear to be hybrids of *A. trachycaulum* and *A. repens*. The plants have the rhizomatous habit and long anthers (4–5 mm) of the latter species; the rachilla is glabrous or villous, disarticulating as in *A. trachycaulum*, and the glumes and lemmas are prominently awned. Such plants may be the basis for reports from the state of *A. pseudorepens* Scribner & J. G. Smith (see Pohl, 1962).

3. **A. dasystachyum** (Hooker) Scribner Fig. 85

Map 165. Restricted to sandy shores and dunes of the Great Lakes; frequent on Lakes Michigan and Huron, but known from Lake Superior only from two collections from Keweenaw County (*Farwell 794*, Aug. 18, 1890, BLH, MICH, MSC, NY; *794A*, June 27, 1895, BLH, MICH).

A conspicuous blue-green glaucous grass. The lemmas are usually nearly or quite awnless, although some plants from all three of the Great Lakes have prominently awned lemmas. Plants from the sandy shores of the Great Lakes constitute a fairly distinct endemic variety, var. *psammophilum* (Gillett & Senn) E. Voss. Typical *A. dasystachyum* of the plains to the west tends to have less glaucous foliage, less villous lemmas, and less attenuate glumes. It has also been suggested that American plants belong to the same species as the earlier described *A. dasyanthemum* Ledeb., an endemic of Russia.

4. **A. smithii** Rydb.

Map 166. Adventive along railroads, roadsides, on shores, etc. – much less common than the next species. Native west of Michigan in prairies and plains, whence it has spread eastward.

The glume differences often emphasized between this species and the next are rather subtle and difficult to describe. In *A. smithii*, the glumes tend to start tapering from below the middle to a narrow or awned tip; in *A. repens*, they taper more abruptly from about or above the middle. Plants with slightly pubescent lemmas rarely occur. A few plants have broader and flatter leaf blades than usual (though still rather deeply grooved), possibly as a result of hybridization with *A. repens*.

5. **A. repens** (L.) Beauv.   Figs. 80, 81   Quack Grass

Map 167. A too familiar weed of roadsides, clearings, fields, gardens, and waste ground generally; spreading into woods and on shores and dunes, where often appearing as if native. While American populations may in part be of native origin, the species is considered to be chiefly introduced from Eurasia.

Extremely variable, with as many as 8 named forms often recognized in the northeastern United States. These are based on the nature of awns, glumes, and pubescence of rachis; they are so intergradient that it does not seem worthwhile to recognize them here. They often grow together, at least 4 of the forms being known throughout the state, from Isle Royale to Detroit. Rarely a robust plant will have 2 spikelets at most nodes of the spike, and with wide leaves will thus

166. Agropyron smithii   167. Agropyron repens   168. Trisetum melicoides

closely resemble *Elymus glaucus*; however, the anthers of *E. glaucus* are shorter than the distinctive long ones of *A. repens* (3–6 mm).

3. AVENEAE

The awns on the lemmas of many species in this tribe are very distinctive in being tightly twisted or coiled, and often darker in color, on the lower portion.

Danthonia is included by Gould (1968) in a tribe Danthonieae, which with Arundineae (*Phragmites*) is in a distinct subfamily, Arundinoideae. All of our other genera are retained by Gould in the Aveneae, in addition to *Beckmannia*, traditionally in the Chlorideae; *Agrostis*, *Apera*, *Alopecurus*, *Ammophila*, *Calamagrostis*, *Cinna*, *Milium*, *Phleum*, and *Polypogon* – all from the former Agrostideae; and all of the Phalarideae.

KEY TO THE GENERA OF AVENEAE

1. Lemmas all awnless; larger glumes ± obovate (broadest above the middle), generally shorter than the lowest floret*
  2. Rachilla and callus prominently bearded with long straight hairs . . . .
    . . . . 1. **Trisetum (melicoides)**
  2. Rachilla and callus glabrous or at most with short hairs (under 0.5 mm long)
    3. Axis and branches of inflorescence glabrous, at most scabrous; larger glumes not over 3 (3.2) mm long . . . . 2. **Sphenopholis**
    3. Axis and branches of inflorescence densely short-pubescent; larger glumes 3–4.2 (4.7) mm long . . . . 3. **Koeleria**
1. Lemmas with distinct twisted or curved awn (sometimes largely hidden by the glumes or absent on some florets of a spikelet); glumes mostly ovate to lanceolate, at least one of them longer than the lowest floret
  4. Larger glumes 6–27 mm long
    5. Ligule a fringe of short hairs with a long tuft at each side; lemma with awn arising between terminal teeth . . . . 4. **Danthonia**
    5. Ligule membranous, hairless; lemma with awn arising dorsally
      6. Spikelets less than 10 mm long (excluding awns), the lower floret staminate with strong awn and the upper floret perfect with (usually) weak awn . . . . .
      . . . . 5. **Arrhenatherum**
      6. Spikelets ca. 20–27 mm long, the florets all perfect or the upper rudimentary; awns various . . . . 6. **Avena**
  4. Larger glume less than 6 mm long
    7. Perfect (lowermost) floret awnless; awn ± incurved or recurved at tip, claw-like, subterminal, on a reduced staminate floret (fig. 89); foliage and inflorescence pilose . . . . 7. **Holcus**
    7. Perfect florets awned; awn twisted or spreading; foliage and inflorescence glabrous or (sometimes in *Trisetum*) pilose
      8. Awn arising above middle of lemma; panicle ± crowded and spike-like . . . . .
      . . . . 1. **Trisetum (spicatum)**
      8. Awn arising well below middle of lemma; panicle at maturity very open and diffuse . . . . 8. **Deschampsia**

*These genera are all included, in addition, in the key to genera of Poeae because of the tendency to short glumes.

1. Trisetum

KEY TO THE SPECIES

1. Lemmas awnless; panicle ± lax and open . . . . 1. **T. melicoides**
1. Lemmas with prominent spreading dorsal awn; panicle erect and ± crowded or spike-like . . . . 2. **T. spicatum**

1. **T. melicoides** (Michaux) Scribner

Map 168. Gravelly, often marly or rocky shores, swales, cedar swamps, riverbanks.

2. **T. spicatum** (L.) Richter Fig. 86

Map 169. Rock crevices and shores, including dolomite on Drummond Island.

An extremely variable species of arctic affinity, found around the northern portion of the globe. Some of our specimens, including the one from Beaver Island (*H. Gillman* in 1871, MICH) have pilose glumes and would hence be referred to var. *pilosiglume* Fern.

2. Sphenopholis "Wedgegrass"

KEY TO THE SPECIES

1. First glume 0.3–0.7 mm wide on a side (midrib to margin), about half as wide as second glume; back of upper lemmas scabrous, at least toward tip; anthers (0.5) 0.7–1.5 mm long . . . . 1. **S. nitida**
1. First glume less than 0.3 mm wide on a side, much less than half as wide as second glume; back of upper lemmas smooth (midnerve often slightly scabrous); anthers not over 0.5 (0.8) mm long
   2. Larger (second) glume distinctly swollen or distended, abruptly truncate and usually shallowly 2-lobed at apex . . . . 2. **S. obtusata**
   2. Larger glume not swollen or distended (fig. 87), ± obtuse to acute at apex . . . . 3. **S. intermedia**

1. **S. nitida** (Biehler) Scribner

Map 170. Dry woods (oak, etc.); bluffs above streams.

The leaf sheaths and blades are rarely glabrous, though usually ± densely short-pilose as in *Koeleria*.

2. **S. obtusata** (Michaux) Scribner

Map 171. Dry woods (e. g., oak–hickory), shores, prairie-like areas. The Keweenaw County record, labeled as from Copper Harbor (*Farwell 7748½* in 1926, MICH) is suspicious (cited in Am. Midl. Nat. 10: 315. 1927).

The inflorescence is dense and contracted, while in the next species it is more open and lax.

3. S. intermedia (Rydb.) Rydb.   Fig. 87

Map 172. In a diversity of usually moist situations, including gravelly or marly shores, depressions and clearings in hardwoods, tamarack swamps, marshy and swampy borders and thickets; occasionally in dry woods.

An occasional specimen may have glumes as large as in *Koeleria*, from which this is readily distinguished by the essentially glabrous foliage and panicle and relatively open, lax inflorescence, as well as by the very narrow first glume.

3. Koeleria

1. K. macrantha (Ledeb.) Schultes   Fig. 88   June Grass

Map 173. Jack pine and oak woods and woodlands, sand dunes, and dry prairies. A specimen labeled as coming from Keweenaw County (*Farwell 7748½* in 1926, BLH) is of such questionable reliability that it is not mapped; it was apparently intended as a duplicate of *Sphenopholis obtusata* (see above) and was so identified.

Occasionally confused with *Sphenopholis*, but most easily distinguished by the densely short-pilose rachis and branches of the ± crowded panicle. In *Koeleria*, the glumes tend to remain on the pedicels after the rest of the spikelet has fallen; in *Sphenopholis*, the glumes and pedicel fall with the rest of the spikelet. The anthers of *Koeleria* tend to be longer (1 or usually 1.3–1.8 mm) than in any of our species of *Sphenopholis* except *S. nitida*.

This species appears in most manuals under the illegitimate name *K. cristata* Pers.

4. Danthonia

KEY TO THE SPECIES

1. Lemmas pilose only on margins and at base (on callus), ca. 6 mm or more in length (including teeth but not awn); spikelets strongly purple or bronze in color . . . . .
. . . . **1. D. intermedia**
1. Lemmas at least sparsely pilose across the back, less than 5.5 mm long (very rarely 7 mm if teeth are as long as 2–3 mm); spikelets ± greenish or at most purplish at their tips. . . . . **2. D. spicata**

169. Trisetum spicatum   170. Sphenopholis nitida   171. Sphenopholis
                                                                     obtusata

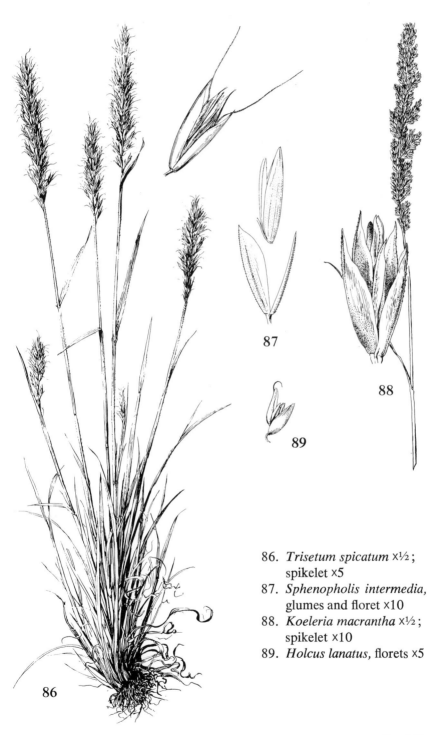

86. *Trisetum spicatum* ×½;
    spikelet ×5
87. *Sphenopholis intermedia,*
    glumes and floret ×10
88. *Koeleria macrantha* ×½;
    spikelet ×10
89. *Holcus lanatus,* florets ×5

1. D. intermedia Vasey

Map 174. Keweenaw County plants collected in meadows near Cliff Mine (*Farwell 709* in 1899, MICH, BLH) and determined by Vasey may be distinguished from the common *D. spicata* as indicated in the key. The longest lemmas are only about 6–6.5 (7) mm long, although in this species they are said normally to be 7–8 mm. Three additional collections also have the lemmas glabrous on the back and purplish spikelets, and may be *D. intermedia*, although the largest lemmas are about 4.5–5 mm long: two Keweenaw County collections (*Farwell 12532* in 1940, BLH, from Lake Superior shore near Eagle Harbor; *C. D. Richards 2161* in 1949, MICH, from rocky shore of Porter's Island) and one from calcareous sandy or stony beach of Lake Michigan east of Manistique, Schoolcraft County (*Fernald & Pease 3100* in 1934, MICH).

2. D. spicata (L.) R. & S.   Fig. 90   Poverty Grass; Oatgrass

Map 175. Chiefly in sandy or rocky, more or less open ground, including aspen, oak, and pine woodlands on plains or dunes; particularly common on jack pine plains, where it may form a solid carpet after disturbance; occasionally found in marshy or boggy places.

A distinctive grass vegetatively, with ± strongly curled lower leaves and a prominent tuft of white hairs on each side of the summit of the sheath – but an extremely variable species in pubescence and size of organs. It is here treated in the broadest sense, including var. *pinetorum* Piper and at least Michigan plants

172. Sphenopholis
intermedia

173. Koeleria macrantha

174. Danthonia intermedia

175. Danthonia spicata

176. Arrhenatherum elatius

177. Avena fatua

referred to *D. allenii* Austin. (For discussions of the differences, see Fernald in Rhodora 45: 239–246. 1943; and Butters & Abbe in Rhodora 55: 122–124. 1953.) The dubious *D. allenii* is often said to approach *D. compressa*, frequently reported from Michigan in the past but now generally understood to have a more eastern range. Some of our specimens which seem best to be referred to the variable *D. spicata* on their overall assemblage of characters would on one particular character or another (especially long narrow teeth on the lemma) run to *D. compressa* in some keys. A very few specimens (e. g., *Farwell 8529*, BLH, from Keweenaw County and *5892*, BLH but not MICH sheet, from Macomb County) have more characters of *D. compressa* (teeth 2–3 mm long, awn pale at base, panicle tending to be compound with lower branches spreading, leaves elongate); however, since other collections vary so greatly in some or all of these characters, the status of the *"compressa-*like" form in Michigan is unclear — as it apparently is elsewhere as well. Plants rarely occur with purplish spikelets and small, pilose lemmas; these, too, have been referred to *D. spicata*. See also comments under the preceding species.

5. Arrhenatherum

1. **A. elatius** (L.) Presl   Fig. 91   Tall Oatgrass
    Map 176. Railroads, roadsides, and other disturbed places; sometimes in adjacent woods. A native of Europe, escaped from cultivation as a forage grass in North America.
    In f. *biaristatum* (Peterm.) Holmb., both lemmas have strong awns; it is known from Ingham and Marquette counties.

6. Avena

REFERENCE

Huskins, C. Leonard. 1946. Fatuoid, Speltoid, and Related Mutations of Oats and Wheat. Bot. Rev. 12: 457–514.

KEY TO THE SPECIES

1. Lemmas with a stout, strongly twisted awn and often with stiff hairs on the back (fig. 93); florets falling from the spikelet by a distinct oval disarticulation surface
. . . . 1. **A. fatua**
1. Lemmas with the awn usually straight, weak, or absent, and the back glabrous (fig. 94); florets falling by fracture of rachilla at base of spikelet . . . . 2. **A. sativa**

1. **A. fatua** L.   Fig. 93   Wild Oats
    Map 177. Railroads, roadsides, beaches, and other disturbed places. A native of Europe.
    The typical form, with stiff hairs on the back of the lemma, has been collected in Michigan only rarely and in the northern part of the state, but is to

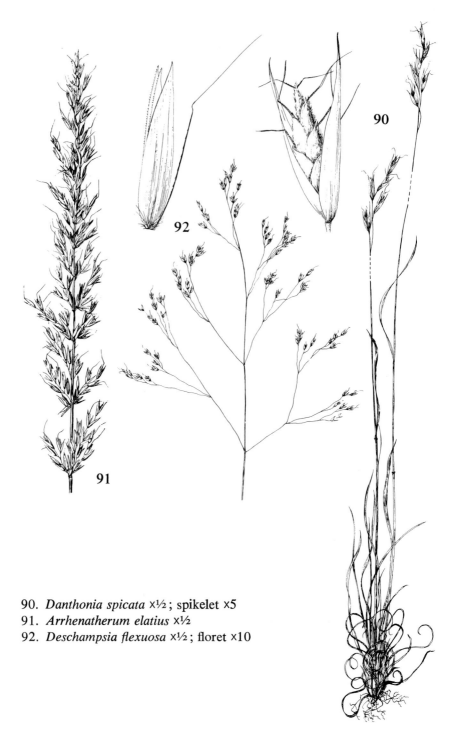

90. *Danthonia spicata* ×½ ; spikelet ×5
91. *Arrhenatherum elatius* ×½
92. *Deschampsia flexuosa* ×½ ; floret ×10

be expected elsewhere (it is known from a Canadian island in Lake St. Clair). Plants referred here from southern Michigan are mostly the form with glabrous lemmas; it can be distinguished from *A. sativa* by the stout, strongly twisted awn and prominent articulation surface in the rachilla — although "fatuoid" *A. sativa* would also display these characters. (As with many plants long in cultivation, the full genetic and taxonomic story of the oats is complex.)

2. **A. sativa** L.   Fig. 94   Oats
   Map 178. Roadsides, fields, railroads, and other waste places; generally a temporary waif, escaped from cultivation.

   Perhaps better treated as a cultivated variety of the preceding, as *A. fatua* var. *sativa* (L.) Haussk.

7. Holcus

1. **H. lanatus** L.   Fig. 89   Velvet Grass
   Map 179. Roadsides and waste ground, usually in more or less shaded situations and spreading to woods and bogs. A native of Europe.

8. Deschampsia   Hair Grass

KEY TO THE SPECIES

1. Awn scarcely if at all exserted beyond tip of glumes, ± straight; lemmas smooth; leaf blades involute or often flat (1.5–5 mm wide); ligule 3–10 (17) mm long . .
   . . . . 1. **D. cespitosa**
1. Awn conspicuously exserted, becoming bent at middle; lemma scabrous or minutely hispidulous; leaf blades involute-filiform; ligule 0.5–3 (5) mm long . . .
   . . . . 2. **D. flexuosa**

1. **D. cespitosa** (L.) Beauv.
   Map 180. Mostly along the shores of the Great Lakes, in gravelly, sandy, or marly places and thriving in crevices of rocks; occasionally inland on shores and riverbanks and in bogs.

178. Avena sativa   179. Holcus lanatus   180. Deschampsia
                                                              cespitosa

2. **D. flexuosa** (L.) Beauv. Fig. 92

Map 181. Open generally dry sandy or rocky shores, dunes, and plains, usually associated with pine, oak, and aspen; particularly characteristic of jack pine plains and of the lumbered and burned-over sand ridges with *Cladonia* lichens and scattered pines along Lake Superior from Whitefish Bay west to the Pictured Rocks.

4. AGROSTIDEAE

The traditional concept of the tribe as characterized by one-flowered spikelets lends itself well to identification but is admittedly an unnatural one. Gould (1968) does not recognize the tribe, but places its type genus, *Agrostis*, along with *Alopecurus*, *Ammophila*, *Apera*, *Calamagrostis*, *Cinna*, *Milium*, *Phleum*, and *Polypogon*, in the Aveneae; *Stipa* and *Oryzopsis* in the Stipeae (where some authors have placed *Aristida* and *Milium* also); *Brachyelytrum* as the sole representative of the Brachyelytreae; *Muhlenbergia*, *Calamovilfa*, *Sporobolus*, and *Heleochloa* [included in *Crypsis*] in the Eragrosteae; and *Aristida* as the only genus of the Aristideae, also in the subfamily Eragrostoideae.

KEY TO THE GENERA OF AGROSTIDEAE

1. Lemma with awn (or awns) strictly terminal
 2. Awns of lemma 3 (lateral ones sometimes very short) . . . . 1. **Aristida**
 2. Awn of lemma solitary
  3. Body of lemma 8–23 mm long
   4. Glumes 9.5–45 mm long . . . . 2. **Stipa**
   4. Glumes rudimentary or one of them up to 5 mm long . . . . 3. **Brachyelytrum**
  3. Body of lemma less than 7 mm long
   5. Glumes acute to obtuse, more than 1 mm wide, scarcely if at all keeled, the spikelets nearly terete; lemma rounded on the back, ± indurated . . . . . 4. **Oryzopsis**
   5. Glumes acuminate, not over 1 mm wide, keeled, the spikelets somewhat compressed (or glumes obsolete in *M. schreberi*); lemma ± keeled, membranous or thin
    6. Inflorescence various (if spike-like, the plants with scaly rhizomes); ligules up to 2 mm long; glumes gradually tapered into awn or awnless . . . . 5. **Muhlenbergia**
    6. Inflorescence a dense thick spike-like panicle, the plants annual, without rhizome; ligules at least 3 mm long; glumes ± 2-lobed or rounded at apex, not tapered into the very slender awn (ca. 5–8 mm long) . . . . 8. **Polypogon**
1. Lemma with awn absent or dorsal or subterminal
 7. Spikelets 10–15 mm long; anthers (4) 5–8 mm long; panicle crowded and ± spike-like, (10) 12–20 (28) mm across at the middle . . . . 6. **Ammophila**
 7. Spikelets less than 8 mm long (excluding awns); anthers up to 4.5 mm long (usually much shorter); panicle various
  8. Spikelets sessile or nearly so, crowded in a very dense spike-like panicle (branches of panicle suppressed, scarcely if at all visible without dissection of panicle)
   9. Glumes awned
    10. Plants from scaly rhizomes; glumes gradually tapered into awn; ligule up to ca. 1 mm long . . . . 5. **Muhlenbergia** (couplet 6)

10. Plants without scaly rhizomes; glumes abruptly rounded or truncate, the awn distinct; ligule over 1 mm long
  11. Awn of glume rather stout and stiff, not over ca. 3 mm long; anthers ca. 1–2 mm long; spikelets articulated above the glumes; glumes prominently pectinate-ciliate on keel basally, otherwise glabrous or variously (but not so prominently) pubescent or ciliate; lemmas awnless . . . . 7. **Phleum**
  11. Awn of glume very slender, ca. 5–8 mm long; anthers less than 1 mm long; spikelets articulated below the glumes; glumes ± evenly hispidulous basally; lemmas often with delicate awn . . . . 8. **Polypogon**
9. Glumes awnless
  12. Panicle ± ovoid, ca. 2–3.5 times as long as wide; spikelets articulated above the glumes; ligule a fringe of hairs; lemmas awnless [if ligule membranous, try *Phalaris* – see text] . . . . 9. **Heleochloa**
  12. Panicle cylindrical (in common species slender and pencil-like), 3–15 times as long as wide; spikelets articulated below the glumes; ligule membranous; lemma with slender awn attached near the middle or base of keel (in common species, often very inconspicuous, shorter than the glumes; see fig. 112) [if spikelets articulated above the glumes and lemmas stoutly awned, try *Anthoxanthum* – see text] . . . . 10. **Alopecurus**
8. Spikelets in ± open or contracted (but not densely spike-like) inflorescences, with evident pedicels and/or panicle branches
13. Spikelets rounded on back, not keeled (neither glumes nor lemma with a midvein more prominent than other nerves), at least 2.5 mm long; lemma ± shiny, distinctly firmer in texture than the glumes
  14. Lemmas with appressed pubescence; leaves with blades usually involute; upper ligules not over 3 mm long . . . . 4. **Oryzopsis**
  14. Lemmas glabrous; leaves with blades broad and flat; upper ligules mostly 4–6 (8) mm long . . . . 11. **Milium**
13. Spikelets keeled (glumes and/or lemmas with midvein more prominent than other nerves) or less than 2.5 mm long; lemma no firmer in texture than the glumes
15. Ligule a fringe of short hairs; lemmas awnless
  16. Lemma (4.7) 5–6.7 mm long, surrounded with a tuft of long hairs (more than half its length) at its base . . . . 12. **Calamovilfa**
  16. Lemma (3) 3.5–5.5 mm long, without long hairs at its base . . . . 13. **Sporobolus**
15. Ligule membranous (at most minutely ciliate at summit of membrane); lemmas awned or awnless
17. Lemma with long hairs at base (on or near callus)
  18. Long hairs at least in part arising from lower portion of lemma; glumes (excluding awn-tips if present) shorter than lemma . . . . 5. **Muhlenbergia**
  18. Long hairs restricted to callus at base of floret; glumes slightly exceeding lemma . . . . 14. **Calamagrostis**
17. Lemma without long hairs at base (at most with hairs on callus less than 0.5 mm long)
19. Glumes both distinctly shorter than lemma; lemma awnless . . . . 5. **Muhlenbergia** (couplet 2)
19. Glumes (one or both of them) equalling or exceeding the lemma and/or the lemma awned
20. Floret raised above base of glumes on a short stipe; spikelet articulated below the glumes; lemma with a small subterminal awn; stamen 1 . . 15. **Cinna**
20. Floret not stipitate; spikelets articulated above the glumes; lemma awnless or with long subterminal awn or with dorsal awn; stamens 3

21. Lemma with a long subterminal awn, much exceeding the body in length; rachilla prolonged (scarcely 0.5 mm) behind the palea . . . 16. **Apera**
21. Lemma awnless (or very rarely with mid-dorsal awn); rachilla not prolonged . . . . 17. **Agrostis**

1. Aristida   Three-awned Grass

Except for *A. purpurascens*, the species are all quite uncommon in Michigan and all are at the edge of their range in the southern part of the state; in fact there may be doubt as to whether any of them are native here. Our plants are fairly easily distinguished, but the measurements in the key may fail completely to apply to forms of the same species farther south and west, where there is apparently much more overlapping of characters and even greater variability in size of awns, glumes, and lemmas.

KEY TO THE SPECIES

1. Awns tightly twisted and ± connate, forming a column ca. 5–9 mm long at summit of lemma before diverging into 3 ± equal and much longer free portions (fig. 95) .
. . . . 1. **A. tuberculosa**
1. Awns not forming a column, separate from their bases
   2. First glume with 3–5 distinct nerves, ca. 16–24 mm long (plus awn if present); body of lemma ca. (12) 15–17 (21) mm long; awns ca. 3.5–5 cm long . . . . . .
   . . . . 2. **A. oligantha**
   2. First glume with 1 distinct nerve, 2.5–12.5 mm long; body of lemma 3.5–11 mm long; awns less than 3.5 cm long
      3. Middle awn on most lemmas loosely spiraled (at least when dry) in 1 or 2 loops toward its base (fig. 96)
         4. Body of lemma ca. 5–7 mm long, with middle awn mostly 4–8 mm long and lateral awns less than 2 mm long; glumes mostly subequal, both longer than body of lemma . . . . 3. **A. dichotoma**
         4. Body of lemma ca. 7–11 mm long, with middle awn 9–18 mm long and lateral awns 7–12 mm long; glumes clearly unequal, the first usually equalling or shorter than the body of the lemma . . . . 4. **A. basiramea**
      3. Middle awn on most lemmas bent, slightly twisted, or straight, without spiraled loops at base (fig. 97)
         5. Middle awn of lemma ca. 7–12 mm long, strongly divergent or somewhat reflexed; lateral awns 1–4 mm long, ± erect or slightly spreading; first glume 2.5–4 (6) mm long (excluding awn-tip if present); body of lemma ca. 3.5–5.5 mm long . . . . 5. **A. longispica**
         5. Middle awn of lemma ca. 15–33 mm long; lateral awns ca. 9–26 mm long; all awns somewhat spreading or divergent; first glume (4) 5–12.5 mm long; body of lemma ca. 5–8 mm long
            6. First glume ca. (4) 5–7 (8.5) mm long, slightly shorter than second glume; lower sheaths essentially glabrous, the nodes mostly exposed; plants annual .
            . . . . 6. **A. necopina**
            6. First glume (8) 8.5–12.5 mm long, slightly longer than second glume; lower sheaths usually ± pilose, covering the nodes; plants perennial . . . 7. **A. purpurascens**

1. **A. tuberculosa** Nutt.   Fig. 95

Map 182. The only Michigan collection seen was made from sandy barrens, Keeler, in August of 1908, by H. S. Pepoon (MSC).

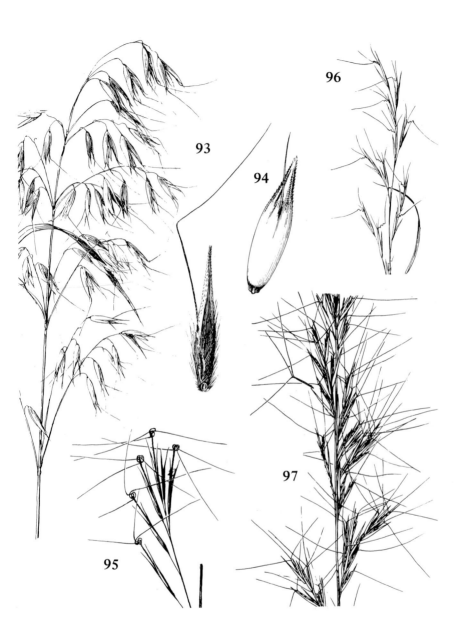

93. *Avena fatua* ×½ ; floret ×2
94. *A. sativa,* floret ×2
95. *Aristida tuberculosa* ×⅘
96. *A. basiramea* ×1
97. *A. purpurascens* ×1

This species has longer awns and larger spikelets than any other of our species except *A. oligantha*. The distinctive twisted column of awns is shorter in material from near the head of Lake Michigan than in specimens from the east coast of the United States.

2. **A. oligantha** Michaux

Map 183. Dry open ground in fields and along railroads.

3. **A. dichotoma** Michaux

Map 184. Sterile field in Charleston Tp., Kalamazoo County (*Hanes 335*, Sept. 22, 1935, WMU, MICH, GH).

4. **A. basiramea** Vasey   Fig. 96

Map 185. Dry sandy open ground. The collection from Keweenaw County reported by Farwell (Pap. Mich. Acad. 23: 125. 1938) is far enough out of range to be suspicious (as is the case with many of his "½" numbers): "Dry foothills of the bluffs at Cliff Mine" (*Farwell 3911½*, Oct. 1, 1914, BLH).

The measurements in the key for awns of this and the preceding species are based directly on the distance from base of awn to apex, without any allowance for "straightening out" the spiral loops.

5. **A. longispica** Poiret

Map 186. Only two collections have been seen from Michigan: Dry sandy open ground near Algonac (*Dodge*, Sept. 15, 1900, MICH, MSC); small prairie-like area at junction of Telegraph Road and Detroit Industrial Expressway (*Rogers 12123*, Sept. 25, 1959, WUD).

In var. *geniculata* (Raf.) Fern., found south of Michigan, the glumes are up to 9 mm long and the lateral awns (sometimes also middle one) are longer than in the typical variety, although the lateral awns are still distinctly shorter than the middle awn.

6. **A. necopina** Shinners

Map 187. Dry sterile fields.

181. Deschampsia flexuosa   182. Aristida tuberculosa   183. Aristida oligantha

This is the plant which, in our area, has been referred to *A. intermedia* Scribner & Ball; true *A. intermedia* is the same as var. *geniculata* of the preceding species, according to Shinners (Rhodora 56: 30. 1954).

7. A. purpurascens Poiret   Fig. 97

Map 188. Dry (rarely moist) usually sandy soil, prairies, sand barrens.

2. Stipa   Needle Grass

A distinctive characteristic of this genus and the preceding is the very firm lemma, with elongate, hard, sharp-pointed, usually bearded callus at the base (included in measurements of lemma). (See fig. 98.)

The awns in *Stipa* are striking enough for their length, but are even more remarkable in aiding the burial of the grain (which is tightly enclosed in the lemma). The lower portion of the awn is very tightly twisted when dry, and untwists with an increase in moisture. This hygroscopic action of twisting and untwisting with changes in moisture drives the fruit into the ground with the aid of the sharp-pointed callus, which is upwardly bearded. In some ways the action could perhaps be compared to that of a carpenter's brace and bit.

KEY TO THE SPECIES

1. Glumes 9.5–12 (14) mm long; awn of lemma ca. 5–6.5 cm long; body of lemma 8–10 mm long . . . . 1. S. avenacea
1. Glumes 17–45 mm long; awn of lemma ca. 9–20 cm long; body of lemma 8–23 mm long
  2. Body of lemma (8) 10–13 mm long, whitish or very pale brown at maturity; glumes (17) 22–28 mm long, including long-attenuate awned tip . . . . . . 2. S. comata
  2. Body of lemma (15) 17–23 mm long, usually dark brown at maturity; glumes (27) 30–45 mm long, including attenuate tip . . . . 3. S. spartea

1. S. avenacea L.

Map 189. Usually in dry ground, especially oak woods, but rarely on marshy shores.

2. S. comata Trin. & Rupr.

Map 190. Dry soil along railroads in Kalamazoo County; the only other

184. Aristida dichotoma

185. Aristida basiramea

186. Aristida longispica

Michigan collection, and the first for the state, is from the banks of the Saginaw River in Bay County (*Dreisbach 6972* in 1930, MICH). The species is probably adventive in Michigan although considered native not far to the south.

The awn tends to be slender, deciduous, and conspicuously curly in this species, while in the next it is stouter, persistent, and merely bent once or twice above the twisted base. The base of the panicle is generally enclosed in the inflated upper sheath in *S. comata*, while in *S. spartea* it is usually exserted.

3. **S. spartea** Trin.   Fig. 98

Map 191. Open sandy often calcareous ground, dune ridges, oak woodland, dry prairies, and along railroads.

See comments under *S. comata*.

3. Brachyelytrum

REFERENCE

Stephenson, Stephen N. 1971. The Biosystematics and Ecology of the Genus Brachyelytrum (Gramineae) in Michigan. Mich. Bot. 10: 19–33.

1. **B. erectum** (Roth) Beauv.   Fig. 99

Map 192. Both varieties occur in rich deciduous woods and oak or oak—

187. Aristida necopina

188. Aristida purpurascens

189. Stipa avenacea

190. Stipa comata

191. Stipa spartea

192. Brachyelytrum erectum

98. *Stipa spartea,* glumes and floret ×2
99. *Brachyelytrum erectum* ×½ ; floret ×5

99

98

hickory stands; the northern var. *septentrionale* is also in lowland woods, damp thickets, sandy pine woods, bogs, and coniferous swamps. This is the only species in the genus and grows in eastern North America and eastern Asia.

A very distinctive grass, with slender, long-awned spikelets. The rachilla is prolonged as a rather conspicuous bristle back of the palea (fig. 99). The slender culms arise from a very scaly, knotty rhizome with buds and several strongly nerved scales at the culm bases. Sterile plants are easily recognized by these underground parts and the broad leaf blades, which are ± pilose beneath (and sometimes also above).

Most of our plants have nearly glabrous to scabrous or hispidulous 3–5-nerved lemmas, and are var. *septentrionale* Babel, which has florets ca. 8–10 mm long (excluding awn), anthers not over 4 mm long, and more than 15 cilia per 5 mm of leaf margin. Many of the collections from the southern part of the Lower Peninsula, however, have strongly hispid (5) 7–9-nerved lemmas and are the typical variety, which has at least the larger florets ca. 10–12 mm long, anthers more than 5 mm long, and fewer than 10 cilia per 5 mm of leaf margin. Intermediates between these well marked varieties are not common.

4. Oryzopsis   Rice-grass

REFERENCE

Voss, Edward G. 1961. Which Side is Up? A Look at the Leaves of Oryzopsis. Rhodora 63: 285–287.

KEY TO THE SPECIES

1. Leaf blades involute, less than 2 (2.3) mm wide; body of lemma 2.5–4 mm long; ligules of upper leaves ca. (1) 1.5–3 mm long
   2. Awn 6–9 mm long, ± twisted; glumes perfectly smooth . . . . 1. **O. canadensis**
   2. Awn absent or less than 2 (3) mm long, nearly straight; glumes very minutely scabrous toward apex [use 20X lens!] . . . . 2. **O. pungens**
1. Leaf blades flat (or margins involute in *O. asperifolia*), the larger ones (4) 5–18 mm wide; body of lemma 5.5–7 mm long; ligules ca. 0.5 mm long or obsolete
   3. Principal leaf blades cauline, not evergreen, the upper surface pale or darker green, ± short-pilose, with many fine veins (basal leaves reduced or mere sheaths); mature lemma dark brown or blackish . . . . 3. **O. racemosa**
   3. Principal leaf blades basal or nearly so, evergreen, the upper surface glaucous (very rarely green), densely and very finely rough-puberulent, with strong closely spaced veins (cauline leaves with blades less than 3 cm long or reduced to sheaths); mature lemma pale or yellowish . . . . 4. **O. asperifolia**

1. O. canadensis (Poiret) Torrey

Map 193. A northern species, often common north of Lake Superior, ranging south locally in sandy open ground with jack pine and white spruce.

2. O. pungens (Sprengel) Hitchc.

Map 194. Sandy woods and woodlands on dunes and plains, usually with aspen, oak, jack pine, and/or red pine; rocky woods and summits in the western Upper Peninsula.

3. O. racemosa (Sm.) Hitchc.   Fig. 100

Map 195. Usually in rich deciduous woods and wooded dunes, sometimes in disturbed places; less often associated with jack pine and oak.

The awn of the lemma may be as long as 23 mm in this species, although usually less than 20 mm. The widest leaf blades are usually 1 cm or more broad.

4. O. asperifolia Michaux

Map 196. In a great variety of woods, including rich deciduous forests and hemlock–hardwoods, but most characteristic of rather dry open woods associated with jack pine, oak, and/or aspen; on wooded dunes with pine and in rocky mixed woods.

The awn may be up to 13.5 or rarely 15 mm long. The leaf blades are all less than 1 cm broad. Although manuals uniformly state that the leaves are glaucous beneath, an examination of the leaves will show that it is the upper surface which is glaucous. I have collected plants on Isle Royale in which both surfaces are green.

5. Muhlenbergia   "Muhly"

REFERENCES

Fernald, M. L. 1943. Five Common Rhizomatous Species of Muhlenbergia. Rhodora 45: 221–239 (also Contr. Gray Herb. 148).
Pohl, Richard W. 1969. Muhlenbergia, Subgenus Muhlenbergia (Gramineae) in North America. Am. Midl. Nat. 82: 512–542.
Shinners, L. H. 1941. Notes on Wisconsin Grasses–II. Muhlenbergia and Sporobolus. Am. Midl. Nat. 26: 69–73.

193. Oryzopsis canadensis

194. Oryzopsis pungens

195. Oryzopsis racemosa

KEY TO THE SPECIES

1. Lemmas not pilose at the base, glabrous (or with minute, even pubescence on back), awnless; culms loosely or densely tufted or matted, without elongate scaly rhizomes
    2. Spikelets less than 2 mm long, mostly on pedicels more than twice as long, in an open panicle . . . . **1. M. uniflora**
    2. Spikelets ca. 2.4–3.5 mm long, mostly on pedicels less than twice as long, in a slender contracted panicle
        3. Ligules less than 0.5 mm long; lemmas with a little minute pubescence on back
        . . . . **2. M. cuspidata**
        3. Ligules ca. (1.2) 1.5–2.5 mm long; lemmas glabrous across back . . **3. M. richardsonis**
1. Lemmas pilose at base, glabrous or short-pubescent on back, awned or awnless; culms arising from elongate scaly rhizomes (except in *M. schreberi*)
    4. Glumes minute, the larger one less than 0.5 mm long, the other obsolete or absent; culms often rooting at nodes of decumbent bases, but without elongate scaly rhizomes . . . . **4. M. schreberi**
    4. Glumes at least half as long as body of lemma; culms from elongate scaly rhizomes
        5. Glumes (including prominent awn-tip) (3) 3.5–6.5 (7.5) mm long, mostly distinctly longer than the body of the lemma; lemma at most short-awned; anthers 0.5–1.3 mm long
            6. Internodes of culm smooth and glabrous over most of their surface; ligule 0.7–1 mm long; anthers ca. 0.5–0.8 mm long . . . . **5. M. racemosa**
            6. Internodes minutely puberulent or roughened over much of their surface (rarely nearly glabrous); ligule (excluding cilia) 0.5–0.7 mm long or shorter; anthers 0.8–1.3 mm long . . . . **6. M. glomerata**
        5. Glumes generally less than 3.6 mm long (rarely, especially on lower spikelets of panicle, up to 4 mm), mostly about equalling or shorter than the body of the lemma; lemma awnless to long-awned; anthers not over 0.5 mm long (except in *M. tenuiflora* with distinctive short, broad glumes)
            7. Larger glumes 0.6–1 mm wide, less than 4 times as long, hence ovate and usually ± abruptly tapered at the tip; culms puberulent below the nodes
                8. Anthers 1–1.5 mm long; broader leaf blades (4) 6–13 mm wide; ligules 1 (1.2) mm long or shorter; sheaths (at least some of them) usually ± pubescent
                . . . . **7. M. tenuiflora**
                8. Anthers not over 0.5 mm long; broader leaf blades 3–5 mm wide; ligules (1) 1.3–2 mm long; sheaths all glabrous . . . . **8. M. sylvatica**
            7. Larger glumes not over 0.6 mm wide, more than 4 times as long, hence narrowly lanceolate and usually ± attenuate at the tip; culms puberulent or glabrous below the nodes
                9. Culm smooth and glabrous throughout, sometimes decumbent at base and rooting at nodes, generally much branched and bushy above; inflorescences (except terminal one) with base often enclosed in upper leaf sheath . . . .
                . . . . **9. M. frondosa**
                9. Culm puberulent below the nodes, ± erect, simple or branched; inflorescences all generally exserted
                    10. Ligules, at least the longest, (1) 1.3–2 mm long; spikelets all on distinct (though sometimes rather short) pedicels . . . . **8. M. sylvatica**
                    10. Ligules ca. 1 (1.3) mm long or shorter; some spikelets in panicle sessile or subsessile . . . . **10. M. mexicana**

1. **M. uniflora** (Muhl.) Fern.   Fig. 101

Map 197. Local, damp sandy lake shores, often becoming abundant following a lowering of water levels; meadows and swamp borders; margins of rock
pools at Isle Royale (Passage Island).

A slender, delicate plant with diffuse inflorescence, in aspect quite unlike our
other Muhlenbergias. Unique in the development of buds just below the nodes,
those on old decumbent culms producing new erect flowering culms (these
therefore *not* axillary). Some of the spikelets are frequently 2-flowered.

2. **M. cuspidata** (Hooker) Rydb.

Map 198. Collected on rocky edges of bluffs at Cliff Mine (*Farwell 848* in
1895, BLH) and at Eagle Harbor (*T. E. Boyce 3375* in 1885, GH).

The culms are erect in this species, with bulb-like offshoots at the base.

3. **M. richardsonis** (Trin.) Rydb.

Map 199. Marshy ground and boggy meadows, said to be common on anthills
(Hanes). Farwell's Keweenaw County collection (*849½*, bluffs at Clifton in
1895, BLH) is suspicious.

The culms are erect or arise from old decumbent culms, sometimes with hard
and knotty but usually not bulb-like thickenings.

196. Oryzopsis asperifolia

197. Muhlenbergia uniflora

198. Muhlenbergia
cuspidata

199. Muhlenbergia
richardsonis

200. Muhlenbergia
schreberi

201. Muhlenbergia
racemosa

4. **M. schreberi** J. F. Gmelin   Fig. 102   Nimblewill

Map 200. In woods, both moist and dry, especially in disturbed areas, but more often a weed of roadsides, fields, and gardens.

5. **M. racemosa** (Michaux) BSP.

Map 201. Dry soil along railroads, collected at two places in Michigan: Schoolcraft Tp., Kalamazoo County (*Hanes 4316* in 1944, WMU, MICH) and city of Menominee (*Grassl 3227* in 1933, MICH). Generally considered adventive in Michigan from farther west.

The internodes may be puberulent near the summit and in a narrow band covered by the center of the sheath, but are otherwise glabrous and shining.

6. **M. glomerata** (Willd.) Trin.   Fig. 103   Marsh Wild-timothy

Map 202. Widespread in both marshy and swampy places, bogs (open mats and older tamarack), boggy meadows and calcareous shores, springy places, moist clearings in woods.

The culms are seldom branched above the base, while in the preceding species there are many branches. The panicle is often strongly tinged with purple.

7. **M. tenuiflora** (Willd.) BSP.

Map 203. Usually found on wooded dunes, hillsides, and riverbanks, whether in oak or beech—maple woods. This is another of the large group of species found in eastern North America and eastern Asia.

8. **M. sylvatica** Torrey

Map 204. Moist often boggy ground, creek and pond borders, etc., usually in shaded places.

9. **M. frondosa** (Poiret) Fern.   Fig. 104

Map 205. Usually in thickets along creeks and rivers on banks, mudflats, and floodplains; also on shores and in disturbed ground.

Many of our plants have awned lemmas and are f. *commutata* (Scribner)

202. Muhlenbergia   203. Muhlenbergia   204. Muhlenbergia
      glomerata   tenuiflora   sylvatica

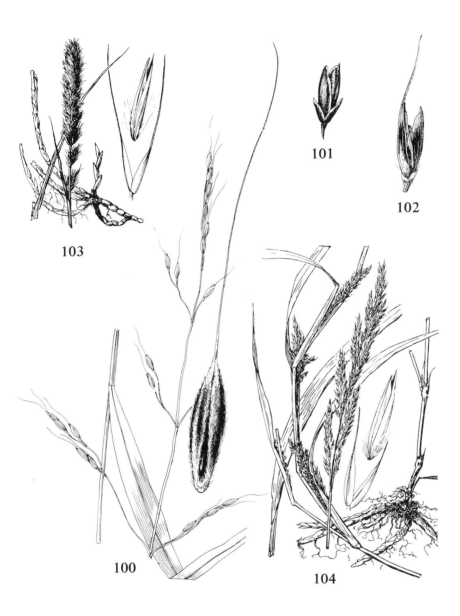

100. *Oryzopsis racemosa* ×½ ; floret ×5
101. *Muhlenbergia uniflora,* spikelet ×10
102. *M. schreberi,* spikelet ×10
103. *M. glomerata* ×1; glumes and floret ×8
104. *M. frondosa* ×1; glumes and floret ×10

Fern. The ligules on our specimens are usually about 0.5–1 mm long, rarely as long as 1.3 mm.

10. **M. mexicana** (L.) Trin.

Map 206. Widespread, as in *M. glomerata*, in marshy, swampy, and boggy habitats, stream border thickets, moist hardwoods; also on shores and beaches and in disturbed ground along roads and trails.

Some plants approach *M. sylvatica*, but can usually be distinguished by the more slender glumes (5–8 times as long as wide in *M. mexicana*, not over 6 times as long as wide in *M. sylvatica*). Also, the spikelets in *M. mexicana* are often tinged with purple and form denser, thicker, more closely spaced panicle branches than in *M. sylvatica*. Most of our plants have essentially awnless lemmas, but some are f. *ambigua* (Torrey) Fern., with long-awned lemmas.

The use of older literature on *Muhlenbergia* is complicated by some involved problems of synonymy. What was long known as *M. mexicana* is now called *M. frondosa*, and *M. mexicana* is now applied to what was formerly called *M. foliosa*.

6. Ammophila

1. **A. breviligulata** Fern.   Fig. 105   Beach Grass

Map 207. Common to abundant on the sandy shores and dunes of Lakes Michigan, Huron, and Superior. Collected inland only on the east shore of Douglas Lake, Cheboygan County (*Ehlers 1217* in 1920, MICH).

This species of the Great Lakes and Atlantic region is closely related to the European *A. arenaria* (L.) Link. Both are excellent sand binders. Observations on *A. breviligulata* at the Sleeping Bear dunes, Leelanau County, by Frank C. Gates showed that the rhizomes may grow as much as 8 feet in a single year as sand accumulates. In possessing both vertical and horizontal rhizomes, this plant is unusually well adapted for survival on sand dunes and it is often the species to "capture" and stabilize them.

205. Muhlenbergia
     frondosa

206. Muhlenbergia
     mexicana

207. Ammophila
     breviligulata

7. Phleum

KEY TO THE SPECIES

1. Awns of glumes (at least the longest ones) 2.5−3 mm long (fig. 106); spike-like
    panicle 2−4 cm long . . . . 1. **P. alpinum**
1. Awns rarely as long as 2 mm (fig. 107); spike-like panicle (1) 4−14 (31) cm long .
    . . . . 2. **P. pratense**

1. **P. alpinum** L.   Fig. 106   Mountain Timothy
Map 208. A plant of damp meadows and shores, but none of Farwell's several collections from Keweenaw County have habitat data.

A native boreal and alpine species with a short and relatively thick spike-like panicle. Occasional specimens of *P. pratense* with short panicles are often misidentified as this species, but differ in having shorter awns and in most or all other characters which tend to distinguish *P. pratense*: base of culm ± enlarged and bulbous, summit of culm very minutely roughened [visible with 20× lens], ligule usually with distinct notch on each side, sheaths not inflated. In *P. alpinum*, the base of the culm is not bulbous and the summit is perfectly smooth, the ligule is not notched on both sides, and the upper leaf sheaths are ordinarily somewhat inflated.

2. **P. pratense** L.   Figs. 107, 108   Timothy
Map 209. A thoroughly naturalized species, widely cultivated for hay and originally a native of Eurasia; established on roadsides, fields, and waste places generally; spreading to woods and shores.

The spike-like panicle is normally cylindrical and elongate, but may be shorter in depauperate plants; see comments under the preceding species.

8. Polypogon

1. **P. monspeliensis** (L.) Desf.   Fig. 109   Rabbitfoot Grass
Map 210. A European introduction, more common in the western United

208. Phleum alpinum   209. Phleum pratense   210. Polypogon
                                                                    monspeliensis

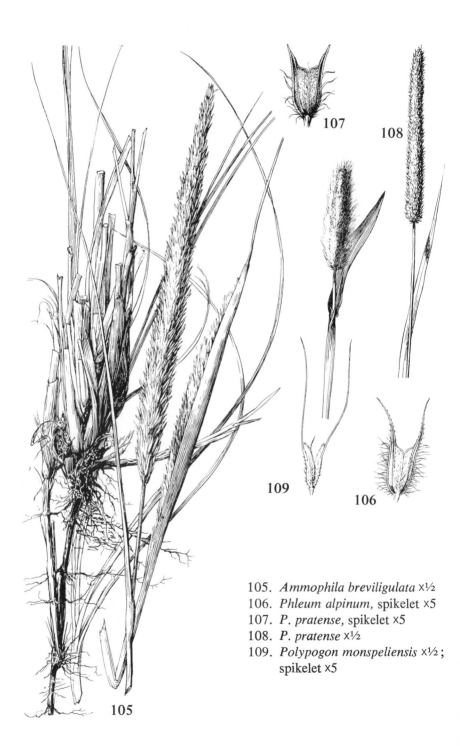

107

108

109

106

105. *Ammophila breviligulata* ×½
106. *Phleum alpinum*, spikelet ×5
107. *P. pratense*, spikelet ×5
108. *P. pratense* ×½
109. *Polypogon monspeliensis* ×½ ; spikelet ×5

105

States than in the eastern. A single Michigan collection has been seen: on waste heap from tannery at Kegomic on Little Traverse Bay, Emmet County (*Fallass* in 1896, ALBC, BLH, AQC, MSC).

9. Heleochloa

Phalaris might be keyed here if not recognized as belonging to the tribe Phalarideae. It is a genus of tall, erect plants with membranous ligules, in contrast to the nearly prostrate *Heleochloa*, with ligule of hairs. The small scales or tiny hairy flaps representing the lemmas of reduced lateral florets at the base of the perfect floret in the spikelet of *Phalaris* are absolutely distinctive.

1. H. schoenoides (L.) Roemer   Fig. 116

Map 211. A native of the Mediterranean region, collected in 1930 at River Rouge on newly made ground from dredging of the river (*Farwell 8762*, MICH, BLH, WUD, GH).

10. Alopecurus   Foxtail

Anthoxanthum might be keyed here if not recognized as belonging to the tribe Phalarideae. The lemma and palea of the fertile floret so closely envelop the grain that they may not be recognized in a superficial examination, in contrast to the pair of prominent, hairy, awned, dark brown sterile lemmas which might be mistaken for a lemma and palea. The spikelets are articulated above the glumes, unlike *Alopecurus*, and the glumes are very unequal. Note also the comments on *Phalaris* under the previous genus.

KEY TO THE SPECIES

1. Spikelets (excluding awns) ca. 4–6.5 mm long; awns mostly exserted ca. 3.5–6 mm beyond tips of glumes; anthers ca. 2.4–3.5 mm long . . . . . . 1. **A. pratensis**
1. Spikelets not over 3 mm long; awns at most exserted ca. 2–3 mm; anthers less than 2 mm long
  2. Awn exserted at most about 1 mm beyond tips of glumes, usually included, inserted about a third or half the distance from base of lemma . . . . 2. **A. aequalis**
  2. Awn of most lemmas exserted ca. 2–3 mm, inserted near the base of lemma (on lower fifth to fourth) . . . . 3. **A. carolinianus**

1. A. pratensis L.

Map 212. A Eurasian species, locally naturalized in North America. Our few collections are from "meadows" (Bay County, *Bradford* in 1897, MSC), roadsides, lawns, and fields.

A. myosuroides Hudson, another European species, has been reported from Michigan but no specimens have been located. It is similar to *A. pratensis* but differs in the glumes at most hispidulous or scabrous toward their tips; in *A. pratensis* the glumes are conspicuously hispid-ciliate on the keel, especially toward their tips. In both species the glumes are connate up to about a third of their length.

110. *Milium effusum* ×½
111. *Sporobolus cryptandrus* ×½ ; glumes and floret ×10

2. **A. aequalis** Sobol   Fig. 112

Map 213. In shallow water and wet places: shores, ponds, ditches, marshes, moist depressions and low meadows, stream margins, swamps and bogs, sometimes in very calcareous places.

Similar to the next species in size of spikelets and short anthers (up to 1 mm long), but with the awns very inconspicuous.

3. **A. carolinianus** Walter

Map 214. Marshy ground, perhaps adventive in Michigan from a native range to the south — at least those plants in wet fields and gardens.

A. geniculatus L., by some considered introduced from Eurasia and by others a native species, has been reported from Michigan but no specimens have been found. It has spikelets a little larger than the preceding two species (2.6–3 mm rather than 2–2.5 mm) and anthers ca. 1 mm or more in length. The awns are exserted as in *A. carolinianus*.

11. Milium

1. **M. effusum** L.   Fig. 110

Map 215. Rich often moist deciduous and mixed woods, common along trails and clearings; rarely in cedar–hemlock stands.

211. Heleochloa
schoenoides

212. Alopecurus pratensis

213. Alopecurus aequalis

214. Alopecurus
carolinianus

215. Milium effusum

216. Calamovilfa longifolia

A very striking plant when seen in the woods, rather strongly glaucous in aspect, with prominent pale ligules. The mature panicle branches are widely spreading or somewhat deflexed as in other large woodland grasses with which it may be growing, *Melica smithii* and *Festuca obtusa*. The glumes are very minutely scabrous. This is the only species of the genus occurring in North America, and plants here have been segregated from typical Eurasian ones as var. *cisatlanticum* Fern.

12. Calamovilfa

REFERENCE

Thieret, John W. 1960. Calamovilfa longifolia and Its Variety magna. Am. Midl. Nat. 63: 169–176.

1. C. longifolia (Hooker) Scribner   Fig. 113

Map 216. Almost entirely, in our region, a plant of the sand dunes and beaches of the Great Lakes; rarely spreading along sandy roadsides and railroads, in jack pines and sand barrens. The only collections seen from Lake Superior are from Houghton (*W. M. Canby* in 1868, NY) and the shore at the Huron Mountain Club in Marquette County (*M. T. Bingham* in 1940, BLH).

Michigan specimens are almost all the local var. *magna* Scribner & Merrill (TL: lake shore at the mouth of the Kalamazoo River [Allegan County]). This has a ± open, spreading panicle, compared with the typical variety, which ranges west of Michigan and has a narrowly contracted panicle. It is apparently adventive along railroads in Kalamazoo County. *C. longifolia* var. *magna* is readily distinguished from *Ammophila*, our other common dune grass, by the smaller spikelets in a more open panicle and, vegetatively, by the usually villous lower sheaths and often greater height (sometimes over 2 m). The characteristic tough scaly rhizomes (fig. 113) attest its powers as a sand binder.

13. Sporobolus   Dropseed

This is one of the genera of grasses in which the fruit differs from a true grain or caryopsis because the pericarp is not adnate to the seed coat.

217. Sporobolus
     cryptandrus

218. Sporobolus neglectus

219. Sporobolus
     vaginiflorus

KEY TO THE SPECIES

1. Summit of leaf sheath with conspicuous dense beard of long hairs on the outside
. . . . 1. **S. cryptandrus**
1. Summit of leaf sheath glabrous or at most with a few hairs (except inside)
  2. Spikelets mostly 1.7–3 mm long . . . . 2. **S. neglectus**
  2. Spikelets 3.5–6 (6.5) mm long
    3. Lemma and palea pubescent with short appressed hairs; glumes nearly equal;
      palea usually prolonged . . . . 3. **S. vaginiflorus**
    3. Lemma and palea glabrous; glumes distinctly unequal; palea not prolonged
      4. Panicle exserted, ± open with spreading branches at maturity; larger glume
        equalling or exceeding lemma . . . . 4. **S. heterolepis**
      4. Panicle partly or mostly included in upper leaf sheath, narrow and contracted;
        glumes both shorter than lemma . . . . 5. **S. asper**

1. S. cryptandrus (Torrey) Gray   Fig. 111

Map 217. Sandy shores, oak woodland, barrens, and dunes; locally common in disturbed areas (roadsides, railroads, fields). At some stations, particularly in the northern Lower Peninsula, the species is undoubtedly adventive, although it is presumably indigenous in the southern part of the state.

The base of the inflorescence and sometimes the entire panicle is enclosed in the upper leaf sheath. This is our commonest species of the genus, easily distinguished by the hairy summits of the sheaths and small spikelets (nearly or quite as small as in the next species, but with very unequal glumes, the smaller one less than 1.5, usually 1, mm long).

2. S. neglectus Nash

Map 218. Dry sandy roadsides and fields; shores and mudflats.

See comments under the next species, of which this is sometimes treated as var. *neglectus* (Nash) Scribner.

3. S. vaginiflorus (Torrey) Wood   Fig. 114

Map 219. Dry roadsides and fields. Farwell's collection labeled as from bluffs at Cliff Mine, Keweenaw County (*849 1/3* in 1895, MICH, BLH) is, like so many such geographic anomalies, the result of either assiduous collecting or mixture of specimens.

Our specimens have the tip of the palea ± beak-like, distinctly prolonged beyond the lemma and glumes (fig. 114) and are thus the so-called var. *inaequalis* Fern. This species and the preceding are very similar, being impossible to distinguish without spikelets. Both differ from our other species in being rather delicate annuals with their glumes nearly equal. The panicle is nearly or completely enclosed in the subtending leaf sheath in both species.

4. S. heterolepis (Gray) Gray   Fig. 115

Map 220. Shallow soil on dolomite pavement, Drummond Island; "boggy

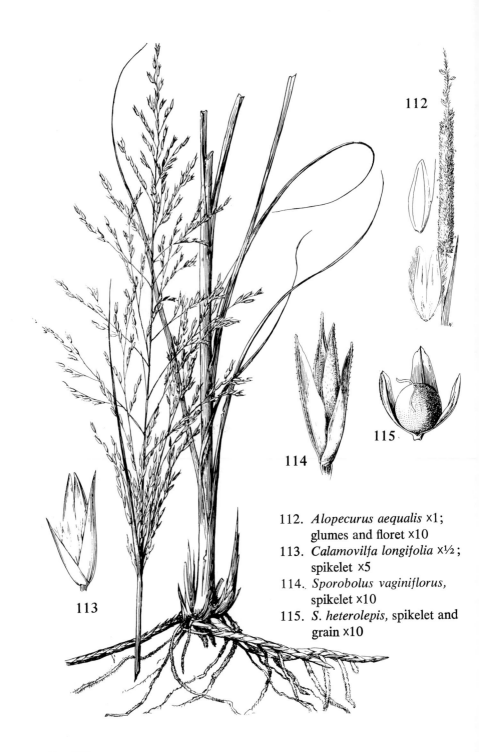

112. *Alopecurus aequalis* ×1;
     glumes and floret ×10
113. *Calamovilfa longifolia* ×½;
     spikelet ×5
114. *Sporobolus vaginiflorus*,
     spikelet ×10
115. *S. heterolepis*, spikelet and
     grain ×10

meadows" in Livingston County; shrubby banks and bridle path in Kalamazoo County.

The mature grain is globose, spreading the parts of the spikelet and splitting the palea (fig. 115).

5. S. asper (Michaux) Kunth

Map 221. Native east and south of Michigan, presumably adventive here along railroads.

14. Calamagrostis   Reedgrass

REFERENCE

Stebbins, G. L., Jr. 1930. A Revision of Some North American Species of Calamagrostis. Rhodora 32: 35–57 (also Contr. Gray Herb. 87).

KEY TO THE SPECIES

1. Callus hairs and glumes at least 1½ times as long as the lemma . . . . . 1. **C. epigeios**
1. Callus hairs and glumes barely if at all exceeding lemma
   2. Panicle mostly open with rather loosely ascending to spreading branches at flowering time; lemma nearly or quite smooth, membranous and translucent for at least the apical half; awn nearly or quite smooth, at least on basal half; callus hairs about as long as lemma (occasionally shorter), ± uniform in length and distribution; palea not over 2 mm long . . . . 2. **C. canadensis**
   2. Panicle mostly narrow and contracted with strongly ascending branches at flowering time; lemma usually firm and prominently scabrous, colorless and translucent only toward the tip; awn distinctly but minutely antrorsely scabrous its entire length [use 20X lens!]; callus hairs generally shorter than lemma, ± unequal in length or distribution (those immediately below the middle of the lemma shorter than those at the side, or absent; do not confuse the hairy prolongation of the rachilla behind the palea); palea various (longer in common species)
      3. Upper ligules 1.5–2.5 (3) mm long; leaf blades mostly smooth below and involute; palea less than 2 mm long; longer callus hairs ca. 1.7–2.2 mm long . .
         . . . . 3. **C. stricta**
      3. Upper ligules mostly 3.5–7.5 (8) mm long; leaf blades ± antrorsely scabrous below, involute to flat; palea (1.8) 2–3.2 mm long; longer callus hairs ca. 2.4–3.2 (3.4) mm long
         4. Awn straight, at most very obscurely twisted, usually inserted on middle half of lemma (fig. 118); palea mostly about 0.5 mm (or more) shorter than the lemma; callus with hairs immediately below the middle of the lemma shorter than those of the lateral tufts . . . . 4. **C. inexpansa**
         4. Awn twisted at base, mostly bent at middle (the tip therefore protruding from the sides of some of the spikelets), inserted on lower third of lemma (fig. 119); palea nearly or quite as long as lemma; callus lacking hairs immediately below the middle of the lemma . . . . 5. **C. lacustris**

1. C. epigeios (L.) Roth

Map 222. Locally adventive in North America from the Old World, collected thus far only once in Michigan: well established in moist lowland area in Rose Lake State Game Area, Clinton County (*S. N. Stephenson* in 1968, MSC, MICH).

The glumes are very narrow, linear-lanceolate and attenuate in this species.

2. C. canadensis (Michaux) Beauv.   Fig. 117   Blue-joint

Map 223. Marshes, shores, swales, bogs, wet prairies, and moist places generally; thickets and openings along streams.

A rather variable species, sometimes confused with the others, especially *C. inexpansa*, but distinguishable once one is familiar with the very thin texture of the lemma. The sheaths may be either glabrous or pubescent.

3. C. stricta (Timm) Koeler

Map 224. Several collections from wet places in Kalamazoo County, and one (*Farwell 12563½* in 1940, BLH) said to have come from conglomerate shores of Lake Superior east of Eagle Harbor.

This is the grass long known as *C. neglecta* Gaertner, Meyer, & Scherb.

220. Sporobolus
heterolepis

221. Sporobolus asper

222. Calamagrostis
epigeios

223. Calamagrostis
canadensis

224. Calamagrostis stricta

225. Calamagrostis
inexpansa

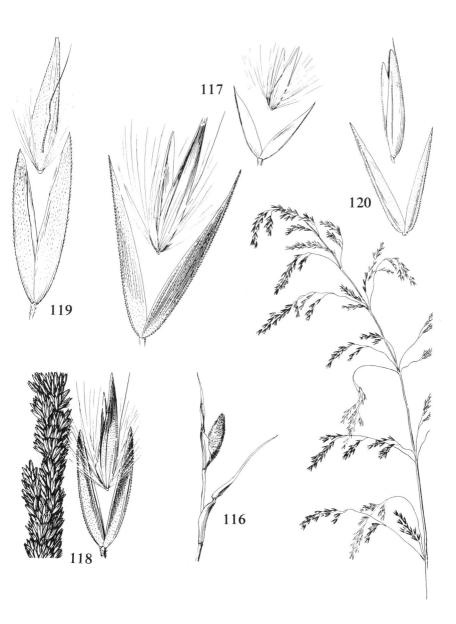

116. *Heleochloa schoenoides* ×½
117. *Calamagrostis canadensis,* two extremes of glumes and florets ×10
118. *C. inexpansa* ×1; glumes and floret ×10
119. *C. lacustris,* glumes and floret ×10
120. *Cinna latifolia* ×1; glumes and floret ×10

4. **C. inexpansa** Gray   Fig. 118

Map 225. Often in the same places as *C. canadensis*, but more characteristic of shores, interdunal flats, and rock crevices along Lake Superior; also on sand dunes.

Our plants are apparently most if not all var. *brevior* (Vasey) Stebbins. A few specimens with all other characteristics of this species have the short ligules of the preceding.

5. **C. lacustris** (Kearney) Nash   Fig. 119

Map 226. Lake shores and rock crevices in the Lake Superior region.

Reports of *C. pickeringii* Gray from Isle Royale should be referred to this species, which is sometimes considered a variety of it. *C. pickeringii* differs in having very short and rather sparse callus hairs, less than 1 mm long.

Several ambiguous collections from Houghton and Keweenaw counties resemble this species in having a slightly bent awn, inserted on the lower third of the lemma; but in the presence of callus hairs below the middle of the lemma and in the short palea (distinctly shorter than the lemma) they resemble *C. inexpansa*, to which they are tentatively referred.

15. **Cinna**   Wood Reedgrass

KEY TO THE SPECIES

1. Spikelets 2.5–3.8 (4) mm long (excluding awn), the second (longer) glume glistening and with lateral nerves very obscure or obsolete; anther 0.4–0.9 mm long . . . . 1. **C. latifolia**
1. Spikelets (4.2) 4.5–6.5 mm long, the second glume rather dull, with a well developed lateral nerve on each side; anther 0.8–1.8 mm long . . . . . . 2. **C. arundinacea**

1. **C. latifolia** (Goepp.) Griseb.   Fig. 120

Map 227. In cedar swamps and other coniferous woods and boggy places, as well as deciduous and mixed woods, especially on wet seepy slopes, in depressions and clearings, and along woods roads.

226. Calamagrostis
       lacustris

227. Cinna latifolia

228. Cinna arundinacea

The panicle is normally green. A few specimens (especially from the Keweenaw Peninsula and Isle Royale) have the panicle rather conspicuously tinged with purplish.

2. C. arundinacea L.

Map 228. Swampy woods, floodplain forests, depressions in deciduous woods, thickets, very rarely on dry wooded sites. Not mapped because it is such a dubious record is a specimen labeled as coming from Keweenaw County (*Farwell 749* in 1890, MICH, BLH).

The panicle tends to be stiffer and less open than in the preceding species, in which it is typically lax and drooping. The ligules of *C. arundinacea* are more deeply tinged with reddish brown than in *C. latifolia*, in which they are generally colorless or nearly so.

16. Apera

1. A. spica-venti (L.) Beauv.   Fig. 121

Map 229. Sparingly introduced in North America from Europe. The Michigan collections seen were both made before 1900: "near Lansing" (*L. H. Bailey* in 1886, GH); John Ball Park, Grand Rapids (*M. B. Fallass* in 1896, ALBC).

Often included in the next genus as *Agrostis spica-venti* L.

17. Agrostis   Bentgrass

Immature specimens of *Poa palustris* are sometimes hastily identified as an *Agrostis* or are included in mixed collections. They may readily be distinguished by having at least the rudiments of a second floret in the spikelet and a cobwebby tuft at the base of the villous-margined lemma.

A difficult genus, the lines between several of the species rather vague. The treatment here follows that of Philipson and is similar to that of Gleason except that his *A. stolonifera* var. *major* (Gaudin) Farw. is here tentatively accorded specific rank as *A. gigantea*. This is the species called *A. alba* L. in many American manuals, but that long misinterpreted name actually applies to a *Poa*. Gleason's var. *compacta* Hartman of *A. stolonifera* should be called var. *palustris*, as the latter is a much older epithet in varietal rank. Fernald's treatment of this group in *Gray's Manual* is unique and cannot be correlated with other works.

REFERENCES

Philipson, W. R. 1937. A Revision of the British Species of the Genus Agrostis Linn. Jour. Linn. Soc. Bot. 51: 73–151.
Shinners, L. H. 1943. Notes on Wisconsin Grasses—III Agrostis, Calamagrostis, Calamovilfa. Am. Midl. Nat. 29: 779–782.

KEY TO THE SPECIES

1. Palea present, about half as long as the lemma or longer; anthers ca. 0.8–1.5 mm long
  2. Larger ligules less than 2.2 mm long; leaf blades not over 4 mm wide; panicle ± open (the branches diverging from their bases) . . . . 1. **A. tenuis**
  2. Larger (upper) ligules ca. 2.5–6 (7) mm long; leaf blades usually more than 4 mm wide *or* the panicle branches ± strongly ascending at their bases
    3. Plants rhizomatous but not stoloniferous – i.e., culms arising from underground rhizomes, straight or curved at the very base, otherwise erect and nearly or quite straight; larger leaf blades mostly 3–7 (10) mm wide; spikelets usually flushed with red or purplish; bases of middle panicle branches mostly meeting the axis of the panicle at an angle of 30–45° (except when very immature) . . . 2. **A. gigantea**
    3. Plants stoloniferous but not rhizomatous – i.e., culms usually decumbent at their bases, the lower nodes often strongly bent and/or rooting, but underground rhizomes absent; larger leaf blades 1.7–3 mm wide; spikelets pale, greenish; bases of middle panicle branches usually strongly ascending or appressed to axis of panicle, at most diverging about 15° (but panicle branches often spreading distally) . . . . 3. **A. stolonifera**
1. Palea absent or obsolete; anthers ca. 0.6 mm long or shorter (except in *A. canina*)
  4. Anthers ca. 0.8 mm or more long; lemma usually with a stiffish bent awn . . 4. **A. canina**
  4. Anthers ca. 0.6 mm long or shorter; lemma awnless (very rarely awned)
    5. Longest panicle branches less than 6 (12) cm long *and* the uppermost leaf blade more than 5 cm long; leaf blades flat, the wider ones 1.5–3.5 (6) mm broad; panicle branches forked about or below the middle, often smooth or only sparingly hispidulous-scabrous; panicle pale, greenish (very rarely tinged with reddish) . . . . 5. **A. perennans**
    5. Longest panicle branches more than 6 cm long *or* uppermost leaf blade less than 5 cm long (or both conditions); leaf blades usually ± involute, the widest up to 1.5 (rarely 3) mm broad; panicle branches often not forked until beyond the middle, copiously hispidulous-scabrous; panicle ± flushed with reddish . 6. **A. hyemalis**

1. **A. tenuis** Sibth.   Colonial Bent; Rhode Island Bent

Map 230. Generally considered introduced from Europe. Widely planted as a lawn grass in the northeastern United States, but only rarely collected as a presumed escape in Michigan, in fields and dry woods.

Some of our specimens seem to have ligules a little long for this species, in which they are normally about 1 mm. Some other specimens from the Lake

229. Apera spica-venti   230. Agrostis tenuis   231. Agrostis gigantea

Superior region may be this species, but are more likely depauperate *A. gigantea*. Ordinarily, small specimens of *A. gigantea* will differ in their longer ligules, leaf blades wider and usually longer on the upper part of the culm than in *A. tenuis*, and a tendency for the panicle branches to have spikelets near their base, while in *A. tenuis* the panicle branches have no spikelets toward their base and there is no rhizome. From occasional specimens of *A. stolonifera* which may have shorter ligules than usual, *A. tenuis* tends to differ in its erect culms, reddish panicle, and more spreading panicle branches.

2. **A. gigantea** Roth   Fig. 122   Redtop

Map 231. A common grass throughout the state, in open to wooded, wet to dry places, whether disturbed (roadsides, fields, etc.) or relatively undisturbed (bogs, shores, woods, dunes). Both this species and the next are usually considered introductions from Europe, although so well established that they often appear as if indigenous. Both are commonly cultivated as forage grasses and form a good turf.

Extremely variable and, together with *A. stolonifera*, the source of much taxonomic as well as nomenclatural contention. *A. gigantea* is characteristically a distinctive large plant with a large, open, decidedly red inflorescence. However, plants growing in the shade, as in woods, may have greenish and often more delicate looking panicles, somewhat like those of the next species which, however, usually grows in open situations, where *A. gigantea* develops its red color. The panicles of *A. gigantea* are normally at least 10 cm long. Depauperate specimens with smaller panicles may, if the panicle branches are ascending (as when young), resemble the next species, but usually can be distinguished by the reddish color, wider leaf blades, and of course the rhizome.

Some collections from Stony Creek near Rochester, Oakland County, are ambiguous. The culms are prominently decumbent, rooting at the nodes, as in the most decumbent forms of *A. stolonifera*, but the leaf blades are very broad (up to 7 mm). The panicles are immature, with ascending branches, green, about 10–12 cm long, the bases included in the uppermost leaf sheath. These would apparently be what the *Manual of Grasses* refers to *A. nigra* With., which is included by Philipson and current British manuals in the synonymy of *A. gigantea*. A very few other collections are intermediate between these and most of our specimens of *A. stolonifera*. Our collections of *A. gigantea* are otherwise of the erect form; decumbent plants are more common in Europe, and the Rochester specimens presumably belong here, although they might be considered robust forms of *A. stolonifera* if the habit were considered more significant than the size.

In a collection made by Dodge near Port Huron in 1892, many of the lemmas have a delicate slender dorsal awn conspicuously exceeding the spikelet; in all of our other collections the lemmas are awnless. A pathological form with greatly elongate lemmas and glumes occasionally is found and is likely to cause confusion unless some normal spikelets are located for identification.

3. **A. stolonifera** L. Creeping Bent

Map 232. Shores and marshy places, trails, and waste ground. Perhaps partly native although at least in part introduced.

Sometimes ± densely tufted [var. *stolonifera*, with leaf blades mostly less than 5 cm long], more often the culms few and loosely stoloniferous with longer leaves [var. *palustris* (Hudson) Farw., often recognized as *A. palustris* Hudson, a name which has also been misapplied to the preceding species]. The panicle is short, up to 10 or occasionally 12 cm long, and its branches are ± contracted, at least after flowering. See also discussion under the preceding species.

4. **A. canina** L. Velvet Bent

Map 233. A lawn and sod grass, reported by several authors as escaped in Michigan, but the only collection seen was recorded as "rare" at Detroit in 1895 (*Farwell 940½*, BLH, US).

5. **A. perennans** (Walter) Tuckerman Autumn or Upland Bent

Map 234. Usually in wooded places (oak, beech–maple), often on the damp side but sometimes in dry open woods and adjacent fields.

Unusually robust plants with panicle branches as long as 12 cm will have the wider leaf blades much broader than in *A. hyemalis* and thus may be readily distinguished. See also comments under the next species.

All Michigan specimens examined have been awnless, although a form with awned lemmas is known.

6. **A. hyemalis** (Walter) BSP. Fig. 123 Ticklegrass

Map 235. Occurring in a diversity of habitats, especially in open sandy or rocky areas: on dunes, shores, and rock outcrops; often becoming abundant after fire or other disturbance on shores, jack pine plains, bog borders, etc.; under oak and jack pine; in bogs and on sand barrens.

In this species, the culm and panicle branches give an impression of being much straighter and stiffer than in the more lax appearing *A. perennans*, from which it is quite distinct, although any one of the characters in the key may give

232. Agrostis stolonifera

233. Agrostis canina

234. Agrostis perennans

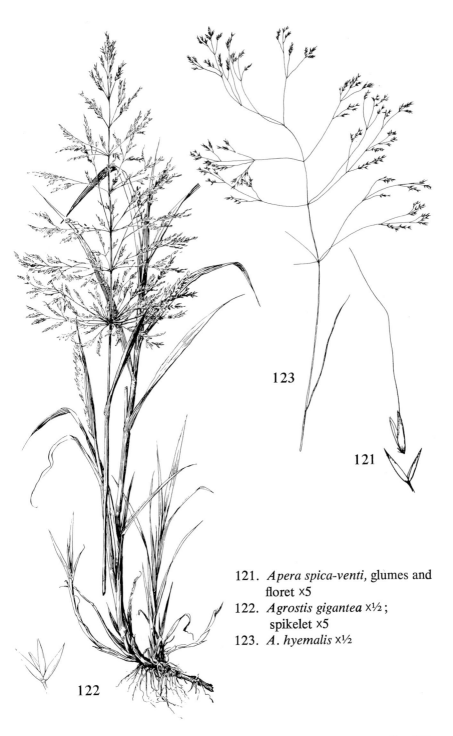

123

121

122

121. *Apera spica-venti*, glumes and
   floret ×5
122. *Agrostis gigantea* ×½;
   spikelet ×5
123. *A. hyemalis* ×½

trouble. As pointed out by Shinners, *A. perennans* has much longer upper leaf blades in relation to the length of the panicle branches than does *A. hyemalis*. Therefore, small specimens of the latter with panicle branches as short as in *A. perennans* will have the leaves too short to be that species.

A few collections from the southernmost Lower Peninsula seem to be typical var. *hyemalis*; the spikelets are 1.2–1.5 (2) mm long, subsessile or on pedicels up to 1.5 (2) mm long, crowded in spikelike clusters at the ends of the panicle branches. All the rest of our material, although extremely variable, is var. *tenuis* (Tuckerman) Gleason (*A. scabra* Willd., as treated by many authors), with spikelets (1.5) 1.8–2.5 (3) mm long, not so crowded, at least some of the pedicels exceeding 2 mm in length.

Plants with many of the lemmas bearing a slender awn from the middle or above the middle of the back rarely occur [f. *setigera* (Fern.) E. Voss]. This form is known in Michigan from Isle Royale and Keweenaw, Mecosta, and Marquette counties.

5. CHLORIDEAE

All of our representatives of this tribe may be easily distinguished at a glance. *Beckmannia* is transferred to the Aveneae in Gould (1968), but the rest of our genera are retained in the Chlorideae, which are included in the subfamily Eragrostoideae.

KEY TO THE GENERA OF CHLORIDEAE

1. Spikes radiating from summit of culm (i.e., umbellate or nearly so) or at least the lower ones whorled (solitary in depauperate individuals); anthers not over 1.5 mm long
    2. Lemmas conspicuously awned; glumes short-awned to narrowly acuminate; ligule white, membranous with ciliate edge . . . . 1. **Chloris**
    2. Lemmas and glumes awnless; ligule various
        3. Glumes ca. 2.5–4.5 mm long; spikelets at least 2–3-flowered; larger spikes 4–6 mm broad; plant a tufted annual; ligule membranous . . . . 2. **Eleusine**
        3. Glumes less than 2.5 mm long; spikelets 1-flowered; larger spikes 1.5–3 mm broad; plant a creeping perennial; ligule a fringe of white hairs . . . . 3. **Cynodon**
1. Spikes all racemose or panicled; anthers various
    4. Glumes equal, ca. 2–3 mm long, deeply pouch-like and largely covering the floret, the spikelet strongly flattened and about as wide as long; ligule membranous, eciliate; anthers ca. 0.5–1 mm long . . . . 4. **Beckmannia**
    4. Glumes unequal, the longer ones ca. 3.5–11 mm long, not pouch-like, equalling or shorter than florets, the spikelet not strongly flattened, much narrower than long; ligules and anthers various
        5. Ligule membranous, eciliate, ca. (2.5) 4–6 mm long, becoming shreddy; anthers ca. 0.5 mm long or shorter; end of rachis of spike spikelet-bearing, not prolonged; spikelets short-pedicelled . . . . 5. **Leptochloa**
        5. Ligule ciliate (entirely of hairs, or hairs longer than any membranous portion), up to 3.5 mm long (a few hairs rarely to 4.5 mm); anthers 2.5–6 mm long; end of rachis of spike prolonged into a stout naked projection 3–14 mm beyond the last spikelet; spikelets sessile or subsessile

6. Spikes 20–40 (55), 0.6–2.2 cm long, on closely hispidulous nodding
peduncles up to 3 mm long . . . . 6. **Bouteloua**
6. Spikes (1) 2–20 (occasionally more), 1.5–12 cm long, on glabrous to hispid
peduncles 0.5–20 [reported to 40] mm long [Plants with spikes as short as in
Bouteloua have them few and on short but smooth peduncles.] . . . . . . .7. **Spartina**

1. Chloris

1. **C. verticillata** Nutt.   Fig. 128   Windmill Grass
   Map 236. Adventive from the Southwest; collected in a pasture in Franklin
Tp., Lenawee County (*W. S. Benninghoff* in 1962, MICH).
   The spikelets consist of one perfect floret and one sterile floret. The mature
inflorescence acts as a tumbleweed.

2. Eleusine

1. **E. indica** (L.) Gaertner   Fig. 124   Goose Grass
   Map 237. Naturalized from the Old World and locally established as a weed
of waste ground in southern Michigan. Collected as early as 1838 in Berrien
County by the First Survey.

3. Cynodon

1. **C. dactylon** (L.) Pers.   Fig. 125   Bermuda Grass
   Map 238. Roadsides, railroads, and other waste places, apparently not well
established this far north. An introduced species, more common as a pasture and
lawn grass farther south.

4. Beckmannia

1. **B. syzigachne** (Steudel) Fern.   Fig. 129   Slough Grass
   Map 239. A plant of wet places, important to waterfowl where more
common but collected only three times in Michigan: ditch ca. 3 miles north of
Pinconning (*Hebert* in 1950, ND); border of cold spring, Alanson (*Ehlers 5112*

235. Agrostis hyemalis

236. Chloris verticillata

237. Eleusine indica

in 1932, MICH, UMBS); McCargo Cove, Isle Royale (*Brown 3537* in 1930, MICH).

This species also occurs in Asia, and American plants belong to one of the two subspecies sometimes recognized there: ssp. *baicalensis* (Kusnez.) Koyama & Kawano.

5. Leptochloa

1. L. fascicularis (Lam.) Gray   Fig. 126   Sprangletop; Salt Meadow Grass
Map 240. Collected by Walpole at Ypsilanti in 1920 (BLH) and found again in Michigan in 1966 (*S. N. Stephenson 1319*, MSC) in a plantation in the Kellogg Forest, Kalamazoo County (Mich. Bot. 6: 24–25. 1967). Probably adventive from farther south.

Sometimes segregated as *Diplachne fascicularis* (Lam.) Beauv. and placed in the tribe Poeae (where, because of the long inflorescence branches, pedicelled spikelets, and obscurely one-sided spikes, it is likely to run in keys).

6. Bouteloua

1. B. curtipendula (Michaux) Torrey   Fig. 127   Grama Grass
Map 241. Prairies and dry open places, especially on banks and hillsides. The

238. Cynodon dactylon

239. Beckmannia
syzigachne

240. Leptochloa
fascicularis

241. Bouteloua
curtipendula

242. Spartina patens

243. Spartina pectinata

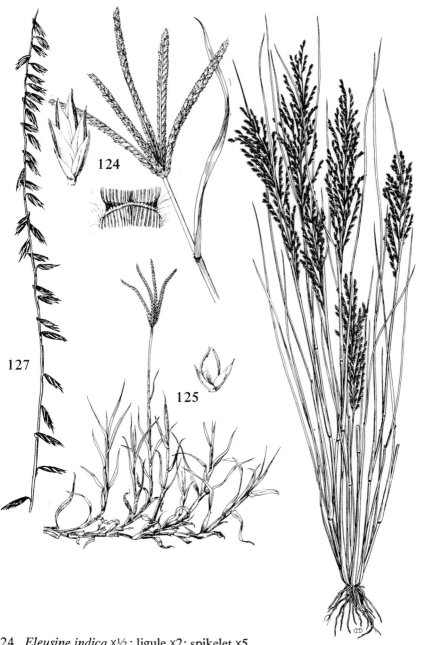

124. *Eleusine indica* ×½; ligule ×2; spikelet ×5
125. *Cynodon dactylon* ×½; spikelet ×5
126. *Leptochloa fascicularis* ×¼
127. *Bouteloua curtipendula* ×¾

genus is predominantly one of prairies and grasslands and only this species ranges into Michigan.

7. Spartina

KEY TO THE SPECIES

1. Glumes awnless; plants wiry, with slender culms bearing (1) 2–4 spikes 1.5–3.5 (4.5) cm long; leaves tightly involute, up to 3 mm wide (when flattened); rhizomes 1.5–3 mm thick; anthers ca. 2.5–3.5 mm long . . . . 1. **S. patens**
1. Glumes (at least the longer) distinctly awned; plants robust, with stout culms bearing 6–20 (or more) spikes (3) 4–12 (15) cm long; leaves with flat blades (often involute when dried), the larger 5–10 (15) mm broad; rhizomes 3–8 mm thick; anthers (3) 3.5–6 mm long . . . . 2. **S. pectinata**

1. **S. patens** (Aiton) Muhl.   Salt-meadow Cordgrass

Map 242. Apparently native, at least formerly, in salt marshes on the southwest side of Detroit; and established in freight yards near Port Huron, where collected by Dodge in 1916 and possibly introduced. Characteristically in salt marshes along the coast from Quebec to Texas, inland only in New York and Michigan.

2. **S. pectinata** Link   Fig. 130   Cordgrass

Map 243. Most characteristic of marshes, wet prairies, and sandy shores, but sometimes in dry sand.

A tall grass (up to 6 feet or more) with rough foliage and abundant distinctive strong rhizomes. These are very effective soil binders. Plants with the spikes longer than usual and more slender, with longer peduncles, have been segregated as var. *suttiei* (Farw.) Fern. (TL: Island Lake [Livingston County]), but seem scarcely distinct.

6. PHALARIDEAE

The spikelets in this tribe basically have 3 florets, but the lower pair may be only staminate (*Hierochloë*) or reduced to tiny scales (*Phalaris*) and hence easily

244. Hierochloë odorata   245. Anthoxanthum   246. Phalaris canariensis
odoratum

overlooked or misinterpreted. Only the upper (terminal) floret is perfect. This tribe is included in the Aveneae by Gould (1968).

KEY TO THE GENERA OF PHALARIDEAE

1. Panicle open, pyramidal, the branches spreading or drooping; glumes nearly equal in length, with lateral nerves obscure or prominent only on basal half; lower florets staminate, at least as large as perfect floret, awnless . . . . 1. **Hierochloë**
1. Panicle contracted, the branches ascending or suppressed (the lower ones sometimes spreading-ascending only during flowering); glumes equal or not, with lateral nerves (at least on larger glumes) prominent beyond the middle; lower florets sterile, either vestigial or large and awned
  2. Glumes very unequal; lower lemmas with prominent dorsal awns, concealing the awnless perfect floret . . . . 2. **Anthoxanthum**
  2. Glumes nearly or quite equal; lower lemmas awnless, small and inconspicuous, only the awnless perfect floret evident . . . . 3. **Phalaris**

1. Hierochloë

1. **H. odorata** (L.) Beauv.   Fig. 131   Sweet Grass
    Map 244. Edges of woods, shores, meadows, boggy places; usually in moist ground, locally abundant and spreading.
    An attractive spring-flowering grass, with fragrant foliage used by Indians in making baskets. The long leaves of sterile shoots produced later in the season are used, and when dried retain their vanilla-like fragrance for many years.

2. Anthoxanthum

1. **A. odoratum** L.   Fig. 132   Sweet Vernal Grass
    Map 245. Shores, meadows, borders of woods; roadsides and other disturbed ground. Introduced from Eurasia.
    The glumes are usually at least sparsely hairy. The species is perennial and tetraploid. A ± smaller, diploid, annual species, *A. puelii* Lec. & Lamotte (*A. aristatum* auct.) has been reported from Michigan but the only specimens seen have been from the Grass Garden of the Agricultural College (now Michigan State University). It differs in its slightly smaller flower parts, glabrous glumes, and definite exsertion of both awns; in *A. odoratum*, the awn of at most one of the sterile lemmas is exserted beyond the longer glume.

3. Phalaris

KEY TO THE SPECIES

1. Glumes mostly 6–8 mm long, the keel prominently winged; sterile lemmas glabrous or nearly so, ca. 2.5–3.7 mm long; plant annual, with very dense, compact, ovoid panicle . . . . 1. **P. canariensis**

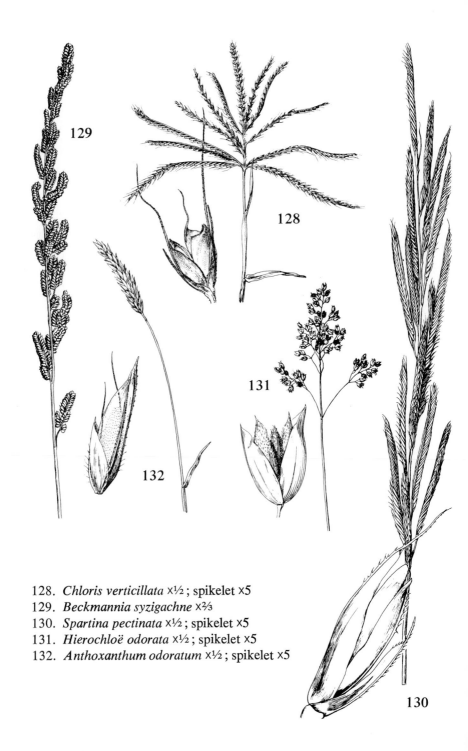

128. *Chloris verticillata* ×½ ; spikelet ×5
129. *Beckmannia syzigachne* ×⅔
130. *Spartina pectinata* ×½ ; spikelet ×5
131. *Hierochloë odorata* ×½ ; spikelet ×5
132. *Anthoxanthum odoratum* ×½ ; spikelet ×5

1. Glumes mostly (3.5) 4–5.7 mm long, the keel not winged; sterile lemmas very
hairy, ca. 0.5–2 (3) mm long; plant a rhizomatous perennial, with elongate lobed
panicle (or the lower branches spreading at anthesis) . . . . 2. **P. arundinacea**

1. **P. canariensis** L. Canary Grass

Map 246. Dumps and waste ground. An introduced species occasionally
escaped from cultivation but not persisting.

Although this grass was named by Linnaeus in recognition of its being native
to the Canary Islands and North Africa, it is also true that the species provides a
main constituent of commercial "bird-seed," so that "canary grass" is a doubly
appropriate common name.

2. **P. arundinacea** L. Fig. 133 Reed Canary Grass

Map 247. Marshes and wet shores, borders of streams and ponds, in ditches.

The glumes are usually glabrous, rarely minutely hispidulous. In f. *variegata*
(Parnell) Druce (Ribbon Grass), frequently cultivated for its ornamental foliage,
the leaf blades are longitudinally striped with white or cream. At least some
colonies of this form are presumably escapes from cultivation and some of our
other plants may be derived from introduction as a pasture grass, for the species
is also native in Eurasia.

7. ORYZEAE

This is the tribe which includes rice (*Oryza sativa* L.), perhaps the oldest of
food crops and eaten by more people than any other grain. Gould (1968)
includes *Zizania* in this tribe, and places it in a distinct subfamily, Oryzoideae.
As traditionally recognized, the tribe includes only the genus *Leersia* in the
United States.

1. Leersia

KEY TO THE SPECIES

1. Spikelets 3–4 mm long, 0.8–1.3 (1.5) mm wide, the tip of one barely overlapping
the base of the next above; panicle branches solitary at all nodes of inflorescence
. . . . 1. **L. virginica**
1. Spikelets (at least larger ones) 4–5.5 mm long, 1.4–1.8 mm wide, mostly
overlapping about the basal half of the next above; panicle branches 2–3 (4) at (or
near) the lowermost node of inflorescence . . . . 2. **L. oryzoides**

1. **L. virginica** Willd. White Grass

Map 248. Swamps, floodplains, creek and river banks, wet depressions in
woods – in more shaded situations than is usual for the next species.

Usually distinct from the next, but some intermediate collections suggest the
possibility of hybridization: e. g., one from Grand Rapids (*Florence Fallass* in

1897, AQC) has short spikelets and solitary panicle branches as in *L. virginica*, but the spikelets are broad and strongly imbricated as in *L. oryzoides* and the foliage is sparsely hispid; another, from near Port Huron (*Dodge* in 1892, MICH), has long spikelets and lower panicle branches paired, but the spikelets are only slightly overlapping, with the aspect of *L. virginica*, and the foliage is only scabrous.

2. **L. oryzoides** (L.) Sw.   Fig. 134   Cut Grass

Map 249. Wet places: ponds, shores, ditches, floodplains, creek and river banks, bogs, pools and wet depressions — often abundant in a distinct zone or band.

A grass with very rough foliage and spikelets, the latter readily adhering to clothing when ripe, much to the annoyance of hikers in moist places where it grows. The margins of the leaf blades are normally roughly hispid-ciliate in this species (rarely only scabrous), while in *L. virginica* they are merely scabrous (rarely with a few rough cilia on some leaves). In both species, the nodes are characteristically bearded with ± retrorse hairs. At least the base of the inflorescence is often hidden by the upper leaf sheath; spikelets developing within leaf sheaths seem more often to produce mature grains than do those of exposed terminal panicles.

8. ZIZANIEAE

Gould (1968) does not separate this tribe from the Oryzeae. We have only one genus, with a single species, in Michigan.

1. Zizania

REFERENCES

Chambliss, Charles E. 1941. The Botany and History of Zizania aquatica L. ("Wild Rice"). Ann. Rep. Smithsonian Inst. 1940, pp. 369–382 + 9 pl.
Dore, William G. 1969. Wild-rice. Canada Dep. Agr. Publ. 1393. 84 pp.
Fassett, Norman C. 1924. A Study of the Genus Zizania. Rhodora 26: 153–160.
Mason, Philip P. 1960. Michigan's First Outdoorsmen. Mich. Conservation 29(2): 40–45.

247. Phalaris arundinacea   248. Leersia virginica   249. Leersia oryzoides

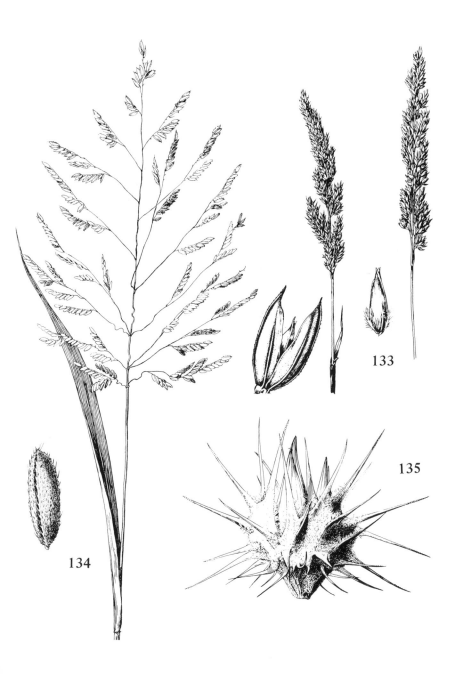

133. *Phalaris arundinacea* ×½ ; spikelet and fertile floret with sterile
    lemmas ×5
134. *Leersia oryzoides* ×½ ; spikelet ×5
135. *Cenchrus longispinus,* bur containing spikelets ×5

1. **Z. aquatica** L.   Fig. 136   Wild-rice

Map 250. Rivers, streams, lakes, and ponds, seldom in water much over 2 feet deep. Wild-rice grows best in shallow (but not stagnant) water where there is at least a slight current over a muddy, mucky, or silty bottom, and where there is little competition from other plants. It is an annual, despite its often large size, and seed is sometimes sown to grow stands for the benefit of wildlife, especially ducks, for which it is an important food. Our locations thus may not all represent natural occurrences. Wild-rice was also an important food for Indians, in Michigan chiefly in the southern Lower Peninsula and western Upper Peninsula. The Menominee tribe was named for this plant. Among many references, those cited above by Chambliss and Mason may be of interest to those concerned with Indian lore.

The pistillate spikelets disarticulate from the distinctive concave summit of the clavate pedicels. The strong lateral nerves of the pistillate lemma closely clasp the palea, producing the illusion of a single cylindrical sheath around the rod-like grain. Three varieties occur in Michigan, but are not always distinct. Typical var. *aquatica* is normally a tall plant (up to about 3 meters) with broad leaves (1–4.5 cm); the pistillate lemma is thin and membranous and at least sparsely hispid-scabrous between the strong nerves; aborted pistillate spikelets are less than 1 mm broad. While var. *aquatica* is apparently restricted to the southern Lower Peninsula (north to Muskegon and Iosco counties), var. *angustifolia* Hitchc. occurs throughout the state and is of smaller stature, narrower-leaved; the pistillate lemma is firm and tough, scabrous-hispid only on the nerves and at most at the base and apex; aborted pistillate spikelets are slightly more than 1 mm broad. A few plants (var. *interior* Fassett) have the large stature of var. *aquatica* and ligules over 1 cm long, but the tough pistillate spikelets of var. *angustifolia*. The latter characteristically has ligules less than 1 cm long and has fewer spikelets on the panicle branches than var. *aquatica* or var. *interior* (fewer than 10 spikelets on lower pistillate branches, fewer than 20 on lower staminate branches, in var. *angustifolia*). Plants with the thin lemmas of var. *aquatica* but narrow-leaved and not so tall have been collected very rarely. Dore (1969) recognizes *Z. palustris* L. as a good species, characterized by the tough pistillate lemmas; var. *palustris* iș equivalent to *Z. aquatica* var. *angustifolia* and var. *interior* (Fassett) Dore is the other variety.

9. PANICEAE

This tribe is a natural assemblage, and these genera are all included in it by Gould (1968).

KEY TO THE GENERA OF PANICEAE

1. Spikelets with an involucre consisting of a spiny bur or of long subtending bristles
  2. Involucre a spiny bur enclosing much or all of the spikelets, the whole readily
     disarticulating (fig. 135) . . . . 1. **Cenchrus**

2. Involucre of long slender bristles subtending but not concealing the spikelet, remaining attached to pedicels when spikelets disarticulate (fig. 138) . . . . . 2. **Setaria**
1. Spikelets without an involucre (although glumes or lemmas may be awned)
   3. Spikelets ± spiny-hispid and usually also awned; ligule none . . . . 3. **Echinochloa**
   3. Spikelets glabrous or pubescent but not coarsely hispid and not awned; ligule present, distinct or nearly obsolete (of hairs or membranous)
      4. Inflorescence composed of 1-sided spikes or spike-like racemes, the rachis of each winged or at least flat on the side opposite the spikelets
         5. Ligule membranous, conspicuous (0.8–2.2 mm long), without a dense row of hairs; fertile lemma rather leathery in texture, acute, about twice as long as wide (or longer), with thin flat translucent margins (fig. 141) . . . . 4. **Digitaria**
         5. Ligule a short (less than 1 mm, usually ca. 0.5 mm) membranous band behind which is a dense row of much longer hairs; fertile lemma very hard, broadly rounded, less than 1½ times as long as wide, with thickened, ± inrolled, indurate margins (fig. 142) . . . . 5. **Paspalum**
      4. Inflorescence an open panicle, not spike-like nor distinctly one-sided
         6. Ligule a membranous collar ca. 1–1.5 mm high, without hairs; base of leaf blade and very summit of sheath without special zone of short pubescence or long hairs or cilia; first glume minute (scarcely 0.5 mm) or obsolete; fertile lemma leathery in texture, with thin flat translucent margins . . . . . . 6. **Leptoloma**
         6. Ligule usually partly or entirely of short or long hairs; (if ligule membranous, plants not otherwise as above: ligule ca. 0.5 mm long or obsolete; or summit of sheath or basal margin of blade pubescent or ciliate; and/or first glume more than 0.5 mm long) . . . . 7. **Panicum**

1. Cenchrus

REFERENCE

DeLisle, Donald G. 1963. Taxonomy and Distribution of the Genus Cenchrus. Iowa St. Jour. Sci. 37: 259–351.

1. **C. longispinus** (Hackel) Fern.   Fig. 135   Sandbur; Sandspur

Map 251. Roadsides, railroads, fields, and waste places generally; an obnoxious weed, especially frequent in dry sandy places but mercifully not yet common on beaches. A most unpleasant species to man (especially if barefoot where *Cenchrus* grows) and animals (the readily deciduous burs rendering an otherwise good forage grass unpalatable when mature). Our plants are presum-

250. Zizania aquatica   251. Cenchrus longispinus   252. Setaria verticillata

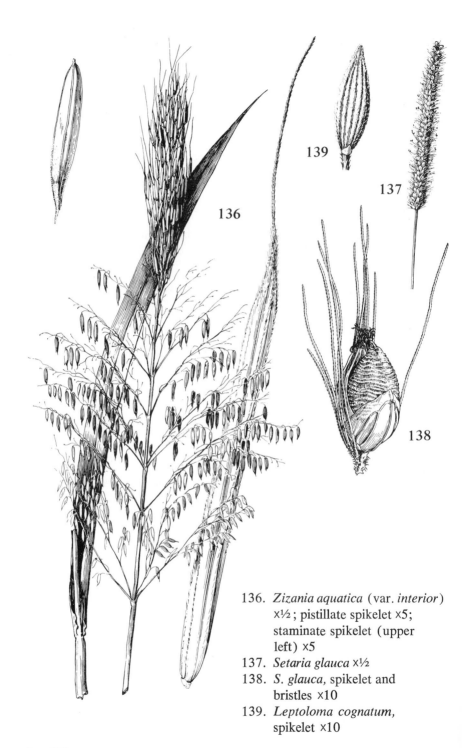

136. *Zizania aquatica* (var. *interior*)
×½; pistillate spikelet ×5;
staminate spikelet (upper
left) ×5
137. *Setaria glauca* ×½
138. *S. glauca,* spikelet and
bristles ×10
139. *Leptoloma cognatum,*
spikelet ×10

ably largely adventive, although perhaps originally native in the southwest part of the state, where known since the First Survey in 1838.

By some authors not distinguished from *C. incertus* M. A. Curtis [*C. pauciflorus* Benth.], a species of the southern states, Mexico, and southward.

2. Setaria   Foxtail Grass

Specimens of *Pennisetum* would run here in the key. *P. villosum* R. Br., an ornamental grass, was collected by Dodge in a cemetery in Port Huron in 1907, where it was planted; there is no definite evidence of this African species escaping in Michigan. The lanceolate spikelets are twice (or more) as long as those of any of our *Setaria*; the bristles are villous basally and fall with the spikelets (in *Setaria* the spikelets fall free and the bristles remain attached to the panicle).

REFERENCES

Pohl, Richard W. 1962. Notes on Setaria viridis and S. faberi (Gramineae). Brittonia 14: 210–213.
Reeder, John R. 1951. Setaria lutescens an Untenable Name. Rhodora 53: 27–30.
Rominger, James M. 1962. Taxonomy of Setaria in North America. Illinois Biol. Monogr. 29. 132 pp.

KEY TO THE SPECIES

1. Bristles, summit of culm, and axis of panicle scabrous with retrorse barbs; panicle branches tending to appear whorled, the panicle ± interrupted toward its base . . .
. . . . 1. **S. verticillata**
1. Bristles, summit of culm, and axis of panicle scabrous or pubescent with antrorse barbs or hairs; panicle very compact throughout
  2. Fertile lemmas mostly ca. (2.7–) 3 (–3.4) mm long, rugose with distinctly transverse ridges, the upper half exposed at maturity (fig. 138); bristles 5 or more per spikelet, becoming orange or golden-brown; sheaths glabrous . . . . . . 2. **S. glauca**
  2. Fertile lemmas less than 3 mm long, evenly and finely rugose or reticulate or smooth (without transverse ridges), the upper half largely or entirely concealed at maturity; bristles fewer than 5 per spikelet, pale greenish or purple (rarely yellow) at maturity; sheaths ciliate with long hairs on the margins
    3. Spikelet articulated above the glumes and sterile lemma; fertile lemma distinctly yellow or darker at maturity; panicle very dense, often ± lobed in appearance .
. . . . 3. **S. italica**
    3. Spikelet articulated below the glumes; fertile lemma pale green or brown; panicle not lobed
      4. Panicle straight and erect or rarely slightly nodding; spikelets not over 2.5 mm long, the blunt fertile lemma nearly or quite concealed by the second glume; leaf blades glabrous above . . . . 4. **S. viridis**
      4. Panicle strongly nodding, bent below the middle; spikelets mostly over 2.5 mm long, the fertile lemma ± tapering to a distinctly exposed tip; leaf blades ± hairy above . . . . 5. **S. faberi**

1. **S. verticillata** (L.) Beauv.

Map 252. A weed in gardens, dumps, vacant lots, and waste places. Introduced from Europe.

2. **S. glauca** (L.) Beauv.   Figs. 137, 138   Yellow Foxtail

Map 253. A common weed of roadsides, railroads, fields, and waste places generally. Introduced from Europe.

Widely known as *S. lutescens* (see Reeder, 1951).

3. **S. italica** (L.) Beauv.   Foxtail or Hungarian Millet

Map 254. Fields, dumps, and waste ground, basically an escape from cultivation. An Old World species, long cultivated; one of the grain crops grown by Indians in Michigan.

4. **S. viridis** (L.) Beauv.   Green Foxtail

Map 255. Roadsides, gardens, fields, railroads, and waste places; spreading into woods and on shores. Another Eurasian species, widely naturalized in North America.

Large plants with broad leaves, long (over 8 cm) and somewhat nodding panicles have been called var. *major* (Gaudin) Posp. Superficially they resemble the next species, but the panicle is less nodding and the small spikelets are identical with those of typical *S. viridis*.

5. **S. faberi** Herrm.   Giant Foxtail

Map 256. To be expected elsewhere in southern Michigan, as the species (a recent invader from Asia) is spreading aggressively in the United States, but only one collection has been seen, from Westphalia Tp., Clinton County (*George McQueen* in 1965, MICH, MSC, BLH).

A tetraploid, this species is ordinarily more robust than the preceding, which is diploid. The lemmas of *S. viridis* are sometimes spotted or mottled dark brown; those of *S. faberi* are unspotted. In its pubescent leaf blades and larger, somewhat exposed grains, this species might be thought to resemble *S. glauca*,

253. Setaria glauca

254. Setaria italica

255. Setaria viridis

but is readily distinguished by its ciliate sheath margins, less rugose fertile lemma, and smaller first glume. (The first glume in *S. glauca* is more than half as long as the second glume.)

3. Echinochloa

REFERENCE

Fassett, Norman C. 1949. Some Notes on Echinochloa. Rhodora 51: 1–3.

KEY TO THE SPECIES

1. Leaf sheaths (at least the lower ones) ± rough-hairy (very rarely smooth); spikelets each with 2 awns (one on sterile lemma, one on second [larger] glume) . . . 1. **E. walteri**
1. Leaf sheaths smooth, essentially glabrous; spikelets with 0 or usually 1 awn (on sterile lemma, the second glume ± acuminate, rarely short-awned)
    2. Polished body of fertile lemma sharply demarcated from withered green tip by a fringe of very minute pubescence [use 20X lens]; stiff bristles of second glume and sterile lemma seldom with pustulate bases . . . . 2. **E. crusgalli**
    2. Polished body of fertile lemma tapered into the firm tip without sharp demarcation or pubescence; stiff bristles of second glume and sterile lemma often with pustulate bases . . . . 3. **E. muricata**

1. **E. walteri** (Pursh) Heller   Fig. 140

Map 257. Banks of rivers and ponds, creek bottoms and ditches, marshes and wet shores — locally common in the marshes at the western end of Lake Erie.

The awn on the second glume is sometimes much shorter than the one on the sterile lemma; but often it is as long, and the spikes have a very dense, bristly look compared to the other species. The sheaths are very rarely smooth [f. *laevigata* Wieg.].

2. **E. crusgalli** (L.) Beauv.   Barnyard Grass

Map 258. Roadsides, fields, gardens, and waste ground; often in damp meadows, on exposed shores and riverbanks, etc. A variable species, naturalized from the Old World.

One of the more distinctive variants is var. *frumentacea* (Link) Wight

256. Setaria faberi   257. Echinochloa walteri   258. Echinochloa crusgalli

("Billion Dollar Grass"), with dense, plump, awnless, chocolate-brown to purplish spikelets at maturity, the palea of the sterile floret purple-tinged.

3. E. muricata (Beauv.) Fern.

Map 259. Occasionally in waste places, but more generally in damp places than the preceding — sometimes even in shallow water: wet shores, ditches, riverbanks, and floodplains. Apparently native in America but variable and with weedy tendencies like the preceding.

Appears in many manuals under the incorrect name *E. pungens* (Poiret) Rydb.

4. Digitaria   Crab Grass
The first glume is minute, sometimes membranous or apparently lacking.

KEY TO THE SPECIES

1. Rachis of racemes scarcely if at all winged; culms erect, not rooting at lower nodes
. . . . 1. **D. filiformis**
1. Rachis of racemes distinctly winged (wing broader than rachis proper); culms usually ± decumbent at base, often rooting at lower nodes
  2. Spikelets ca. (1.7) 2–2.3 mm long, the fertile lemma dark brown; second glume nearly or fully as long as the floret; sheaths and blades usually nearly or quite glabrous (except around summit of sheath) . . . . 2. **D. ischaemum**
  2. Spikelets ca. 2.5–3 mm long, the fertile lemma light or dark grayish; second glume only about half as long as the floret (fig. 141); sheaths and usually blades ± pilose, at least toward base of plant . . . . 3. **D. sanguinalis**

1. D. filiformis (L.) Koeler

Map 260. Apparently not found in Michigan since it was collected by the First Survey in 1838 from a dry prairie at Edwardsburg (MICH, GH, NY).

The sheaths are pilose, as in *D. sanguinalis*; the spikelets are similar in size to those of *D. ischaemum* (or slightly smaller) and likewise have the second glume and sterile lemma of about equal length, minutely pubescent with knob-tipped hairs.

259. Echinochloa muricata   260. Digitaria filiformis   261. Digitaria ischaemum

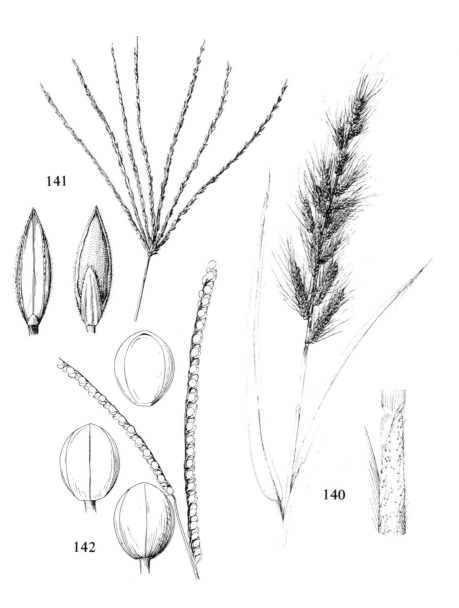

141

140

142

140. *Echinochloa walteri* x½ ; lower leaf sheath x1
141. *Digitaria sanguinalis* x½ ; spikelets x10
142. *Paspalum ciliatifolium* x1; spikelets and floret x10

2. **D. ischaemum** (Schreber) Muhl.

Map 261. Dry roadsides, fields, waste places, and sand barrens; lawns and gardens; sometimes spread into woodlands. Naturalized from Eurasia or perhaps in part also native.

The second glume (and, more sparsely, the sterile lemma) is softly pubescent, especially between the nerves, with very fine, minutely gland-tipped hairs.

3. **D. sanguinalis** (L.) Scop.   Fig. 141

Map 262. A weed in similar situations as the preceding; also found occasionally in low ground. More common in the southern Lower Peninsula; not found north as far as the Straits of Mackinac since 1928 when last collected in a millyard in Pellston, Emmet County. Naturalized from Europe.

The spikelets are usually glabrous or nearly so, especially between the nerves; if ciliate, the hairs are mostly not gland-tipped.

5. Paspalum

1. **P. ciliatifolium** Michaux   Fig. 142

Map 263. Sandy open ground, fields, and oak woodland.

The only species of this large genus thus far found in Michigan, but represented by two varieties, each of which is sometimes recognized as a species distinct from *P. ciliatifolium*: in var. *stramineum* (Nash) Fern. [*P. stramineum* Nash], the leaf is finely puberulent on the upper side of the blade, with or without longer hairs; in var. *muhlenbergii* (Nash) Fern. [*P. muhlenbergii* Nash], the leaf blade is not puberulent, but has only long hairs.

6. Leptoloma

1. **L. cognatum** (Schultes) Chase   Fig. 139

Map 264. Dry prairies, old fields, sandy open ground.

The spikelets are narrowly elliptic (unlike most *Panicum*), and often have stripes of silky appressed pubescence.

262. Digitaria sanguinalis   263. Paspalum ciliatifolium   264. Leptoloma cognatum

7. Panicum   Panic Grass

A large and difficult genus. A more conservative view of the species than that of Hitchcock and Chase is here followed, as advised by Fernald. The present treatment follows in general the concepts of Shinners and Pohl, but with the key owing much to Deam's *Flora of Indiana*.

Many of the species flower twice, with terminal ("vernal") panicles in the early summer and lateral ("autumnal") panicles in late summer or fall. The key is designed for use with early panicles. The autumnal form may appear quite different, and autumnal foliage and spikelets are less satisfactory for determination.

REFERENCES

Fernald, M. L. 1934. Realignments in the Genus Panicum. Rhodora 36: 61–87 (also Contr. Gray Herb. 103).

Hitchcock, A. S., & Agnes Chase. 1910. The North American Species of Panicum. Contr. U. S. Natl. Herb. 15: 1–396.

Pohl, Richard Walter. 1947. A Taxonomic Study on the Grasses of Pennsylvania. Am. Midl. Nat. 38: 513–600. [*Panicum* on pp. 575–590]

Shinners, L. H. 1944. Notes on Wisconsin Grasses—IV Leptoloma and Panicum. Am. Midl. Nat. 32: 164–180.

KEY TO THE SPECIES

1. Spikelets glabrous
  2. Spikelets all or mostly 3 mm or more in length, strongly nerved
    3. First glume more than half as long as second glume; lower floret staminate; plants over 5 dm tall, essentially glabrous (except for margin and throat of leaf sheath), from strong scaly rhizome, with panicle terminal and over 15 cm tall .
     . . . . 1. **P. virgatum**
    3. First glume not over half as long as second glume (except in *P. miliaceum*); lower floret with neither stamens nor pistil; plants shorter, usually ± pubescent, not rhizomatous, with panicles usually shorter or several
      4. Leaves (blades & sheaths) essentially glabrous
        5. Plant a branching annual with axillary (but not basal) panicles (often enclosed basally by the enlarged sheaths) in addition to terminal one; stem usually ± bent at nodes, hence zigzag, with distinctly alternate leaves, the blades usually over 5 mm wide . . . . 2. **P. dichotomiflorum**
        5. Plant a densely tufted perennial (old dead leaves evident), sometimes with basal panicles but not axillary ones along the culm; stem stiffly erect, with very narrow leaves mostly crowded at the base and seldom as wide as 5 mm
         . . . . 14. **P. depauperatum**
      4. Leaves pubescent (at least on sheaths)
        6. Spikelets (4.1) 4.3–5.3 mm long; plant an annual, with dense, somewhat curved or nodding panicle . . . . 3. **P. miliaceum**
        6. Spikelets mostly less than 4.3 mm long; plants annual or perennial, with more open or few-flowered erect panicle
          7. Spikelets plump, nearly half as wide as long; plants perennial, with old dead leaves of previous year among the dense tufts . . . . 14. **P. depauperatum**
          7. Spikelets about a third as wide as long; plants annual, without old dead leaves from previous years . . . . [go to couplet 11]

Page 223

2. Spikelets less than 3 mm long, strongly nerved or not
  8. Spikelets covered with mealy-looking warts, but nearly nerveless (fig. 145) . . .
  . . . . 4. **P. verrucosum**
  8. Spikelets not warty, nerved or not
    9. Sheaths sparsely to heavily pilose on back
      10. Spikelets 2 mm long or shorter; terminal panicle less than half the total
        height of the plant; internodes of stem pilose . . . . 5. **P. philadelphicum**
      10. Spikelets usually more than 2 mm long – if slightly shorter, then the
        terminal panicle half the height of the plant; terminal panicle half or more
        the height of the plant *or* internodes glabrous (or rarely both conditions
        present)
        11. Pulvini in axils of lower primary panicle branches glabrous; internodes
          (except sometimes the lowermost just below the node) glabrous; terminal
          panicle usually less than half the height of the plant, generally longer than
          wide, the branches ascending . . . . 6. **P. flexile**
        11. Pulvini of lower primary panicle branches pilose; internodes often pilose;
          terminal panicle usually half or more the height of the plant, generally
          broader than long at maturity, when branches are widely spreading . . . .
          . . . . 7. **P. capillare**
    9. Sheaths of middle and upper leaves glabrous on back (may be ciliate on
      margins)
      12. Plants with stout culms (1.5) 2.5–5 mm thick at middle internodes;
        terminal panicle at least 10 cm tall; axillary panicles also usually present
        (may be partly included in large leaf sheaths); sheaths equalling or exceeding
        internodes; ligule 0.7–1 mm or more long (erose-membranous in *P.
        rigidulum*)
        13. Spikelets ca. 2 mm long or a little smaller; pedicels all shorter than the
          spikelets (rarely an exception), often with a few long hairs at their summit;
          plants erect, perennial; ligule membranous . . . . 8. **P. rigidulum**
        13. Spikelets ca. 2.5 mm long or longer; pedicels mostly shorter than spikelets
          but some slightly longer, without long hairs; plant usually ± spreading,
          annual; ligule of hairs . . . . 2. **P. dichotomiflorum**
      12. Plants with very slender culms less than 1.5 (usually 0.5–1) mm thick at
        middle internodes; terminal panicle usually smaller; axillary panicles not
        present (later autumnal form is branched); sheaths much shorter than
        internodes; ligule often minute or obsolete (of short hairs)
        14. Spikelets ca. 1.5–1.7 mm long; nodes with prominent retrorse beard . . .
          . . . . 9. **P. microcarpon**
        14. Spikelets ca. 2 (1.8–2.3) mm long at maturity; nodes glabrous or minutely
          pubescent or rarely with spreading hairs . . . . 10. **P. dichotomum**
1. Spikelets at least sparsely pubescent toward their margins
  15. Widest leaf blades over 15 mm broad
    16. Sheaths of all leaves glabrous or with soft pubescence, the hairs not papillose
      (or rarely marginal cilia weakly so) . . . . 11. **P. latifolium**
    16. Sheaths of at least lower and middle leaves (especially on leafy branches) with
      papillose-based hairs (or papillae obvious on backs of sheaths even if hairs are
      gone)
      17. Spikelets 3.5–4 mm long; first glume half as long as spikelet (fig. 148);
        panicle with fewer than 40 (50) spikelets, well exserted . . . . 12. **P. xanthophysum**
      17. Spikelets 2.9–3.4 (3.6) mm long; first glume usually slightly less than half as
        long as spikelet; panicle (at least exserted terminal one) with more than 50
        spikelets, often (especially later & axillary ones) with base included in sheath
        . . . . 13. **P. clandestinum**

15. Widest leaf blades less than 15 mm broad

 18. Longest leaf blades at least 20 times as long as wide (not over 5 or very rarely 6 mm wide); leaves all appearing crowded toward base of plant (internodes few and short), erect, and all elongate, the plants not forming rosettes of short stubby leaves

  19. Second glume and sterile lemma distinctly prolonged (usually 0.5−1 mm) beyond the fertile lemma, forming an acute boat-shaped tip or beak to the spikelet (fig. 150); largest spikelets (3) 3.2−4.1 (4.5) mm long . 14. **P. depauperatum**

  19. Second glume and sterile lemma barely if at all prolonged beyond mature fertile lemma, the mature spikelet ± blunt (figs. 151, 152) (if shortly beaked, then only ca. 2.5 mm long); spikelets (2) 2.2−3.1 (3.3) mm long

   20. Spikelets (2) 2.2−2.6 mm long, 1.2−1.4 mm wide; panicle at maturity ± open, the longest pedicels 8−18 mm long . . . . 15. **P. linearifolium**

   20. Spikelets 2.8−3.1 (3.3) mm long, 1.6−1.8 mm wide at maturity; panicle narrow (at most 1−1.5 cm wide), the longest pedicels seldom over 7 mm long (fig. 153) . . . . 16. **P. perlongum**

 18. Longest leaf blades less than 15 (20) times as long as wide (width various); leaves ± distributed on the culm, spreading or ascending (not all crowded basally and erect), the plants usually forming basal rosettes of short stubby leaves

  21. Largest spikelets (3) 3.1−3.8 (4) mm long

   22. First glume very broadly ovate (fig. 149), not over 1.5 (1.7) mm long, its tip distinctly below the middle of the sterile lemma beneath it; ligule of hairs ca. (0.5) 1−1.5 mm long; spikelets sparsely short-hairy (hairs less than 0.5 mm long) . . . . 17. **P. oligosanthes**

   22. First glume narrowly ovate (figs. 148, 154), 1.6−2.5 mm long, its tip reaching to or beyond the middle of the sterile lemma; ligule obsolete, scarcely 0.5 mm long (longer hairs at sheath margin); spikelets short- or long-hairy

    23. Spikelets minutely hairy; leaf blades glabrous (except for few cilia at base) . . . . 12. **P. xanthophysum**

    23. Spikelets with long soft spreading hairs (the longest ca. 1 mm); leaf blades pubescent. . . . . 18. **P. leibergii**

  21. Largest spikelets less than 3 mm long

   24. Middle and upper internodes and usually sheaths (except for ciliate margin) glabrous; axis of panicle glabrous or nearly so, without long hairs

    25. Longest hairs of ligules ca. 1.5−3 mm or more; spikelets 1.3−1.8 mm long

     26. Panicle branches ± strongly ascending, the panicle half or less as wide as long; pedicels of lateral spikelets mostly equalling or shorter than spikelets. . . . . 19. **P. spretum**

     26. Panicle branches ± spreading, the panicle more than half as wide as long; pedicels of lateral spikelets all or nearly all longer than spikelets . . . . 20. **P. lindheimeri**

    25. Longest hairs of ligules less than 1 (1.5) mm long; spikelets 1.4−3 mm long

     27. Spikelets 1.4−1.7 mm long; ligule, at least on upper leaves, nearly obsolete (rarely a short fringe) . . . . 21. **P. sphaerocarpon**

     27. Spikelets ca. (1.8) 2 mm or more long; ligule (except in *P. calliphyllum*) a short (usually less than 1 mm) but definite fringe of hairs

      28. Spikelets 2.6−3 mm long . . . . 22. **P. calliphyllum**

      28. Spikelets 1.8−2.2 mm long

       29. Leaf blades less than 6 mm wide; spikelets very sparsely puberulent or glabrate. . . . . 10. **P. dichotomum**

29. Leaf blades mostly 6–9 (12) mm wide; spikelets usually conspicuously, though finely, pubescent . . . . 23. **P. boreale**
24. Middle and upper internodes and usually sheaths puberulent to pilose (or both); axis of panicle often with some long hairs
  30. Spikelets (2.1) 2.3–2.9 mm long (pubescence of leaf sheaths of very short hairs, with or without additional long hairs)
    31. Pubescence of leaf sheaths and culms exclusively of sparse to dense fine short hairs (less than 0.5 mm long) except along margin of sheath; ligule nearly or quite obsolete . . . . 24. **P. commutatum**
    31. Pubescence of leaf sheaths, and usually culms, of short puberulence (best seen on upper sheaths and culms – rarely nearly absent) plus ± sparse to copious long hairs intermixed (best seen on lower sheaths); ligule a definite fringe of hairs usually ca. 0.5–4 mm long . . . . . 25. **P. commonsianum**
  30. Spikelets not over 2 mm long, mostly a little less (if rarely up to 2.2 mm in *P. praecocius,* the sheaths with only long spreading hairs)
    32. Sheaths of main (vernal) culm with two types of pubescence: short hairs like those on the spikelets (best seen on upper sheaths) and long hairs (best seen on lower sheaths), the two most clearly intermixed on middle sheaths
      33. Spikelets mostly 1.3–1.6 mm long; leaf blades pilose with ± erect hairs on upper surface, the largest blades on culm leaves less than 4 mm wide and up to 5 cm long . . . . 26. **P. meridionale**
      33. Spikelets mostly (1.6) 1.7–1.9 (2) mm long; leaf blades glabrous to sparsely pilose on upper surface, the largest on culm leaves often more than 4 mm wide or longer than 5 cm . . . . 27. **P. columbianum**
    32. Sheaths of main (vernal) culms with a single type of pubescence: all hairs long (though not necessarily uniform)
      34. Middle and upper sheaths pilose with ± horizontally spreading very fine hairs, many of them ca. 3–4 mm long; spikelets ca. 1.8–2 (2.2) mm long, the first glume a little longer than broad and definitely acute (fig. 159) . . . . 28. **P. praecocius**
      34. Middle and upper sheaths with spreading to ascending hairs less than 3 mm long (except sometimes marginal cilia); spikelets (1.4) 1.5–1.8 (2) mm long, the first glume very short, usually no longer than broad and ± obtuse, rounded, truncate, or erose (fig. 160) . . . . . 29. **P. implicatum**

1. **P. virgatum** L.   Fig. 143   Switch Grass

Map 265. Prairies, dunes, swales, oak woodland, open (often marshy) ground; spreading along roadsides and railroads.

Some Michigan specimens have been referred to var. *cubense* Griseb., but they do not display the characters of that variety fully and may be only variable individuals of the typical variety.

2. **P. dichotomiflorum** Michaux

Map 266. Disturbed ground of shores, banks, roadsides, fields, etc., often in moist places. This species started spreading rapidly in the Detroit area around 1915 according to Farwell (Pap. Mich. Acad. 3: 90. 1924); it is of weedy tendency, although a native plant.

3. **P. miliaceum** L.   Fig. 144   Proso; Broomcorn Millet
Map 267. Roadsides, railroads, cultivated fields, and other waste places. Native of the Old World and occasionally escaped from cultivation.

4. **P. verrucosum** Muhl.   Fig. 145
Map 268. The only Michigan collection seen is from the shore of Cable Lake in Cass County (*L. M. Umbach 7448* in 1915, MICH, BLH). A species of primarily Coastal Plain distribution, disjunct near the head of Lake Michigan.

5. **P. philadelphicum** Trin.
Map 269. Borders of rivers, creeks, marshes, and lakes.

The spikelets are less prolonged at the tip than the distinctive acuminate tips of the second glume and sterile lemma of the next two species, but this is a rather qualitative character. Material from this region is too uncommon to attempt a resolution of the problem of separating *P. tuckermanii* Fern. and *P. gattingeri* Nash (both of which might occur in Michigan) from *P. philadelphicum*. The supposed differences, including presence or absence of hairs in the axils of the lower panicle branches, are not agreed upon by authors, and the complex is here recorded under the oldest name.

265. Panicum virgatum

266. Panicum
dichotomiflorum

267. Panicum miliaceum

268. Panicum verrucosum

269. Panicum
philadelphicum

270. Panicum flexile

6. **P. flexile** (Gatt.) Scribner

Map 270. Damp sandy or gravelly, usually calcareous, shores and marshy places.

7. **P. capillare** L.   Fig. 146   Witch Grass

Map 271. Dry to moist open (and usually disturbed) ground: roadsides and railroads, fields and gardens, waste places, shores and riverbanks. A native species, but weedy in habit.

8. **P. rigidulum** Nees   Fig. 147

Map 272. Swales, shores, and ponds – apparently very local in Michigan. Long known as *P. agrostoides* Sprengel, an illegitimate name.

9. **P. microcarpon** Ell.

Map 273. Moist woods and thickets, our few collections mostly without habitat data.

If a plant of *P. spretum* with glabrous spikelets is found, it might run here in the key but could be readily distinguished by its glabrous nodes and prominent ligule. Some plants of *P. sphaerocarpon* may seem to have spikelets glabrous and hence would run here, but the nodes are glabrous or with appressed-ascending hairs.

10. **P. dichotomum** L.

Map 274. Dry to moist oak, oak–hickory, or mixed woods; stream banks; pine groves.

Included here, following most conservative authors, is *P. barbulatum* Michaux, a form not sharply distinguished by more pubescent nodes and broader leaves. Also included here are old Michigan records of *P. lucidum* Ashe and *P. yadkinense* Ashe. Reports of the latter from Michigan apparently trace back to old collections of Dodge from St. Clair County, labeled by him as *P. maculatum* (a synonym). However, these were re-identified for Dodge by Mrs. Chase about 1911 as *P. lucidum*. The plants are evidently erect, and were collected from a pine grove in mid-July; *P. lucidum* is said to be a species of wet woods and bogs,

271. Panicum capillare

272. Panicum rigidulum

273. Panicum microcarpon

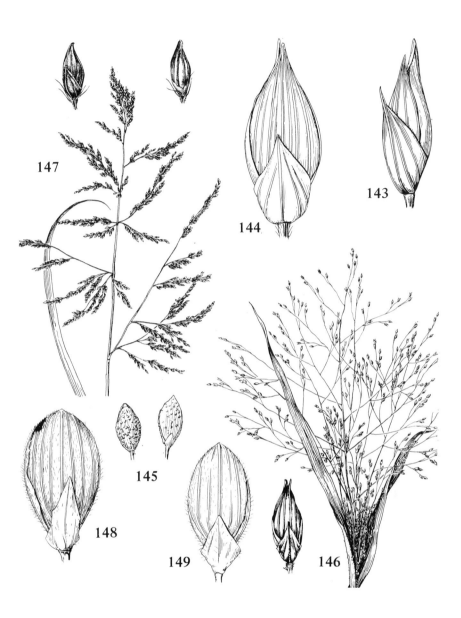

143. *Panicum virgatum,* spikelet ×10
144. *P. miliaceum,* spikelet ×10
145. *P. verrucosum,* spikelets ×7
146. *P. capillare* ×½ ; spikelet ×10
147. *P. rigidulum* ×½ ; spikelets ×5
148. *P. xanthophysum,* spikelet ×10
149. *P. oligosanthes,* spikelet ×10

the culms becoming decumbent. Since in no other characters (including pubescence and leaves) do I find these specimens distinguishable from *P. dichotomum*, I am using the latter name in a broad sense for all Michigan plants with glabrous spikelets in this group (exclusive of *P. microcarpon*). *P. yadkinenese* otherwise ranges well southeast of Michigan. If found in the state, it could be distinguished from plants here referred to *P. dichotomum* by a short prolongation of the second glume and sterile lemma beyond the hard fertile lemma; in *P. dichotomum* the second glume and sterile lemma are barely as long as the fertile lemma. *P. yadkinense* is also distinctive in having numerous whitish, wart-like spots between the veins of the leaf sheaths.

11. P. latifolium L.

Map 275. Woods and thickets, sometimes along borders but not a plant of open ground: in rich deciduous (beech–maple, etc.), oak or oak–hickory, aspen, or mixed woods; on wooded dunes and bluffs and riverbanks.

Some of our specimens have bearded nodes, but the spikelets are too small for *P. boscii* Poiret, which has sometimes been reported from Michigan. See comments under *P. clandestinum*.

12. P. xanthophysum Gray   Fig. 148

Map 276. Usually in dry sandy or rocky open ground or woods (aspen, jack pine, oak), including low dune ridges.

The panicle is very narrow, with stiffly ascending, nearly simple branches. See comments under the next species.

13. P. clandestinum L.

Map 277. Usually in damp and often sandy ground: floodplains and thickets on stream banks; aspen woods, borders, and clearings; marshy ground, ditches, etc.

Quite variable in the length and acuteness of the first glume. Although wide-leaved plants of *P. xanthophysum* may seem, from the key, rather similar to *P. clandestinum*, the species are readily distinguished by several tendencies. *P.*

274. Panicum dichotomum

275. Panicum latifolium

276. Panicum xanthophysum

clandestinum has more strongly nerved spikelets than *P. xanthophysum*, and the leaf blades are more strongly cordate at the base. The panicle, besides having more numerous spikelets, is much more open and lax in *P. clandestinum* — although autumnal panicles may be mostly or entirely included in the leaf sheaths. The ligule in *P. clandestinum* appears to be a membranous collar about 0.5 mm wide, very minutely ciliolate; in *P. xanthophysum* it is also short, but more clearly a fringe of hairs.

Very rarely specimens of *P. latifolium* may have the lowermost sheaths ± papillose, but these can be distinguished from *P. clandestinum* by the fact that the lower internodes of the culm are glabrous, while in *P. clandestinum* they are ± pubescent and papillose. The leaf blades in *P. clandestinum* are very long-tapering with rather straight margins; those of *P. latifolium*, while ± long-tapering, are somewhat more acuminate. This is a relative distinction which is nevertheless very helpful once one becomes familiar with both species.

14. P. depauperatum Muhl.   Fig. 150
Map 278. Dry usually sandy or rocky open ground (rarely in moist places); jack pine, aspen, and oak woods; pine-covered dunes.

This species and the next two are not always easy to separate. It is likely that there is a small incidence of hybridization. A few collections have ± beaked spikelets as in *P. depauperatum* but (although slightly immature) they are nearly as small as in *P. linearifolium*. In the typical variety, the sheaths are pilose; in var. *involutum* (Torrey) Wood [var. *psilophyllum* Fern.] , they are glabrous or nearly so. The spikelets may be glabrous or pubescent.

15. P. linearifolium Britton   Fig. 151
Map 279. Dry often sandy woods (oak, oak–hickory, aspen, jack pine), especially in disturbed areas and clearings; occasional in mixed woods; frequent on dunes under pine, oak, etc.

The typical variety has pilose sheaths; in var. *werneri* (Britton) Fern., the sheaths are glabrous or nearly so — paralleling the variation in *P. depauperatum*. The spikelets apparently always have at least a few fine hairs.

277. Panicum clandestinum

278. Panicum depauperatum

279. Panicum linearifolium

Most Michigan plants referred to *P. bicknellii* Nash are scarcely typical and seem more probably to be unusual specimens of *P. linearifolium* or hybrids. (It has been suggested that *P. bicknellii* itself may be a hybrid.) *P. bicknellii* ordinarily has culm leaves elongate, but not quite so long and narrow as those of *P. linearifolium*; it also has relatively short rosette leaves, unlike *P. linearifolium*, which does not form a winter rosette. *P. bicknellii* branches from the middle nodes in the autumnal form, while in *P. linearifolium* (as in *P. depauperatum* and *P. perlongum*) the reduced autumnal panicles are on short basal branches, inconspicuous among the crowded leaves. All of these species normally have glabrous or very finely pubescent internodes. Some Farwell collections from Dearborn [Wayne County] are branched, but the leaves are typical for *P. linearifolium* and no short rosette leaves are evident on the specimens. The internodes, moreover, are pilose and the spikelets barely 2 mm long, so that altogether the plants suggest possible hybridization of *P. linearifolium* with some other species. Some Hanes collections from Kalamazoo County with elongate culm leaves and stubby rosette leaves, but spikelets only about 2 mm long, were tentatively referred to *P. bicknellii* by Mrs. Chase and J. R. Swallen, with the observation that they differed in some respects. A collection from the E. S. George Reserve, Livingston County, has culm leaves conspicuously elongate, not over 3 mm wide; stubby rosette leaves; and spikelets 2.2–2.4 mm long. It appears to be a hybrid of *P. linearifolium* with some other species, but in many ways it is in accord with descriptions of *P. bicknellii*. Plants with elongate culm leaves and tendency to basal rosettes have also been seen from Algonac, St. Clair County. For the present, the problematic *P. bicknellii* is not admitted to the list of authenticated species known from the state. See also comments under *P. calliphyllum*.

16. **P. perlongum** Nash   Figs. 152, 153

Map 280. Dry prairies and old fields; a collection labeled as from Keweenaw County (*Farwell 840½* in 1894, BLH) is not mapped, as it is suspicious both for the "½" number and the considerable distance out of range.

An uncommon species with us, not always clearly distinguished from the preceding two common species. Some plants have a few of the spikelets longer than usual for *P. perlongum*, and ± beaked, suggesting the possibility of hybridization with *P. depauperatum*, although the broad spikelets and narrow panicle are as in typical *P. perlongum*; others have all blunt spikelets of the right length for *P. perlongum* but these are narrow (not over 1.5 mm wide) and the panicle is open.

17. **P. oligosanthes** Schultes   Fig. 149

Map 281. Dry open usually sandy ground, fields, prairies, oak woods.

Very variable in pubescence of leaf blades and sheaths. Most if not all Michigan specimens may be referred to var. *scribnerianum* (Nash) Fern. [*P. scribnerianum* Nash], which has somewhat more numerous and smaller spikelets

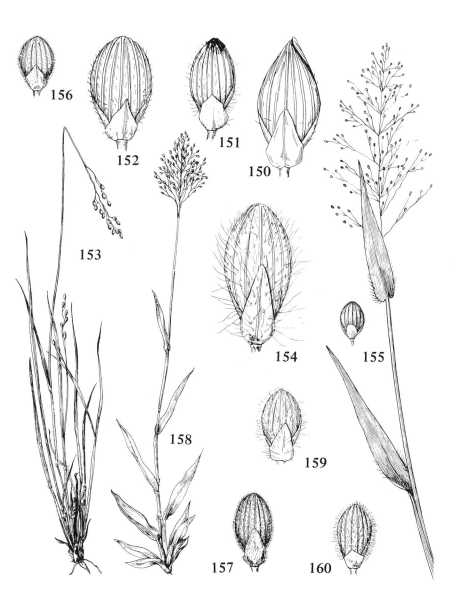

150. *Panicum depauperatum,*
    spikelet ×10
151. *P. linearifolium,* spikelet ×10
152. *P. perlongum,* spikelet ×10
153. *P. perlongum* ×½
154. *P. leibergii,* spikelet ×10

155. *P. sphaerocarpon* ×½;
    spikelet ×5
156. *P. meridionale,* spikelet ×10
157. *P. columbianum,* spikelet ×10
158. *P. columbianum* ×½
159. *P. praecocius,* spikelet ×10
160. *P. implicatum,* spikelet ×10

than the southern var. *oligosanthes*. See also notes under *P. commonsianum*.

The rare and problematic *P. calliphyllum* might run here if the spikelets should exceed 3 mm. It has the short first glume of *P. oligosanthes*, but the ligule is nearly obsolete.

18. **P. leibergii** (Vasey) Scribner   Fig. 154

Map 282. Dry to wet prairies and prairie-like places.

19. **P. spretum** Schultes

Map 283. In Michigan known only from the sandy-mucky drying bed of Lake Sixteen (and nearby) in Presque Isle County and from peaty-marly-sandy lakes in Muskegon County.

20. **P. lindheimeri** Nash

Map 284. Damp sandy to boggy ground, especially on sandy, gravelly, often calcareous shores and marshy flats. The southernmost collections referred here to *P. lindheimeri* (e. g., some from Berrien, Oakland, Washtenaw, and Wayne counties) look somewhat different from the characteristic stiff, glabrate, usually reddish, plants of northern often marly shore meadows along Lakes Michigan and Huron; but they are so glabrous that they are included here. Perhaps these southern specimens include an admixture of something else.

280. Panicum perlongum

281. Panicum oligosanthes

282. Panicum leibergii

283. Panicum spretum

284. Panicum lindheimeri

285. Panicum sphaerocarpon

An ambiguous species, sometimes apparently close to the preceding, and sometimes to *P. implicatum* — which has at times been treated as a variety of it. Fernald (in *Gray's Manual*) includes *implicatum* and *lindheimeri* as two (of several) varieties of the Coastal Plain species *P. lanuginosum* Ell. As here treated, *P. lindheimeri* includes some glabrous collections referred by Hitchcock and Chase and by Swallen to *P. tennesseense*. Authors are not at all agreed on this complex, and recognition of the two species here is strictly tentative. See also comments under *P. implicatum*.

21. **P. sphaerocarpon** Ell.   Fig. 155

Map 285. Dry open ground, fields, and sandy woods.

The leaves, especially those in rosettes, are quite large for a plant with such small spikelets.

22. **P. calliphyllum** Ashe

Map 286. Oak–hickory woods in Richland Tp., Kalamazoo County (*Hanes 1067*, Aug. 5, 1937, WMU, GH, US).

A rare species of uncertain status in eastern North America. Although usually considered related to *P. bicknellii* (see under *P. linearifolium*, above), it has by some been suggested as related to, or a hybrid with, *P. xanthophysum, P. oligosanthes,* or *P. latifolium.* The spikelets may be as much as 3 mm long (see comments under *P. oligosanthes,* above). Plants of this species are said to be light green.

True *P. bicknellii* is said to have spikelets ca. 2.5–2.8 mm long (slightly smaller than in *P. calliphyllum*) and the elongate leaf blades not over 8 or 9 mm wide (in *P. calliphyllum* they tend to be wider — up to 13 mm on the Michigan material).

23. **P. boreale** Nash

Map 287. Usually in damp to marshy sandy or rocky open ground; occasionally in dry aspen or oak woodland.

Usually easily distinguished by glabrous foliage and nodes, at least above, and short ligule. Some plants have the leaves ± pilose and the nodes, especially the lower ones, pilose as in *P. annulum* Ashe, which has been reported from Michigan. Such pubescent plants have been called var. *michiganense* Farw. (TL: Detroit, Wayne County), and can be separated from *P. annulum* by the long slender pedicels (pedicels often shorter than spikelets in *P. annulum*) and the longer, less dense hairs of the leaf blades, which in *P. annulum* are velvety on both surfaces.

24. **P. commutatum** Schultes

Map 288. Sandy open or wooded ground, apparently not common.

Our few collections seem mostly, if not all, to be the somewhat more northern var. *ashei* Fern., which is not always clearly distinguishable.

25. P. commonsianum Ashe

Map 289. Sand barrens, low dunes, open oak and pine woods, shores.

This is typically a species of the Coastal Plain from Massachusetts to Florida, represented inland by var. *euchlamydeum* (Shinners) Pohl (see Pohl in Am. Midl. Nat. 38: 506–509. 1947). Our plants referred here are quite variable in amount of pubescence and length of ligule, but are recognized by the double vestiture (both long and short hairs on sheaths, culms, and lower surfaces of blades) and large spikelets (compared with those of *P. columbianum* and *P. meridionale*, our other related species with double vestiture). The ligule consists of a dense fringe of short hairs (ca. 1 mm or less) and may have long hairs in addition (up to 5 mm). *P. oligosanthes* has sparse pubescence, usually distinctly double, but is quite a different species. Specimens of var. *scribnerianum* with unusually small spikelets might run here in the key, but the spikelets are nearly all green and subrotund (ca. 2 mm or more wide and about 1.5 times as long); in the more pubescent *P. commonsianum* var. *euchlamydeum*, the spikelets are ± strongly flushed with maroon and are ellipsoid (ca. 1–1.2 mm wide, twice as long).

P. deamii Hitchc. & Chase was reported from wooded dunes in Oceana County on the basis of a collection (*Bazuin 4784*, MSC) with spikelets ca. 2.6–2.9 mm long, double vestiture, and ligules ca. 1 mm long; the leaves are much shorter than described for *P. deamii*, and the collection falls easily in the range of variation for *P. commonsianum* var. *euchlamydeum*. The latter, as pointed out by Shinners and by Pohl, includes plants from this region previously referred (in error) to *P. pseudopubescens*, *P. scoparioides*, and *P. villosissimum*. These three are all Coastal Plain species with a single type of pubescence.

26. P. meridionale Ashe   Fig. 156

Map 290. Sandy oak woods, drying shores, and fields.

Many of our specimens have puberulence as well as long pilose hairs on the upper surface of the leaf blades, and thus might be referred to var. *albemarlense* (Ashe) Fern. This species is a puzzling one, and some of the specimens included here may represent unusual states of other species or hybrids.

286. Panicum calliphyllum

287. Panicum boreale

288. Panicum commutatum

27. **P. columbianum** Scribner   Figs. 157, 158

Map 291. Dry sandy open ground, thickets, and woods (jack pine, oak, and/or aspen), dune ridges, pine forests, and less often on moist shores.

This species is quite variable in amount of pubescence, although the double vestiture of sheaths is always characteristic; in shape of the first glume; and in size of spikelets and leaves. Most of our specimens, especially from the northern part of the state, have spikelets 1.8–1.9 mm long and thus would be the very weakly distinguished *P. tsugetorum* Nash. These plants are clearly distinguished from the preceding species by the larger spikelets and generally broader leaves, glabrous or nearly so above. In southern Michigan, some plants (typical *P. columbianum*) with slightly smaller spikelets (mostly 1.6–1.7 mm) and smaller leaves may run close to *P. meridionale*. Some glabrate specimens may have the double vestiture rather obscure, and superficially resemble *P. implicatum*. They may often be distinguished by longer spikelets and longer, more acute first glume.

28. **P. praecocius** Hitchc. & Chase   Fig. 159

Map 292. Dry open, usually sandy ground, oak woodlands, borders and fields.

Apparently hybridizes with *P. implicatum* and perhaps other species.

289. Panicum
    commonsianum

290. Panicum meridionale

291. Panicum
    columbianum

292. Panicum praecocius

293. Panicum implicatum

294. Andropogon
    virginicus

29. P. implicatum Britton   Fig. 160

Map 293. Common and widespread, in a diversity of dry or damp open places and thickets; most often on moist sandy, calcareous, or marshy shores; also in dry oak woods, meadows, sandy barrens and jack pine plains, fields and waste ground, and even bog mats (very rarely).

Very variable in stature and in amount of pubescence, but it does not seem possible to recognize such segregates as *P. huachucae* Ashe and *P. tennesseense* Ashe, which are here included. Fernald includes this species and *P. lindheimeri* with the Coastal Plain *P. lanuginosum* Ell., but the treatment of Pohl is here followed. In the small spikelets, *P. implicatum* may seem close to *P. meridionale*, especially to specimens of the latter in which the fine secondary pubescence is obscure. On the other hand, glabrate specimens of *P. implicatum* run into *P. lindheimeri*. Plants here treated as *P. lindheimeri* have the upper internodes, peduncle, and axis of panicle glabrous; the upper sheaths and blades nearly or quite glabrous; and the lower part of the plant glabrate to moderately pilose. These have sometimes been identified as a robust glabrate form of *P. implicatum*. As here treated, *P. implicatum* has more pubescence in any or all of these parts. Ordinarily, pilosity in the axis of the panicle will distinguish *P. implicatum*.

Plants from Michigan reported as *P. subvillosum* Ashe are referred mostly to *P. implicatum* and *P. columbianum*. The status of Ashe's plant is doubtful (see Shinners, p. 179) and our specimens are easily referable to other species.

10. ANDROPOGONEAE

The spikelets occur in pairs in this tribe, one stalked, the other sessile. The stalked spikelet is staminate, rudimentary, or absent and represented only by a pedicel. The sessile spikelet, as in the Paniceae, contains basically two florets, one sterile and the other fertile. Both the sterile and the fertile lemmas are thin and hyaline; in our species (except sometimes in *Sorghum*) the fertile lemma bears an elongate awn, usually twisted basally. Gould (1968) includes the genera of the Tripsaceae in this tribe.

295. Andropogon scoparius   296. Andropogon gerardii   297. Sorghastrum nutans

KEY TO THE GENERA OF ANDROPOGONEAE

1. Inflorescence of 1–several narrow or spike-like simple racemes (some spikelets
   sessile and some pedicelled) . . . . 1. **Andropogon**
1. Inflorescence an open to contracted panicle
  2. Stalked spikelet absent, represented by a sterile hairy pedicel closely resembling
     the segments of the panicle axis (fig. 165); ligule stiff, ± cartilaginous, glabrous or
     at most with minute hairs . . . . 2. **Sorghastrum**
  2. Stalked spikelet present, sterile or staminate (fig. 166); ligule at least in part of
     evident soft hairs . . . . 3. **Sorghum**

1. **Andropogon**   Beardgrass

KEY TO THE SPECIES

1. Sessile florets with body ca. 3–3.5 (4) mm long and a straight awn; stalked floret
   usually represented by only a hairy pedicel; base of racemes ± included in
   spathe-like leaf base even at maturity; ligule of major culm leaves a minutely ciliate
   scale ca. 0.5–0.6 mm long; stamen 1, the anther not over 1 mm long . . 1. **A. virginicus**
1. Sessile florets with body (5) 6–9 mm long and awn tightly twisted basally; stalked
   floret rudimentary or well developed and staminate; base of at least some racemes
   usually exserted at maturity; ligule of major culm leaves ca. 0.9–2.5 mm long;
   stamens 3, the anthers (2) 2.5–5 (5.5) mm long
  2. Racemes solitary at the ends of long peduncles (fig. 161); stalked floret reduced,
     sterile, the 1–2 glumes usually ± rudimentary (fig. 161) . . . . 2. **A. scoparius**
  2. Racemes 2–6 (10), umbellate or crowded on short stalks at end of long
     peduncles (fig. 162); stalked floret nearly or quite as large as sessile floret,
     staminate (fig. 164) . . . . 3. **A. gerardii**

1. **A. virginicus** L.   Broom-sedge
Map 294. Open sandy fields, juniper savanas, and roadsides; occasionally on
moister shores.

2. **A. scoparius** Michaux   Fig. 161   Little Bluestem
Map 295. A characteristic prairie species, but more often seen in Michigan on
jack pine plains, sand dunes, and shores; dolomite pavements of Drummond
Island; spreading along roadsides and railroads, in sandy old fields, etc.

Several varieties within this variable species have been proposed but are not
widely accepted. The species is sometimes segregated into a separate genus as
Schizachyrium scoparium (Michaux) Nash (see Gould in Brittonia 19: 72–73.
1967).

3. **A. gerardii** Vitman   Figs. 162, 164   Big Bluestem; Turkeyfoot
Map 296. Like the preceding, a characteristic prairie species, but spreading
along roadsides and railroads; in oak woods, jack pine plains, old fields, and bog
borders. May be as tall as 3 m.

The staminate stalked florets are as long as the perfect sessile ones or even a
bit longer, and the anthers are similar in size (usually 2.5–5 mm long – rarely as

small as 2 mm or as large as 5.5 mm). The species was long known as *A. furcatus* Willd.

2. Sorghastrum

1. S. nutans (L.) Nash   Fig. 165   Indian Grass

Map 297. Dry woods (jack pine, oak, etc.), prairies, and moist shores, even sometimes in marshy places; apparently spreading somewhat in disturbed ground, as along roadsides and railroads.

A tall handsome grass with golden panicle in late summer. In lacking even a rudimentary pedicelled spikelet, it resembles *Andropogon virginicus*, from which it conspicuously differs in the larger sessile (fertile) spikelet (twice as long as in *A. virginicus*) and long cartilaginous ligule ± continuous with elongate auricles. Mature disintegrating panicles may seem to have spikelets with *two* sterile hairy pedicels; one is the true sterile pedicel and the other is the very similar internode of the panicle itself, from which the next sessile spikelet and sterile pedicel above have disarticulated.

3. Sorghum

Michigan is at the northern limit of the climate which species of this genus can endure, and we have only a few collections — most somewhat ambiguous as to their identity or to their status in the flora. *S. sudanense* (Piper) Stapf has been collected from a cultivated field in Shiawassee County and apparently also in Allegan, Ingham, and Kalamazoo counties — perhaps never as a true escape. It would run in the key to *S. halepense*, from which it differs in annual habit and spikelets not disarticulating; it is sometimes considered a variety of *S. bicolor*. Immature specimens without basal parts are not easily distinguished from *S. halepense*.

Several species and varieties of *Sorghum* have long been cultivated, especially in the southern states, for silage, oil, wax, juice, and seeds. The foliage may produce enough hydrocyanic acid, especially late in the season or after frost, to be noxious to livestock.

298. Sorghum bicolor

299. Sorghum halepense

300. Tripsacum dactyloides

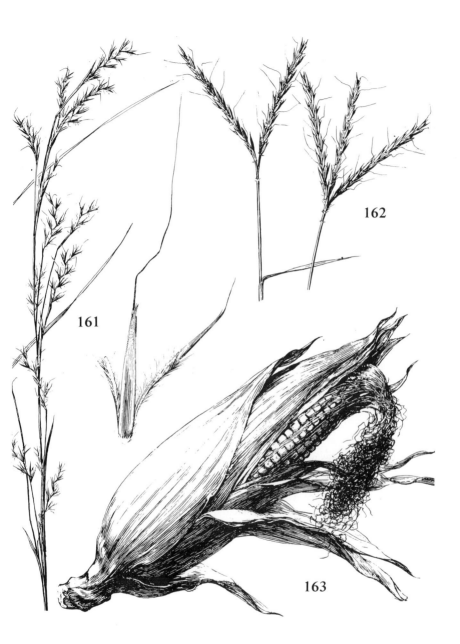

161. *Andropogon scoparius* ×½ ; pair of spikelets ×5
162. *A. gerardii* ×½
163. *Zea mays,* pistillate inflorescence ×½

1. Pedicelled spikelet elliptic-lanceolate, somewhat shorter than sessile spikelet; sessile spikelet usually spread open by the very turgid grain at maturity (fall); spikelets not disarticulating from pedicels at maturity; leaf blades mostly over 2 cm broad; panicle very dense, compact, the main axis usually ± hairy . . . . 1. **S. bicolor**
1. Pedicelled spikelet lanceolate, slightly longer than sessile spikelet; sessile spikelet enclosing grain at maturity; spikelets at maturity disarticulating neatly from cup-shaped summit of pedicels; leaf blades usually not over 2 cm broad; panicle ± open, the main axis not hairy (may be scabrous) . . . . 2. **S. halepense**

1. **S. bicolor** (L.) Moench   Sorghum; Broom-corn

Map 298. Local in waste ground and not persisting. A native of the Old World, cultivated since prehistoric times.

A very stout annual grass, superficially resembling corn vegetatively. Long known as *S. vulgare* Pers., an illegitimate name (see Shinners in Baileya 4: 141–142. 1956).

2. **S. halepense** (L.) Pers.   Fig. 166   Johnson Grass

Map 299. Roadsides, shores, and waste ground. Locally troublesome in the southwest part of the Lower Peninsula. Native to the Old World, especially the Mediterranean area, and grown for forage.

Typically perennial from extensive rhizomes, and often an obnoxious weed very difficult to eradicate, especially where more abundant southward.

11. TRIPSACEAE

Gould (1968) does not separate this tribe from the Andropogoneae.

KEY TO THE GENERA OF TRIPSACEAE

1. Pistillate flowers several, sunken in indurate joints of lower portion of same rachis as the terminal staminate spikelets, at maturity disarticulating into 1-seeded segments. . . . . 1. **Tripsacum**
1. Pistillate flowers aggregated into separate leafy-bracted "ears" in lower axils of plant (fig. 163), not disarticulating into segments . . . . 2. **Zea**

1. Tripsacum

1. **T. dactyloides** (L.) L.   Fig. 167   Gama Grass

Map 300. Collected along railroads 1919–1922 and presumably adventive from somewhat farther south.

2. Zea

The single familiar species of this genus was originally native in North America (probably Mexico) and is widely cultivated as one of the most important crop plants in the world. Corn is apparently a cultivated derivative of

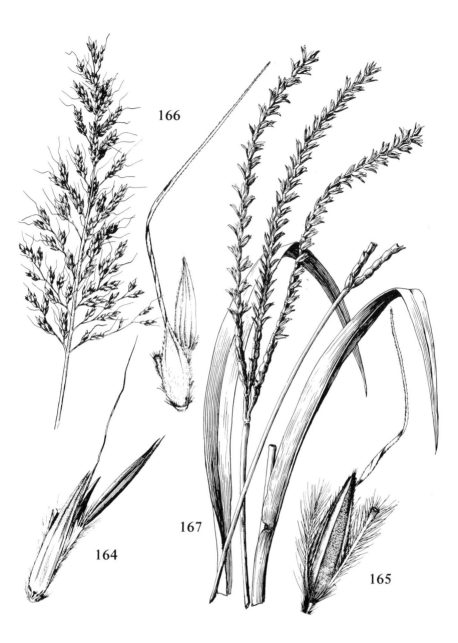

164. *Andropogon gerardii,* pair of spikelets ×5
165. *Sorghastrum nutans,* spikelet and sterile pedicel ×5
166. *Sorghum halepense* ×½; pair of spikelets ×5
167. *Tripsacum dactyloides* ×½

the grass known as teosinte [*Euchlaena mexicana* Schrader] , domesticated and selected for useful qualities of edibility and harvestability by Mexicans of some 8,000 years ago.

REFERENCE

Galinat, Walton C. 1971. The Origin of Maize. Ann. Rev. Genet. 5: 447–478.

1. **Z. mays** L.   Fig. 163   Indian Corn; Maize
Map 301. An occasional annual waif, rarely appearing spontaneously established along roadsides and railroads in the southernmost part of the state, not withstanding winters northward. Doubtless of more frequent (though temporary) occurrence out of cultivation than the few collections would indicate.

Except for relatively late sites in North Dakota, maize from a Late Woodland site on Bois Blanc Island, dating from at least a thousand years ago, represents the northernmost archaeological occurrence known for this plant (Mus. Anthr. Univ. Mich. Anthr. Pap. 30: 189. 1967).

CYPERACEAE   Sedge Family

The genera in the Cyperaceae are in many cases a matter for some controversy. The generic limits (and hence names) familiar in this region are here used, as much because of tradition — and hence utility and lack of confusion for the reader — as anything else. *Scirpus hudsonianus* has sometimes been placed as *Eriophorum alpinum*, and may belong in that genus. *Eleocharis pauciflora* and *E. rostellata*, in lacking a distinct tubercle, are transitional to *Scirpus* — and indeed some authors include all of *Eleocharis* in the genus *Scirpus*. On the other hand, some divide *Scirpus* into several genera. In his monograph of *Rhynchospora* (1949–1951), Kükenthal includes the species traditionally assigned to *Psilocarya*. Critical solution of such problems as generic definition is beyond the scope of a local flora. One may assign our species to whatever genus his taxonomic inclinations prefer, utilizing the appropriate name, without affecting the distribution maps, habitats, or key characters which are the features of this Flora.

KEY TO THE GENERA

1. Flowers all imperfect; achene either enclosed in a sac (perigynium) or exposed and bony (± spherical and white)
  2. Achene (and pistillate flower, except protruding style) enclosed in a sac-like structure (perigynium) . . . . 1. **Carex**
  2. Achene exposed, hard, bony, ± spherical, white or whitish . . . . 2. **Scleria**
1. Flowers (or at least one of them when spikelets are few-flowered) perfect; achene neither enclosed in a sac nor bony, spherical, and white
  3. Scales of spikelets 2-ranked and usually keeled or laterally compressed; spikelets several to many (sometimes crowded in a head)

4. Stems terete, hollow, appearing jointed (fig. 252); inflorescences axillary; achenes beaked, subtended by bristles . . . . 3. **Dulichium**

4. Stems ± 3-angled, solid, not appearing jointed; inflorescences terminal; achenes beakless, not subtended by bristles . . . . 4. **Cyperus**

3. Scales of spikelets spirally imbricated (or if 2-ranked, the spikelet solitary); spikelets solitary to many

5. Inflorescence a single erect terminal spikelet, not surpassed by any involucral bract (at most, the callous tip of the lowermost scale slightly exceeding summit of spikelet)

6. Achene with a tubercle formed by the persistent style base, this either sharply differentiated as a broad to narrow cap, or conical and apparently a continuation of the tapered summit of the achene; bristles present and at most slightly surpassing achene or obsolete, not silky; scales of spikelet essentially similar; plants leafless, the cauline sheaths without blades . . . . 5. **Eleocharis**

6. Achene without a tubercle (at most blunt with a short apiculus not differentiated in color or texture); bristles few to many, either silky and much surpassing the achene at maturity or shorter and inconspicuous in 2 species with lowermost scale of spikelet ± prolonged into a callous tip; plants often with short leaf blades

7. Scales blackish or lead color, the several lower ones sterile; achene ca. 3 mm long, subtended by numerous long silky bristles . . . . . . 14. **Eriophorum (spissum)**

7. Scales brownish or yellowish (often green or light in center), at most the lowest 1 or 2 sterile; achene less than 2 mm long, subtended by 6 bristles (silky or not) . . . . 11. **Scirpus** (couplet 2)

5. Inflorescence of several spikelets, or if single this surpassed by one or more involucral bracts

8. Basal empty scales of spikelet (2) 3 or more; perfect flowers 1–3 (4) (some staminate flowers may also be present)

9. Styles 2-cleft (at least at very tip), the basal portion persistent as a distinct tubercle or beak on the achene; bristles often present . . . . 6. **Rhynchospora**

9. Styles 3-cleft, deciduous; bristles none . . . . 7. **Cladium**

8. Basal empty scales of spikelet 0–2, or perfect flowers few to many, or (usually) both these conditions present

10. Style or at least the expanded base of it persistent as a minute tubercle or beak on the achene; perianth bristles none [if bristles present, try *Rhynchospora*]; body of achene ca. 0.7–0.9 mm long

11. Leaves capillary; base of style persistent as a minute brown bulbous tubercle on the 3-sided achene (fig. 266); anthers ca. 0.3 mm long; scales minutely strigose . . . . 8. **Bulbostylis**

11. Leaves flat; most of style persistent (but fragile), the base almost as broad as body of the 2-sided achene (fig. 267); anthers 0.5–0.9 mm long; scales glabrous . . . . 9. **Psilocarya**

10. Style deciduous and achenes beakless, or *if* the style base ± persistent, then not expanded; perianth bristles present or none; achene various

12. Inflorescence of 1–several crowded spikelets appearing to be lateral, an erect (or curved or somewhat spreading) involucral bract appearing to be a continuation of the culm

13. Plants very slender, mostly less than 10 cm tall (often less than 5 cm) and the culms less than 0.5 mm thick (fig. 268); perianth bristles none, but a tiny translucent (easily overlooked) scale sometimes present between achene and rachilla; achene cylindrical, papillate, ca. 0.6–0.7 mm long. . . . . 10. **Hemicarpha**

13. Plants with culms thicker or taller or (usually) both; perianth bristles often present, with no additional scale; achene much larger, 2–3-sided . . . . 11. **Scirpus**

12. Inflorescence appearing terminal
14. Perianth of 3 slender bristles alternating with 3 which include an expanded scale-like portion (fig. 275) . . . . 12. **Fuirena**
14. Perianth entirely of slender bristles or none
15. Base of style slightly enlarged (but readily deciduous); bristles none; achenes white, the surface ± reticulate or papillate . . . . 13. **Fimbristylis**
15. Base of style slender, not enlarged; bristles present; achenes various
16. Perianth bristles not over 8, not silky, usually short (if much exserted, then strongly curled) . . . . 11. **Scirpus**
16. Perianth bristles apparently numerous, long-exserted, silky, and straightish at maturity . . . . 14. **Eriophorum**

1. Carex
Sedge

This genus, by far the largest in our flora as in most temperate regions, is composed of somewhat grass-like plants, but usually with clearly 3-ranked leaves. As in other Cyperaceae, the flowers are each in the axil of a *single* scale. The flowers, however, are unisexual. The staminate flowers consist solely of 3 stamens, and may be in separate spikelets or in the same spikelets as the pistillate flowers. The latter consist solely of a pistil, with the style forked into 2 or 3 stigmas. Except for the stigmas, which protrude at flowering time, the pistil is surrounded by a sac-like or flask-shaped structure called the *perigynium*. This is absolutely diagnostic for the genus *Carex* and is the source of many of the characters used in identification and classification.

Accurate identification requires mature perigynia and often the basal parts of the plant. Perigynia from the apex or base of a spikelet are often less characteristic than those from the middle, and the latter should be used in comparisons. Except when expressly stated to the contrary, the term "body" refers to all of the perigynium except the beak (i. e., including any narrowed base which may be present); in some groups, however, determination of where the body ends and the beak begins is somewhat difficult without a definition which is too circular – the beak beginning at that point at which the perigynium is definitely narrowed and prolonged beyond what might be expected were the body of the perigynium beakless.

As in the Gramineae, examination of the ligule is sometimes helpful, although careful study suggests that in certain cases ligule characters are more ambiguous than some keys have implied. When ligule measurements or proportions are given, the "width" is really the same as the width of the free leaf blade at the summit of the sheath. The "length" is the total length, so that if the ligule has the appearance of an inverted V, the length is the entire depth of the V, not just the width of the free V-shaped band.

The genus is divided into a number of traditional groups, which have in recent years been called "sections" by some authors. However, many of the group names are not valid ones in the technical rank of section or if so, should be ascribed to authors other than those usually given. Hence, they are here treated as group names for convenience, without formal taxonomic standing – although the group including the type species of the genus, *C. hirta*, is called *Carex* in accord with the spirit of the Code.

The serious student of this large and difficult genus in our region will find the important references listed below to be useful, together with the standard manuals. For many years K. K. Mackenzie was the leading student of *Carex* in this country, and his monograph (1931–1935) has been widely followed, although the number of species has generally been slightly reduced by other authors. Mackenzie's work was the basis of Hermann's thorough study of *Carex* in Michigan. The traditional species concepts in *Carex* have, for the most part, been followed in the present treatment, although some innovations and simplifications have been introduced into the keys, which are based largely on the several thousand Michigan specimens critically identified by F. J. Hermann during the past third of a century.

REFERENCES

Hermann, Frederick J. 1941. The Genus Carex in Michigan. Am. Midl. Nat. 25: 1–72. ["Additions" in Am. Midl. Nat. 46: 482–492. 1951.]
Mackenzie, Kenneth Kent. 1931–1935. Cariceae. N. Am. Flora 18: 1–478.
Mackenzie, Kenneth Kent. 1940. North American Cariceae. Illustrated by Harry Charles Creutzburg. N. Y. Bot. Gard. 2 vol. (539 pl.).

KEY TO THE GROUPS

1. Spikelet solitary, terminal (entirely staminate, entirely pistillate, or mixed)
  2. Styles 2-cleft; achenes 2-sided (lenticular)
    3. Perigynia very obscurely or not at all serrulate, plump (usually at least as convex on upper face as on the lower), the lowermost tending to be remote (often as much as 0.5–1 mm distant at points of attachment); plants producing slender rhizomes (fig. 168); spikelets without empty basal scales; anthers ca. 1.5–2.5 (3) mm long . . . . 1. DIOICAE (p. 256)
    3. Perigynia minutely but strongly and regularly serrulate on apical portion and beak, ± flattened (less convex ventrally, or even plane), crowded; plants densely cespitose, not rhizomatous; spikelets usually with 1–2 empty basal scales; anthers 2–3.5 mm long . . . . 11. STELLULATAE (*C. exilis*, p. 270)
  2. Styles 3-cleft; achenes 3-sided (or nearly terete)
    4. Spikelets unisexual (either staminate or pistillate); perigynia pubescent
      5. Spikelets mostly (1.2) 1.5–3.5 cm long; culms mostly exceeding leaves of current year; scales ciliate; perigynia very hairy . . . . 17. SCIRPINAE (p. 289)
      5. Spikelets shorter, often on short culms; scales eciliate; perigynia minutely (and sometimes sparsely) pubescent . . . . 16. MONTANAE (p. 285)
    4. Spikelets containing both staminate and pistillate flowers; perigynia usually glabrous
      6. Perigynia minutely pubescent . . . . 16. MONTANAE (p. 285)
      6. Perigynia glabrous
        7. Spikelets staminate at base, pistillate toward apex, densely flowered, mostly 1 cm or more thick; perigynia inflated (i.e., much larger than the included

achene), abruptly contracted to a long, very slender beak . . . .
. . . . 44. SQUARROSAE (p. 322)
7. Spikelets pistillate at base, staminate above, more slender and sparsely
flowered (fewer than 10 perigynia); perigynia various but not as above
8. Lower pistillate scale leaf-like, at least on most spikelets, much exceeding
the perigynium; perigynia distinctly beaked, the body plumply filled by the
mature achene . . . . 15. PHYLLOSTACHYAE (p. 284)
8. Lower pistillate scale not leaf-like, scarcely if at all exceeding perigynium;
perigynia essentially beakless or linear-lanceolate (tapering into indistinct
beak)
9. Perigynia slender, linear-lanceolate (more than 5 times as long as thick),
spreading or (usually) strongly reflexed at maturity (fig. 232) . . . .
. . . . 40. ORTHOCERATES (p. 316)
9. Perigynia broader, less than 5 times as long as thick, appressed-ascending
(fig. 194) . . . . 14. POLYTRICHOIDEAE (p. 284)
1. Spikelets 2 or more (except sometimes in depauperate individuals) – sometimes
crowded but distinguishable by lobate appearance of inflorescence or protruding
bracts or visible short segments of rachis between spikelets
10. Styles 2-cleft; achenes 2-sided
11. Lateral spikelets penduncled, or if sessile, then elongate; terminal spikelet often
entirely staminate
12. Plants slender, the culms up to 2.8 (or rarely 3.7) dm tall and less than 1 mm
thick (excluding leaf bases) even toward the base; terminal (staminate or
sometimes mixed) spikelet solitary, (0.3) 0.6–1.1 (1.8) cm long; lowermost
bract usually with a short sheath ca. 1.5–7 (rarely 10–30) mm long;
perigynia white-pulverulent or golden yellow at maturity (except in the small
specimens of *C. lenticularis* which will run here)
13. Lowermost pistillate spikelet (except rarely one arising from near base of
plant) sessile or nearly so; terminal spikelet staminate; perigynia green or
slightly glaucous, crowded . . . . 38. ACUTAE (couplet 5, p. 314)
13. Lowermost pistillate spikelet nearly always penduncled; terminal spikelet
often pistillate near apex, or the pistillate spikelets ± loosely flowered; fresh
perigynia white-pulverulent or golden yellow . . . . . . 21. BICOLORES (p. 292)
12. Plants coarser, the culms over (3) 5 dm tall and usually over 1 mm thick, at
least toward base; staminate spikelets often 2 or more, (2) 2.5–7 (8) cm long;
lowermost bract essentially sheathless (rarely with very short sheath);
perigynia neither white-pulverulent nor golden yellow
14. Pistillate spikelets on ± lax peduncles, at length drooping (fig. 229), the
scales prominently awned; body of achene with an irregular notch,
constriction, or wrinkle on one side . . . . 39. CRYPTOCARPAE (p. 316)
14. Pistillate spikelets erect or strongly ascending, often sessile (fig. 227), the
scales acute or acuminate, not awned; body of achene smooth and ± regular
. . . . 38. ACUTAE (p. 314)
11. Lateral spikelets sessile, short, often crowded; terminal spikelet at least partly
pistillate (rarely staminate in an occasional individual)
15. Culms arising mostly solitarily from prominent rhizome or decumbent stolon
(figs. 169, 170); anthers 2–3.5 (4) mm long (except in certain species of
Heleonastes keyed here for added convenience but related elsewhere)
16. Perigynia plumply plano-convex to subterete in cross section, not winged or
sharply margined; plants of sphagnum bogs, cedar swamps, etc.
17. Scales pale-hyaline with green midrib; perigynia apiculate or with very
small beak; at least the lower few-flowered spikelets ± separated; plants
loosely cespitose, producing slender rhizomes . . . . 10. HELEONASTES (p. 267)

17. Scales rich brown; perigynia with distinct beak ca. 0.5 mm long; spikelets crowded as if in a single head; culms arising from axils of old decumbent culms or stolons (fig. 170) . . . . 5. CHORDORRHIZAE (p. 258)

16. Perigynia strongly flattened, with distinctly winged or sharply edged margins; plants mostly of wet or dry open habitats

    18. Perigynia mostly over 2 mm wide; staminate flowers only at the base of some or all spikelets . . . . 13. OVALES (couplet 8, p. 274)

    18. Perigynia mostly not over 2 mm wide; staminate flowers not restricted to bases of spikelets

        19. Mature perigynia with the body ± narrowly wing-margined above and the beak bidentate (firm teeth 0.5 mm long); rhizome slender (ca. 1−1.5 mm in diameter), with brownish fibrous sheaths; spikelets often dissimilar, some largely or entirely staminate or pistillate, others mixed . . . .
. . . . 4. ARENARIAE (p. 258)

        19. Mature perigynia distinctly 2-edged but not winged, the beak with short weak teeth; rhizome stout (ca. 2−3 mm in diameter), with black fibrous sheaths; spikelets mostly similar (each one staminate apically and pistillate basally; in Intermediae the upper sometimes largely staminate)

           20. Sheaths of upper leaves green-nerved ventrally, usually not covering the inconspicuous nodes . . . . 3. INTERMEDIAE (p. 256)

           20. Sheaths of upper leaves with broad white-hyaline stripe on ventral side, covering the included nodes . . . . 2. DIVISAE (p. 256)

15. Culms cespitose, the tufts with or without connecting rhizomes; anthers various

    21. Staminate flowers at the bases of some or all spikelets, not at the apex (note especially the terminal spikelet)

        22. Perigynia with thin-winged margins, at least narrowly so along apical part of body and basal part of beak, strongly flattened and scale-like (in some species elongate), ± appressed and overlapping (or in some species, spreading at the tips) . . . . 13. OVALES (p. 272)

        22. Perigynia at most with a ridge along the margin, not winged, the achene plumply filling at least the apical part of the body all the way to the margins

           23. Body of perigynium elliptic or nearly so (except in *C. arcta*) with at most a very short beak, and with rounded or slightly margined edges, nearly or entirely filled by the achene, characteristically greenish-white puncticulate (under strong lens); anthers 0.8−1.5 (very rarely 1.7) mm long . . . . 10. HELEONASTES (p. 267)

           23. Body of perigynium ± ovate or lanceolate or prominently beaked, sharp-edged or margined, only half to two-thirds filled by achene (very spongy around and below base of achene), not greenish-white dotted; anthers 0.8−2.6 mm long

               24. Mature perigynia loosely to strongly appressed-ascending, 4−5.7 mm long; anthers (1.1) 1.3−2.6 mm long . . . . 12. DEWEYANAE (p. 272)

               24. Mature perigynia strongly spreading to reflexed, 2−3.6 mm long; anthers 0.8−2 mm long . . . . 11. STELLULATAE (couplet 2, p. 269)

    21. Staminate flowers at the apex of some or all spikelets (even when anthers have fallen, protruding filaments are usually visible)

        25. Culms stout (often 1.5 mm thick at ca. 3 cm below inflorescence) and very sharply angled (or even narrowly winged), ± soft and easily compressed (flattened in pressing); wider leaves 5−10 mm broad, with rather loose sheaths; perigynia spongy-thickened basally, on short slender stipes; anthers 1.3−2.6 mm long . . . . 9. VULPINAE (p. 266)

        25. Culms slender (not over 1.5 mm thick at ca. 3 cm below inflorescence, or

rarely so in some species), firm, not wing-angled nor easily compressed (hence, not flattened in pressing); leaves, perigynia, and anthers various

26. Spikelets 10 or fewer, usually greenish at maturity, crowded or remote in a simple inflorescence (one spikelet, no branches, at each node of it)

27. Perigynia elliptic, essentially beakless, very plump (nearly terete) and filled by the achene; at least the lower spikelets well separated, containing 1–5 perigynia . . . . . . 10. HELEONASTES (*C. disperma*, p. 267)

27. Perigynia ± ovate, beaked, plano-convex or lenticular; spikelets various

28. Mature perigynia brownish; some spikelets (especially terminal) entirely or mostly staminate or staminate at their bases only [only occasional specimens of some species of this group] . . . . 11. STELLULATAE (p. 269)

28. Mature (not over-ripe) perigynia generally greenish; no spikelets entirely or mostly staminate (a few may have stamens at their base in addition to their apex) . . . . 6. BRACTEOSAE (p. 258)

26. Spikelets numerous (10–many), yellowish or brownish at maturity; inflorescence tending to be compound, at least its lower nodes with 2 or more spikelets crowded on a lateral branch

29. Pistillate scales terminating in a distinct rough awn; bracts, at least lower ones, very slender and exceeding spikelets or branches; ventral surface of leaf sheaths usually transversely wrinkled or puckered (very rarely smooth) . . . . 7. MULTIFLORAE (p. 262)

29. Pistillate scales acute or minutely cuspidate; bracts mostly short, inconspicuous, or absent; leaf sheaths smooth ventrally (or sometimes wrinkled in *C. decomposita*). . . . . 8. PANICULATAE (p. 264)

10. Styles 3-cleft; achenes 3-sided (or nearly terete)

30. Perigynia at least sparsely puberulent, pubescent, hispidulous, or scabrous

31. Style persistent, continuous with mature achene and of similar texture (indurated) at least toward its base; perigynia ca. 4–15 (18) mm long

32. Perigynia ca. 12–18 mm long, in short-oblong to globose spikelets ca. 2–3.5 cm in diameter . . . . 46. LUPULINAE (p. 327)

32. Perigynia ca. 4.5–9 mm long, in elongate, cylindrical spikelets less than 2 cm in diameter

33. Leaves hairy . . . . 33. CAREX (p. 309)

33. Leaves glabrous . . . . 43. PALUDOSAE (couplet 2, p. 321)

31. Style withering, articulated with summit of achene (or above a short apiculus), at length deciduous; perigynia 2–7 mm long

34. Perigynia with distinct and definite slender beak and/or the apex with 2 firm teeth

35. Leaves hairy*

36. Beak of perigynium with minute, scarcely visible teeth; body of perigynium strongly 3-angled, closely enveloping the achene, essentially nerveless, tapered to a stipe-like base (fig. 201); culms pubescent; perigynia mostly slightly less than 5 mm long . . . . 19. TRIQUETRAE (p. 292)

36. Beak of perigynium with strong spreading teeth ca. 1 mm or more long; body of perigynium ± rounded, loosely enveloping achene (especially at summit), strongly ribbed, ± rounded (not cuneate-tapered) at base; culms glabrous; perigynia mostly slightly more than 5 mm long. . . . . 33. CAREX (*C. hirta*, p. 309)

*Rarely, pistillate plants of *C. scirpoidea* (fig. 198), which normally has solitary unisexual spikelets, will have small secondary spikelets. Dioecious plants with strongly ciliate scales and pubescent ventral surface of the leaf sheaths should not be run further here.

35. Leaves glabrous (often rough or scabrous, but not hairy)
  37. Pistillate spikelets not over 10 mm long (occasionally 12 mm in *C. communis*); achenes mostly with very convex or rounded sides (the angles thus obscured), at least apically, very tightly enveloped by the perigynium, especially on the apical half; anthers ca. 1.5–3.7 mm long; plants of dryish habitats . . . . 16. MONTANAE (p. 285)
  37. Pistillate spikelets mostly over 10 mm long; achenes with flattish to slightly concave sides (the angles thus ± evident), the summit (especially around base of style) ± loosely enveloped by the perigynium; anthers ca. 2.5–4.7 mm long; plants of dry to wet habitats
    38. Perigynia scabrous, the beak usually more than half as long as the body, weakly and obscurely toothed (fig. 224); leaf blades and bracts very scabrous, the widest 5–12 mm broad . . . . 34. ANOMALAE (p. 310)
    38. Perigynia ± densely short-hairy, the beak less than half as long as the body, with two firm apical teeth; leaf blades rough or smoothish, not over 4.5 (6) mm broad, often channeled or with revolute or involute margins . . . . 33. CAREX (p. 309)
34. Perigynia beakless or merely apiculate ("beak" not over 0.4 mm long) and the apex not toothed
  39. Spikes entirely pistillate; lowest bract sheathless and with reduced blade (not over 2 cm long, shorter than the spike); leaf sheaths pubescent chiefly on the ventral surface . . . . 17. SCIRPINAE (p. 289)
  39. Spikes at least partly staminate; lowest bract with sheath or with prolonged blade; leaf sheaths glabrous or pubescent
    40. Leaf sheaths (and usually the blades) ± pubescent, especially toward base of plant; terminal spikelet pistillate toward apex, staminate toward base . . . . 32. VIRESCENTES (couplet 2, p. 308)
    40. Leaf sheaths and blades glabrous; terminal spikelet staminate toward apex or entirely staminate
      41. Bract at base of inflorescence sheathless or nearly so, but with well developed blade; foliage glaucous; perigynia nearly terete, in densely flowered cylindrical spikelets over 1 cm long and ca. 4 mm or more thick . . . . 35. PENDULINAE (p. 312)
      41. Bract at base of inflorescence (not counting occasional basal spikelets) usually with sheath (2) 4 mm or more in length, with blade absent or rudimentary (not over 2 (4) cm long); foliage not glaucous; perigynia ± 3-sided, nerved or not, in spikelets less than 1 cm long and/or less than 4 mm thick . . . . 18. DIGITATAE (p. 290)
30. Perigynia glabrous (in some species, papillose or granular, but not even sparsely puberulent or scabrous)
  42. Leaf sheaths and often also the blades pubescent (at least strongly hispidulous), especially toward base of plant*
    43. Beak of perigynium with firm teeth ca. 1.5–3 mm long (fig. 237); style persistent, continuous with summit of mature achene and of similar texture; perigynia ca. 8–10 mm long, in spikelets 4–12 cm long . . . .
      . . . . 43. PALUDOSAE (*C. atherodes,* p. 321)
    43. Beak of perigynium with teeth scarcely 0.5 mm long or absent; style withering, articulated with summit of achene (or above a short apiculus), at length deciduous; perigynia less than 6 mm long, in spikelets less than 3 cm long

*In some specimens of *C. atherodes* from shallow water, the only pubescence evident is a hispidness at the summit of the ventral surface of protected leaf sheaths.

44. Perigynia ca. 4–5.5 (6) mm long, including a beak ca. 1 mm long
  45. Pistillate scales with a rough awn often exceeding the perigynium; perigynia finely and closely many-nerved, in erect or strongly ascending spikelets; sheaths strongly hispidulous, the blades and culms hispidulous or scabrous . . . . 25. OLIGOCARPAE (*C. hitchcockiana*, p. 301)
  45. Pistillate scales not awned or the lowest short-awned; perigynia with 2 strong nerves and often several weak ones, in spikelets ± drooping on slender peduncles; sheaths, blades, and usually culms softly hairy . . . . .
  . . . . .28. SYLVATICAE (*C. castanea*, p. 304)
44. Perigynia ca. 2.4–4.5 mm long (to 6 mm in Gracillimae), beakless or with beak scarcely 0.5 mm long
  46. Pistillate spikelets erect or ascending, sessile or short-peduncled; perigynia ca. 2.4–3 mm long; bract at base of inflorescence sheathless or with sheath at most (and very rarely) as long as 4 mm . . . .
  . . . . 32. VIRESCENTES (couplet 3, p. 308)
  46. Pistillate spikelets on filiform peduncles, usually laxly spreading or drooping; perigynia ca. 3.5–6 mm long; bract at base of inflorescence with prolonged sheath 1–6 cm or more long . . . . 27. GRACILLIMAE (p. 303)
42. Leaf sheaths and blades completely glabrous (though sometimes scabrous)
  47. Style persistent, at least the basal portion continuous with the mature achene and of similar texture; beak of perigynium ± prominent (over 1 mm long) with ± firm, stiff, sharp teeth (in some species these very short)
  48. Body of perigynium obovoid or obconic, ± truncately contracted into a distinct long slender beak (fig. 241); terminal spikelet often mostly pistillate (staminate at base only) . . . . 44. SQUARROSAE (p. 322)
  48. Body of perigynium ovoid to lanceolate or ellipsoid, tapered or contracted into the beak; terminal spikelet normally staminate, at least apically
  49. Pistillate scales subtending at least some of the perigynia terminated by a distinct slender scabrous awn; perigynia ca. 4–9 mm long
  50. Scales toward apex of pistillate spikelets merely acuminate or with awns shorter than their bodies (the latter easily visible, about half as long as perigynia or longer); staminate spikelets 2 or more; body of perigynium rather gradually tapered into a beak ca. 1.5 mm long, including the short (not over ca. 0.8 mm) teeth . . . . 43. PALUDOSAE (p. 321)
  50. Scales toward apex of pistillate spikelets ordinarily with awns (as on the other pistillate scales) nearly or fully as long as their bodies (fig. 234) (the latter small and mostly hidden among the bases of the densely crowded perigynia); staminate spikelet solitary (or very rarely a second smaller one present); body of perigynium tapered or strongly contracted into a beak ca. 1.2–3.5 mm long, including teeth up to 2.2 mm long . .
  . . . . 42. PSEUDO-CYPEREAE (p. 318)
  49. Pistillate scales smooth-margined and awnless or very short-awned, or at most with a scabrous margin toward an acuminate (sometimes inrolled) apex (occasionally a long rough awn in species with perigynia more than 9 mm long); perigynia (4) 4.5–18 mm long
  51. Perigynia very narrowly lanceolate, 4–6.5 times as long as wide and not over 3 mm wide, many-nerved, tapering to apex (not strongly contracted into a beak); staminate spikelet solitary (pistillate spikelets may be staminate at apex) . . . . 41. FOLLICULATAE (p. 318)
  51. Perigynia lanceolate or broader, less than 4 times as long as wide, or more than 3 mm wide, or strongly contracted into a conspicuous beak (or all of these); staminate spikelets solitary or 2 or more

52. Leaves and bracts involute-filiform; perigynia 4–6 mm long, in sessile spikelets few-flowered or short-oblong (not over 2 cm long at the most) (fig. 242); staminate spikelet solitary (rarely a second small one present) . . . . 45. VESICARIAE (*C. oligosperma*, p. 324)
52. Leaves and bracts with definite flat blades; perigynia and spikelets various
  53. Perigynia 4–12 mm long, ca. 6–12(15)-nerved, in cylindrical spikelets; staminate spikelets normally 2–4 (often 1 in *C. retrorsa*). .
  . . . . 45. VESICARIAE (p. 324)
  53. Perigynia (11) 12–17 (18) mm long, ca. 15–20-nerved, in cylindrical, short-oblong, or subglobose spikelets; staminate spikelet solitary . .
  . . . . 46. LUPULINAE (p. 327)
47. Style withering, articulated with summit of achene (or above a very short straight or bent apiculus), at length deciduous; beak of perigynium absent, or without teeth, or at most with short and usually soft teeth
54. Perigynia ± rounded to broadly tapered at summit, beakless or essentially so (the tiny beak or apiculus if present less than 0.5 mm long if distinct or up to 0.8 mm long if vaguely defined, often strongly bent or curved); beak or apiculus (if present) never toothed (or teeth scarcely 0.1 mm long)
  55. Leaf blades linear-filiform, not over 0.5 mm broad; perigynia dark brown or nearly black at maturity, 2 mm or less in length, in few-flowered spikelets, of which at least the upper ones are on peduncles usually surpassing the sessile staminate spikelet . . . . 20. ALBAE (p. 292)
  55. Leaf blades 0.5 mm or more broad; perigynia and spikelets various (but not as above)
    56. Bract of lowest pistillate spikelet sheathless (at most with a thin scarious sheath 1–3 mm long)
      57. Terminal spikelet partly pistillate; pistillate spikelets (except in the very local *C. atratiformis*) nearly or quite sessile and erect or ascending; roots glabrous or nearly so . . . . 37. ATRATAE (p. 313)
      57. Terminal spikelet normally entirely staminate; spikelets and roots various
        58. Pistillate spikelets mostly drooping at maturity on slender peduncles; roots with dense felt-like pubescence; perigynia ± rounded and stipitate at the base . . . . 36. LIMOSAE (p. 312)
        58. Pistillate spikelets erect or ascending, sessile or peduncled; roots glabrous; perigynia strongly tapering at the base . . . . . [go to couplet 62]
    56. Bract of lowest pistillate spikelet with a sheath 4 mm or more in length
      59. Terminal spikelet bearing some perigynia (very rarely a few individuals with one entirely staminate); plants very strongly reddish-tinged at base
        60. Staminate flowers at apex of terminal spikelet, pistillate flowers at base; cauline sheaths bladeless or with rudimentary blades up to 2 (rarely 4) cm long; pistillate spikelets short-cylindric, bearing fewer than 10 perigynia, very long-peduncled, some elongate peduncles usually arising from base of plant . .18. DIGITATAE (*C. pedunculata*, p. 290)
        60. Staminate flowers at base of terminal spikelet, pistillate flowers at apex; cauline sheaths with well developed blades; pistillate spikelets linear-cylindric, bearing more than 10 perigynia, on peduncles about as long as the spikelet or shorter, all arising from the upper part of the culm . . . . 27. GRACILLIMAE (p. 303)
      59. Terminal spikelet entirely staminate; plants reddish or not at the bases
        61. Perigynia concave- or at least cuneate-tapering toward the base, ± 3-angled and often somewhat broadly spindle-shaped

62. Plants with elongate deep or shallow rhizomes and very slender but firm culms; leaf blades ca. 1–4 mm wide . . . . . . 22. PANICEAE (p. 293)
62. Plants without elongate rhizomes, the culms rather weak, easily compressed, sometimes nearly wing-margined, soon shriveling after maturity of the fruit; leaf blades up to 35 mm wide . . . .
. . . . 23. LAXIFLORAE (p. 294)
61. Perigynia convex-rounded toward the base, nearly or quite terete, ellipsoid-cylindric (or very obscurely triangular) to nearly globose
63. Larger perigynia ca. 4–5 mm long, the nerves not raised above the surface at maturity . . . . 26. GRISEAE (*C. amphibola*, p. 303)
63. Larger perigynia ca. 2–3.5 mm long; nerves various
64. Perigynia with the nerves not raised above the surface, usually ± impressed; staminate spikelet usually long-peduncled; plants not strongly rhizomatous nor with any pistillate spikelets on basal peduncles . . . . 26. GRISEAE (*C. conoidea*, p. 303)
64. Perigynia with the nerves slightly raised above the surface; staminate spikelet nearly or quite sessile or, if long-peduncled, the plants strongly rhizomatous and with basal pistillate spikelets . . .
. . . . 24. GRANULARES (p. 300)
54. Perigynium with a definite slender beak 0.5 mm or more in length, or an indistinct tapering beak 1 mm or more in length; beak in some species with short apical teeth
65. Lower pistillate scales leaf-like or bract-like, much exceeding the perigynia (fig. 197); achenes abruptly constricted to a short thick base; body of perigynium nearly terete, essentially nerveless except for 2 ribs; anthers ca. 0.5–1.6 mm long . . . . 15. PHYLLOSTACHYAE (p. 284)
65. Lower pistillate scales scarcely if at all exceeding the perigynia; achenes not abruptly constricted at the base; perigynia and anthers various
66. Perigynia horizontally spreading to deflexed, strongly few-ribbed, densely crowded in subglobose to thick-cylindric spikelets, at least the uppermost pistillate spikelets ± sessile and often crowded, the terminal spikelet (staminate or partly pistillate) sessile or short-peduncled; tip of beak cleft into 2 straight, inconspicuous, but stiff teeth up to ca. 0.5 mm long . . . . 31. EXTENSAE (p. 307)
66. Perigynia not as above (2-ribbed, or finely many-nerved, ascending, and/or in elongate or peduncled spikelets); tip of beak with teeth absent, or very minute, or scarious
67. Bract of lowest pistillate spikelet sheathless (or pistillate spikelets all crowded at base of plant)
68. Body of perigynium strongly obconic or obovoid, very abruptly contracted into a slender long beak over half as long as the body; terminal spikelet staminate at base only [*C. typhina* might run here]
. . . . 44. SQUARROSAE (p. 322)
68. Body of perigynium ovoid to ellipsoid, with short beak; terminal spikelet normally staminate throughout
69. Leaves and bracts involute-filiform, stiff and wiry; perigynia 4–6 mm long, in sessile spikelets; anthers 2.5–4.5 mm long [*C. oligosperma* may run here] . . . . 45. VESICARIAE (p. 324)
69. Leaves and bracts with definite flat blades; perigynia and anthers often a little shorter
70. Pistillate spikelets linear-cylindric, drooping or curving on slender peduncles; perigynia (somewhat twisted, fig. 215) and achenes

strongly angled, the latter with concave sides; tall plants (culms over 3 dm high) with scattered thin leaves . . . .
. . . . 27. GRACILLIMAE (*C. prasina*, p. 304)
70. Pistillate spikelets short, thick, and few-flowered, often crowded at base of plant; perigynia and achenes very convex-sided; low plants (culms less than 1 dm high) with crowded, very stiff leaves . . . . .
. . . . .16. MONTANAE (*C. rugosperma* var. *tonsa*, p. 287)
67. Bract of lowest pistillate spikelet with sheath 4 mm or more in length
71. Perigynia with several to numerous conspicuous fine nerves the full length of each side
72. Nerves of perigynia very numerous and impressed, giving a longitudinally corrugated appearance; awns of pistillate scales rough or even ciliate . . . . 25. OLIGOCARPAE (p. 301)
72. Nerves of perigynia several and slightly raised; awns of pistillate scales absent, smooth, or rough
73. Awns of pistillate scales usually smooth or absent; lower spikelets mostly not drooping; beak not bidentate; plants pale, brown, or reddish at base . . . . 23. LAXIFLORAE (p. 294)
73. Awns rough and/or summit of pistillate scales minutely ciliate; lower spikelets drooping on long capillary peduncles; beak slightly bidentate at maturity; plants strongly reddish at base . . . .
. . . . 28. SYLVATICAE (p. 304)
71. Perigynia with 2−3 main ribs, the sides otherwise nerveless or with less prominent nerves
74. Lowermost pistillate spikelets erect or ascending at maturity
75. Staminate spikelet well peduncled; perigynia ± convex-sided toward the base; bracts with poorly developed blades (fig. 204) . . . . . .
. . . . 22. PANICEAE (*C. vaginata*, p. 293)
75. Staminate spikelet sessile or nearly so; perigynia tapered-cuneate toward the base; bracts with well developed blades . . . .
. . . . 23. LAXIFLORAE (*C. leptonervia*, p. 299)
74. Lowermost pistillate spikelets drooping on long capillary peduncles at maturity
76. Pistillate spikelets not over 15 mm long . . . . 29. CAPILLARES (p. 305)
76. Pistillate spikelets mostly 20 mm or more long
77. Sheaths at base of plant strongly reddened, the lowermost lacking green blades; achenes with rounded angles and slightly convex sides, at least toward apex . . . . 28. SYLVATICAE (p. 304)
77. Sheaths at base of plant pale or brownish, with well developed green blades, or densely fibrous (if occasionally somewhat reddish and bladeless in *C. prasina,* the achenes very sharply triangular with concave sides)
78. Perigynia abruptly contracted into a slender beak ca. 2.5−4.5 mm long (usually slightly longer than the subglobose body, fig. 219); achenes convex-sided, at least apically; base of plant very strongly and densely fibrous . . . .
. . . . 30. LONGIROSTRES (p. 307)
78. Perigynia gradually tapering into a conical beak not over 2 mm long (and about equalling or shorter than the angled body, fig. 215); achenes sharply triangular, with concave sides; base of plant at most slightly fibrous . . . . 27. GRACILLIMAE (*C. prasina*, p. 304)

1. DIOICAE

1. **C. gynocrates** Drejer   Fig. 168
    Map 302. Bogs, boggy shores, and openings in boggy woods (cedar swamps, etc.), often in sphagnum.

2. DIVISAE

2. **C. praegracilis** W. Boott   Fig. 169
    Map 303. A western species known in Michigan only from low sandy depressions among jack pines and junipers near Eagle Harbor (*Fernald & Pease 3158* in 1934, GH, MICH; *Hermann 7763* in 1936, MICH, MSC; *Farwell 11327* in 1936, MICH, BLH; *Farwell 12459a* in 1940, MICH) and wet open boggy ground near Manistique (*Dodge* in 1915, MICH, NY). A collection which may be this species (det. Koyama 1971), although the perigynia are much smaller and narrowed at the base, was made in 1965 near a railroad in Ann Arbor (*Hiltunen 4398*, MICH).

3. INTERMEDIAE

3. **C. sartwellii** Dewey   Fig. 171
    Map 304. Wet open sandy or mucky ground, often calcareous, including marshes and meadows, lake shores, and (Drummond Island) thicket borders on dolomite pavement. Material from Schoolcraft County formerly reported as this species has been reidentified as *C. praegracilis*.
    This species is sometimes included in the Arenariae. The perigynia tend to have slightly thinner margins and firmer teeth at the summit of the beak than in *C. praegracilis* and are not at all stipitate; the leaves are more scattered on the stem (in *C. praegracilis* the perigynium tends to be slightly stipitate and the leaves are crowded at the base of the plant). The ventral surface of the sheath tends to be prolonged, thin, and brownish at the summit, rather than white and slightly thickened as in *C. praegracilis*.

301. Zea mays   302. Carex gynocrates   303. Carex praegracilis

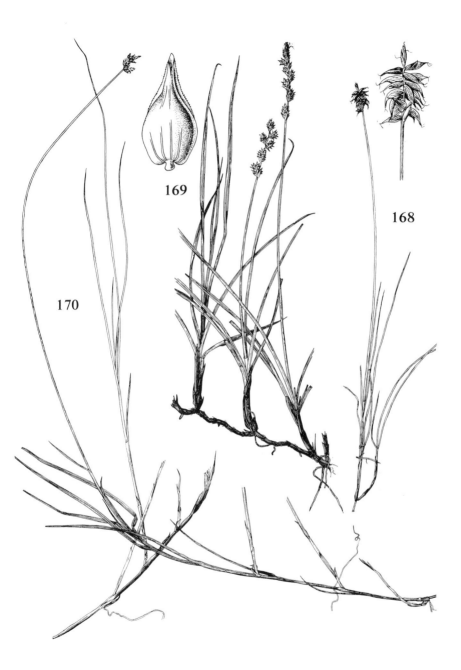

168. *Carex gynocrates* ×⅔; spikelet ×2
169. *C. praegracilis* ×½; perigynium (dorsal) ×8
170. *C. chordorrhiza* ×½

4. C. foenea Willd.

Map 305. Dry sandy or rocky ground, including more or less open jack pine and oak woods.

Long known as *C. siccata* Dewey.

5. CHORDORRHIZAE

5. C. chordorrhiza L. f.   Fig. 170

Map 306. Sphagnum bogs and similar wet sites in interdunal hollows and peaty ground.

6. BRACTEOSAE

1. Leaf sheaths loose, white with green veins or mottled green and white on back; wider blades (4.3) 5–10 mm broad (or rarely only 3 mm in nos. 6 & 7 with very slender elongate stigmas)
  2. Pistillate scales with narrowly acuminate or awned tips reaching over the bases or all the way to the ends of the beaks of the perigynia they subtend; anthers ca. 1.1–2.4 mm long; stigmas quite elongate and slender, when intact and well developed, protruding 1.5 mm or more from the perigynia; spikelets crowded in a dense inflorescence
    3. Sides of mature perigynia mostly straw-colored or yellow-brown; ventral surface of leaf sheaths thin or slightly thickened at the summit . . . . 6. C. gravida
    3. Sides of mature perigynia green; ventral surface of leaf sheath strongly concave and thickened at the summit . . . . 7. C. aggregata
  2. Pistillate scales with short-acuminate, slightly cuspidate, acute, or obtuse tips almost or not at all reaching the bases of the beaks of the perigynia they subtend; anthers 0.7–1.1 (1.3) mm long; stigmas shorter and stouter, protruding slightly from perigynia; spikelets crowded or the lower (in *C. sparganioides*) becoming well separated
    4. Spikelets close together, the lower not separated more than their length, usually ± overlapping; perigynia (3.2) 3.6–4.5 mm long, (1.7) 2–3 times as long as wide, the bodies not wing-margined; widest leaf blades (4.3) 5–7 (8) mm broad
    . . . . 8. C. cephaloidea

304. Carex sartwellii   305. Carex foenea   306. Carex chordorrhiza

4. Spikelets well separated below, the lower ones ± remote; perigynia 3–4.1 mm long, 1.3–1.8 (2) times as long as wide, the bodies ± narrowly thin-winged; widest leaf blades 5.5–10 mm broad . . . . 9. **C. sparganioides**
1. Leaf sheaths ± tight and slender and uniform green or whitish on back (or sometimes mottled in the slender and narrow-leaved *C. leavenworthii*); wider blades 0.9–4.3 (4.5) mm broad
  5. Perigynia mostly ca. 4.5 mm long; ligules, at least on culm leaves, distinctly longer than width of leaf . . . . 10. **C. spicata**
  5. Perigynia rarely as long as 4.1 mm, usually less than 4 mm; ligules about as long as width of leaf, or shorter
    6. Perigynia mostly widely spreading, conspicuously spongy-thickened at their bases and there puckered in drying, the wire-like margin above the base tending to turn inward
      7. Beak of perigynium smooth, only slightly exceeding the tip of the acuminate scale; inflorescences with spikelets rather close together or crowded . . . . . .
      . . . . 11. **C. retroflexa**
      7. Beak of perigynium minutely serrulate, much exceeding the tip of the acute to obtuse or rounded scale; inflorescence interrupted, with ± separated spikelets
        8. Wider leaf blades mostly 0.9–1.8 (very rarely 2.5) mm broad; stigmas reddish to dark brown, slender and elongate (when intact), often protruding 1–1.5 mm or more, often reflexed but otherwise straightish or slightly sinuous . . . . 12. **C. rosea**
        8. Wider leaf blades mostly (1.5) 1.7–2.7 mm broad; stigmas very dark reddish brown, comparatively short and stout, strongly curled . . . . 13. **C. convoluta**
    6. Perigynia mostly ascending and not widely spreading, at most with thin spongy area at base not conspicuously puckered in drying (unless immature), the margin above flat or slightly incurved
      9. Inflorescence crowded to oblong and interrupted (the lower spikelets overlapping but distinct); leaf blades densely papillose above [use 20×–30× lens]; bodies of scales more (often much more) than half as long as bodies of the perigynia they subtend; larger perigynia in a spikelet 3–4.1 mm long, (1.8) 2–2.6 mm wide
        10. Perigynia with at most a few very slender faint nerves on dorsal face; anthers 1.1–1.5 mm long; inflorescence always very densely crowded, ovoid . . . .
        . . . . 14. **C. mesochorea**
        10. Perigynia with several thick (though sometimes rather faint) nerves on dorsal face; anthers 1.5–2.3 mm long; inflorescences crowded to elongate-oblong . . . . 15. **C. muhlenbergii**
      9. Inflorescence densely crowded, ± ovoid, the spikelets in a close head and nearly indistinguishable except by the slightly protruding setaceous bracts; leaf blades smooth above or the cellular outlines conspicuous, but only rarely some leaves papillose; bodies of scales usually about or only slightly more than half as long as bodies of the perigynia; perigynia 2.5–3.2 (very rarely 3.5) mm long, 1.5–1.8 (2) mm wide
        11. Beak of perigynium very short and smooth (often sparsely serrulate at very base, at junction with body); anthers (0.6) 0.8–1.7 mm long; perigynia broadest toward the base of the very broadly ovoid body . . . 16. **C. leavenworthii**
        11. Beak of perigynium serrulate; anthers 0.7–1 (1.3) mm long; perigynia broadest at or near the middle of the orbicular to broadly elliptic body . . .
        . . . . 17. **C. cephalophora**

6. **C. gravida** Bailey

Map 307. Dry open ground.

The Berrien County collection is referred to var. *lunelliana* (Mack.) Hermann, which differs from the typical variety in its perigynia more abruptly beaked and strongly ribbed, with shorter teeth, and its wider leaves.

7. **C. aggregata** Mack.

Map 308. Roadsides and lawns in the village of Schoolcraft, Kalamazoo County, collected 1936–1939 by Mr. and Mrs. Hanes (WMU, MICH, MSC, GH) and F. J. Hermann (NY). Probably adventive from its normal range somewhat to the south.

8. **C. cephaloidea** (Dewey) Dewey

Map 309. Rich deciduous woods, creek banks, thickets, less often in meadows.

See comments under *C. cephalophora* (no. 17).

9. **C. sparganioides** Willd.

Map 310. Rich deciduous woods and borders, usually beech–maple but sometimes in oak–hickory or floodplain woods; rarely in open moist fields.

See comments under *C. alopecoidea* (no. 24) and *C. decomposita* (no. 20).

10. **C. spicata** Hudson

Map 311. An Old World species, one of our few introduced Carices (from Europe) and apparently spreading westward. First collected in Michigan in July of 1965 by John A. Churchill on a dry sandy knoll at Hardwood Point in Huron County (MSC). In 1970, collected in a swale between U. S. Highway 25 and the marshy shore near Hardwood Point (*Hiltunen 4315*, MICH, WUD).

A distinctive species in its large perigynia, nerveless ventrally and weakly nerved dorsally; rich orange-brown scales with green midrib; and long ligules.

11. **C. retroflexa** Willd.

Map 312. Known only from a lawn in Schoolcraft, Kalamazoo County

307. Carex gravida

308. Carex aggregata

309. Carex cephaloidea

(*Hanes 436* in 1943, WMU), and probably only a waif in southern Michigan, which would be at the northern edge of its range if native.

12. **C. rosea** Willd.　　Fig. 172
Map 313. Usually in rich moist deciduous or mixed woods.

This species and the next are frequently confused. The two are quite distinct in well developed individuals, but do have a tendency to intergrade. The distinguishing characters of the stigmas are more comparative than quantitative, so they are difficult to determine from a key unless both are at hand for comparison. The perigynia of *C. rosea* tend to taper into the beak, while in *C. convoluta* the beak is somewhat more abrupt (and often more strongly serrulate). The anthers in both species are 0.8–1.3 mm long.

13. **C. convoluta** Mack.
Map 314. Deciduous woods, usually rich beech–maple or upland oak, seldom in swampy woods.

Those who consider this only a variety of the preceding may call it var. *pusilla* Peck.

14. **C. mesochorea** Mack.
Map 315. Known from a single collection (*Hanes* in 1949, MICH) from a street in Kalamazoo, and presumably only a waif.

310. Carex sparganioides

311. Carex spicata

312. Carex retroflexa

313. Carex rosea

314. Carex convoluta

315. Carex mesochorea

15. **C. muhlenbergii** Willd.

Map 316. Dry sandy fields, banks, and borders of woods; dunes; oak and aspen woods.

This species and *C. cephalophora* are common and easily distinguished in size of perigynia and anthers, as well as the usually more elongate inflorescence of *C. muhlenbergii*. They are less easy to distinguish from the sporadic *C. mesochorea* and *C. leavenworthii*, although the perigynium shape of the latter and the long scales of *C. mesochorea* are distinctive. Plants in which the perigynia are nerveless and flatter on the ventral face have been called var. *enervis* Boott, and have essentially the same range in Michigan as the typical plants, although the very few collections from the Upper Peninsula and northern third of the Lower Peninsula are typical var. *muhlenbergii*.

16. **C. leavenworthii** Dewey

Map 317. Known in Michigan only from lawns, roadsides, and a sandy bank in Kalamazoo County.

17. **C. cephalophora** Willd.

Map 318. Deciduous woods and thickets of all kinds, usually in dry oak, oak–hickory, or aspen, but occasionally in moist beech–maple stands; often on hillsides and banks.

The ventral surface of the sheaths is slightly thickened and concave at the summit. Occasionally one may key a specimen of *C. cephaloidea* here, but the latter species has the ventral surface of the sheaths very fragile, not thickened at the truncate summit, as well as an even more strongly scabrous culm above. See also remarks under *C. muhlenbergii*.

7. MULTIFLORAE

1. Beak of perigynium about a third or less of the total length of the perigynium (fig. 173); leaves mostly shorter than the culms . . . . 18. **C. annectens**
1. Beak of perigynium (at least on most perigynia) nearly or fully half the total length of perigynium (fig. 174); leaves mostly surpassing the culms . . 19. **C. vulpinoidea**

316. Carex muhlenbergii   317. Carex leavenworthii   318. Carex cephalophora

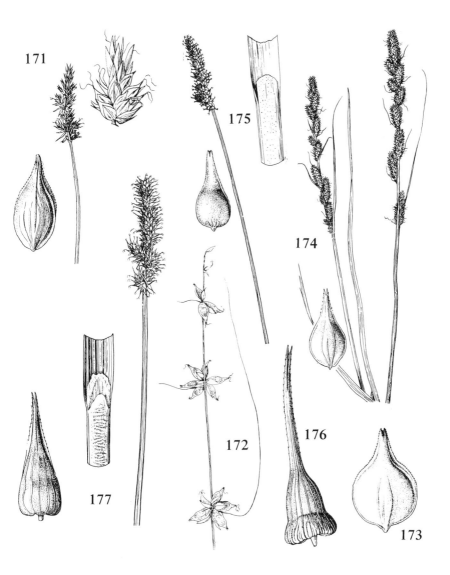

171. *Carex sartwellii* ×½; inflorescence ×2; perigynium (dorsal) ×8
172. *C. rosea* ×2
173. *C. annectens,* perigynium (dorsal) ×8
174. *C. vulpinoidea* ×½; perigynium (dorsal) ×8
175. *C. diandra* ×⅔; sheath ×1; perigynium (ventral) ×8
176. *C. crus-corvi,* perigynium (dorsal) ×8
177. *C. stipata* ×⅔; perigynium (dorsal) ×8; sheath ×1

18. **C. annectens** (Bickn.) Bickn.   Fig. 173

Map 319. Moist to wet fields, swales, clearings, ditches, and borders of swamps.

The scales tend to be somewhat shorter-awned in this species than in the next. The body of the perigynium is broadly ovate or orbicular, abruptly contracted into the short beak. This species is very close to the next, and may be considered a variety of it, as var. *ambigua* Boott. A few specimens with the short beaks characteristic of this species have been referred by Hermann to *C. vulpinoidea*, presumably because the culms are much shorter than the leaves.

19. **C. vulpinoidea** Michaux   Fig. 174

Map 320. Wet, usually open ground of all sorts, especially shores, river and stream margins, ditches, meadows and clearings, and depressions in or bordering woods.

The perigynia tend to taper more into the beak in this species than in the preceding. The extreme of tapering, with ± lanceolate, slender-beaked perigynia, is var. *pycnocephala* Hermann (TL: Big Stone Bay, Emmet County). In f. *segregata* (Farw.) Raymond (TL: Dundee [Monroe County]), the spikelets are more separated than usual.

8. PANICULATAE

Plants without red dots on the leaf sheaths and with longer anthers than in this group should be sought at couplet 19 of the key to groups.

1. Inflorescence 7.5–15 cm long, the branches evident; culms stout, 1.5–2.7 mm thick at about 3 cm below the inflorescence; broadest leaf blades 5–7 mm wide; perigynia deep olive-green at maturity, obovoid, very abruptly beaked; leaf sheaths concave at mouth; anthers ca. 1–1.2 mm long . . . . 20. **C. decomposita**
1. Inflorescence 1.2–6.5 cm long (occasionally 9 cm in *C. prairea*), the branches obscure (spikelets very crowded in them); culms not over 1.5 mm thick at about 3 cm below inflorescence; broadest blades not over 3 mm wide; perigynia golden brown to deep brown (with green beak) at maturity (darker when over-ripe), lanceolate-ovate; leaf sheaths (when intact) ± prolonged ventrally at the mouth; anthers ca. 1.3–2.1 mm long

319. Carex annectens   320. Carex vulpinoidea   321. Carex decomposita

2. Leaf sheaths whitish or pale ventrally except for purplish dots; inflorescence ±
   crowded, the lowermost spikelet (or branch) usually at least slightly overlapping
   the next above it (occasionally separated by a distance no more than its total
   length); perigynia tending to spread at maturity, therefore not concealed by the
   scales . . . . 21. **C. diandra**
2. Leaf sheaths strongly tinged with copper color toward their summits ventrally;
   inflorescence ± interrupted, the lowermost spikelets (or branches) often well
   separated or even peduncled; perigynia ± appressed at maturity, nearly or
   completely concealed by the large scales . . . . 22. **C. prairea**

20. **C. decomposita** Muhl.

Map 321. A species of swamps, barely ranging as far north as Michigan. Of
our very few specimens, the only ones with habitat data were collected in a
"very wet swamp on the River Raisin" in 1832 by N. W. Folwell (NY, GH).

Some robust specimens of *C. sparganioides* with the lower spikelets compound may seem to run here, but may be distinguished by their larger (ca. 3–4.1
mm), thin, ovate, narrowly winged, bright to light green perigynia; and usually
by the absence of purplish dots on the ventral surfaces of the upper leaf sheaths.
(In *C. decomposita*, the perigynia are smaller, biconvex-obovoid, and very dark
green to brown at maturity; and the sheaths are dotted with purplish.)
Furthermore, the anthers of *C. sparganioides* are often less than 1 mm long.

21. **C. diandra** Schrank   Fig. 175

Map 322. Wet marshy ground, shores, bogs (especially calcareous ones),
interdunal swales, springy thickets.

The perigynia in this species are usually smaller than in the next, although the
extremes are about the same. In *C. diandra* they run (2) 2.2–2.5 (3) mm long
and are definitely convex on the ventral face at maturity.

22. **C. prairea** Dewey

Map 323. Usually in wet meadows, swales, marly bogs, marshy shores,
tamarack and cedar swamps, and stream banks.

The perigynia range in length from 2.1 to 3 mm, but are usually 2.6–3 mm,
and flattish on the ventral face at maturity.

322. Carex diandra

323. Carex prairea

324. Carex crus-corvi

9. VULPINAE

1. Perigynia 6.5–8 mm long, enlarged below with a spongy disc-like area much broader than the rest of the body (fig. 176); beak twice as long as body of perigynium, or longer; thin ventral surface of leaf sheaths with copious tiny purplish dots . . . . 23. **C. crus-corvi**
1. Perigynia 3–6.2 mm long, corky below but without so distinct a disc-like area; beak slightly longer than the body of perigynium, or shorter; thin ventral surface of leaf sheaths dotted or not
  2. Perigynia 3–4 (4.2) mm long, rather abruptly contracted into a beak no longer than the body, essentially nerveless ventrally and with only weak nerves dorsally; ventral surface of leaf sheaths sparsely to strongly dotted with purplish, especially toward their summits . . . . 24. **C. alopecoidea**
  2. Perigynia 4–6.2 mm long, somewhat contracted or ± obcuneate and tapered evenly into the beak (this then difficult to define, but about equalling or slightly exceeding the body if the latter is measured from the base of perigynium to summit of achene), strongly several-nerved dorsally and with at least a few nerves ventrally; ventral surface of leaf sheaths not dotted with purplish
    3. Sheaths thickened (or even ± cartilaginous) at the concave or truncate mouth, smooth and unwrinkled ventrally; perigynia 4.7–6.2 mm long . . 25. **C. laevivaginata**
    3. Sheaths thin (usually broken) at the prolonged (when intact) mouth, rather strongly puckered or cross-wrinkled ventrally (fig. 177) – very rarely nearly or quite smooth; perigynia 4–5 (5.5) mm long . . . . 26. **C. stipata**

23. **C. crus-corvi** Kunze   Fig. 176
Map 324. River banks, floodplains, and marshes – very local.

24. **C. alopecoidea** Tuckerman
Map 325. Marshes, swales, and wet meadows; moist woods and clearings.

Some specimens of *C. sparganioides* (no. 9) may run here in the key. They differ from *C. alopecoidea* in nearly always lacking purplish dots on the sheaths, in having generally wider leaves, in the shorter beak of the perigynium (not over about half as long as the body), and in their much shorter anthers.

25. **C. laevivaginata** (Kuek.) Mack.
Map 326. Swampy woods (deciduous, hemlock, or cedar), swales, marshy woodland borders, streamsides in woods, very rarely in bogs.

325. Carex alopecoidea   326. Carex laevivaginata   327. Carex stipata

26. C. stipata Willd.   Fig. 177

Map 327. One of our commonest species throughout the state, in moist, usually shaded ground everywhere, except only very rarely in sphagnum bogs.

10. HELEONASTES

1. Lowest bract bristle-like, several times as long as its spikelet; perigynia mostly 2.8–3.8 (4) mm long, including very short smooth beak; spikelets widely separated, containing 1–5 perigynia each . . . . 27. **C. trisperma**
1. Lowest bract absent or at most about twice as long as its spikelet (if rarely prolonged, the perigynia smaller and often with serrulate beak); perigynia and spikelets various
  2. Perigynia nearly terete or unequally biconvex; spikelets (at least the lower) well separated, containing 1–5 perigynia; staminate flowers at apex of spikelet (but anthers seldom seen on specimens) . . . . 28. **C. disperma**
  2. Perigynia plano-convex; spikelets various; staminate flowers at base of spikelet
    3. Spikelets crowded into a short inflorescence up to 1.3 cm long; perigynia (2.5) 2.8–3.5 mm long, smooth-margined, tapered to apex but not contracted into a beak . . . . 29. **C. tenuiflora**
    3. Spikelets remote or crowded, but total inflorescence over 2 cm long; perigynia 1.7–2.6 mm long, including a tiny beak, this often minutely serrulate or scabrous
      4. Perigynia broadest near the base of the body; spikelets usually ± overlapping or crowded . . . . 30. **C. arcta**
      4. Perigynia broadest at or near the middle of the body; at least the lower spikelets well separated
        5. Perigynia ca. 3–9 per spikelet (occasionally one or two spikelets on a plant, especially terminal one, with as many as 15), loosely spreading, becoming rich brown in age; largest leaves ca. 1–2 mm wide; foliage and perigynia green when fresh . . . . 31. **C. brunnescens**
        5. Perigynia mostly 10–many per spikelet, appressed-ascending, greenish or dull brown in age; largest leaves (1.6) 2–2.7 (3.7) mm wide; foliage and perigynia glaucous or gray-green at least when fresh . . . . 32. **C. canescens**

27. C. trisperma Dewey   Fig. 178

Map 328. Usually in coniferous bogs and swamps, even in dense shade under cedar, spruce, tamarack, fir, and/or hemlock; less often in swampy mixed woods or boggy hollows in sandy soils.

The perigynia are distinctly plano-convex and the staminate flowers basal in this species, which is sometimes confused with *C. disperma*.

28. C. disperma Dewey

Map 329. Sphagnum bogs and cedar swamps (often in sphagnum), boggy clearings and along creeks and pools in low woods, occasionally under hemlock or beech–maple.

Sometimes placed in a separate group, the Dispermae.

29. C. tenuiflora Wahl.

Map 330. Sphagnum bogs, bog forests, and peaty shores; apparently rather local but easily overlooked, especially since it may grow intermixed with *C. disperma* and *C. trisperma*.

30. **C. arcta** Boott
Map 331. Very local in low usually ± open ground.

31. **C. brunnescens** (Pers.) Poiret
Map 332. In a great variety of low or rarely upland woods; borders of bogs and swamps; coniferous swamps and wet hollows; occasionally in bogs and on shores.

Immature specimens of *C. seorsa* (no. 34) may resemble this species, but have smooth (not serrulate) beaks. The perigynia of *C. brunnescens* tend to be more definitely beaked than the merely apiculate ones of *C. canescens*, but the distinction is a subtle one.

32. **C. canescens** L.
Map 333. Usually in sphagnum bogs and older tamarack, cedar, and fir stands; also pond and stream margins, marshy or boggy ground, alder thickets, and rarely elm—maple swamp forests.

The majority of our specimens have been referred by Hermann to var. *disjuncta* Fern., with the spikelets ± cylindrical, many-flowered, and the lowest two spikelets remote (separated by over 1.5 cm). Almost as frequent and

328. Carex trisperma

329. Carex disperma

330. Carex tenuiflora

331. Carex arcta

332. Carex brunnescens

333. Carex canescens

sometimes growing with it is var. *subloliacea* (Laest.) Hartman, with fewer-flowered and hence subglobose spikelets, the lowest two usually not remote (not over 1.5 cm apart). The perigynia tend to be a little smaller in var. *subloliacea* (not over 2.2 mm long), but the two seem to intergrade so that recognition of them seems scarcely warranted here. Typical *C. canescens*, with perigynia and spikelets as in var. *disjuncta* but more crowded, is apparently known only from the Keweenaw Peninsula.

11. STELLULATAE

The distinctions between species in this series are very subtle, as can be seen from the key — and only after familiarity with typical, authentically identified specimens of the species can one name many of his unknowns with any assurance. It is possible that we should return to a more conservative treatment like that of L. H. Bailey (Bull. Torrey Bot. Club 20: 422—426. 1893), in which *C. sterilis* included *C. cephalantha* [as var. *cephalantha* (Bailey) Bailey] and *C. angustior* [as var. *angustata* (Carey) Bailey]. Some authors have, furthermore, even included *C. interior* in this polymorphic species, but the overall problem is too complex for easy solution in a local flora, involving also the question of identity with European species before the correct names could be determined.

1. Spikelet solitary, terminal; anthers 2–3.5 mm long . . . . 33. **C. exilis**
1. Spikelets more than 1; anthers 0.8–2 mm long
  2. Beak of perigynium smooth, the body broadest near the middle, distinctly nerved
   at least on basal half of ventral face . . . . 34. **C. seorsa**
  2. Beak of perigynium minutely serrulate or scabrous, the body usually broadest
   below the middle, nerved or not ventrally
    3. Beak of perigynium about a fourth of its total length, with apical teeth very
     short (rarely more than 0.3 mm long) and inconspicuous (especially from
     ventral view); lower pistillate scales often nearly or quite obtuse (or at least
     blunt) and scarcely more than half as long as the perigynia; radius of mature
     inflorescence (distance at larger spikelets from their farthest point to the main
     axis) 2.5–4.5 (5.5) mm
      4. Broadest leaves (1) 1.2–2.1 (2.4) mm wide, the foliage rather stiffly ascending;
       ventral face of perigynium nerveless or nerved basally . . . . 35. **C. interior**
      4. Broadest leaves less than 1 mm wide, the foliage ± lax and delicate; ventral face
       of perigynium nerved at least on basal half . . . . 36. **C. howei**
    3. Beak of perigynium usually about a third or more of its total length, with apical
     teeth stiffer and generally a little longer, the bidentate aspect usually
     emphasized by a false suture on ventral face below the apex; pistillate scales
     usually ± acute and nearly or quite as long as body of perigynia; radius of
     mature inflorescence, except in depauperate individuals, (3.5) 4.5–7.5 mm
      5. Perigynia with very broadly triangular-ovate bodies, ± strongly nerved
       ventrally, deep or dull green at maturity (brown when over-ripe), with beak
       about a third or a little less the total length of perigynium (fig.180). . 37. **C. atlantica**
      5. Perigynia lanceolate to broadly ovate, nerved or not ventrally, yellow-green to
       straw-colored or brown at maturity, with beak mostly a third or more of the
       total length of perigynium

6. Perigynia ± lanceolate, the beak therefore difficult to distinguish from body, but often nearly or quite equal in length to body, usually essentially nerveless on ventral face, the beak with low and rather sparse serrulations (fig. 181); achene usually 0.5 mm or more longer than wide . . . . 38. **C. angustior**
6. Perigynia ± ovate (or sometimes even deltoid-ovate in *C. sterilis*), the beak about a third of the total length, nerved at least toward the base ventrally, the beak sharply serrulate; achene less than 0.5 mm longer than wide
   7. Beak with sharp but short (scarcely setulose) serrulations, the apical teeth short but distinct and stiff; anthers 0.8–1.4 (1.5) mm long . . . 39. **C. cephalantha**
   7. Beak and uppermost part of body of perigynium copiously setulose-serrulate, the slender apical teeth rather soft, often ± bent or twisted; terminal spikelet occasionally all staminate; anthers (1) 1.2–2 mm long .
   . . . . 40. **C. sterilis**

33. **C. exilis** Dewey

Map 334. Sphagnum bogs, beach-pool bogs, and peaty coniferous swamps of tamarack, spruce, etc.

34. **C. seorsa** Howe

Map 335. Apparently rare (or overlooked), in moist or swampy deciduous woods.

Immature specimens superficially resemble *C. brunnescens*, which differs in having the beak at least sparsely serrulate.

35. **C. interior** Bailey   Fig. 179

Map 336. Widespread in wet places, especially bogs and cedar (or spruce and tamarack) swamps; also on shores, in open meadows, and along streams — sometimes in marly ground.

Plants with the perigynia ± strongly nerved, at least basally, on the ventral face may be referred to f. *keweenawensis* (Hermann) Fern. (TL: Eagle Harbor, Keweenaw County). Such forms occur sporadically throughout the state.

36. **C. howei** Mack.

Map 337. Low woods and borders of marshes, apparently very local in Michigan.

334. Carex exilis   335. Carex seorsa   336. Carex interior

Perhaps better treated as a variety of the preceding, under the name *C. interior* var. *capillacea* Bailey. One of the Kalamazoo County collections referred to this species has many leaves 1–1.7 mm wide, but they are much longer than the culms and the perigynia are strongly nerved ventrally.

37. C. atlantica Bailey   Fig. 180

Map 338. Moist thickets and boggy ground (sometimes in sandy or marly areas).

Our specimens are var. *incomperta* (Bickn.) Hermann (Rhodora 67: 361–362. 1965), previously recognized as a distinct species, *C. incomperta* Bickn. This is a puzzling taxon, in many ways linking the preceding two species to the following ones. The perigynium teeth are quite short and hence specimens may be confused with *C. howei* and *C. interior*. From the former it differs in its broader (the widest 1.6–2.4 mm) and stiffer leaves, and from the latter, in its strongly nerved perigynia; from both it normally differs in having larger spikelets and very broadly *ovate* perigynium bodies with longer, more acute subtending scales. The body of the perigynium in *C. sterilis* is sometimes broadly ovate, but the teeth are longer than in *C. atlantica*.

38. C. angustior Mack.   Fig. 181

Map 339. Bogs, shores, and low wet open ground.

39. C. cephalantha (Bailey) Bickn.

Map 340. Bogs, tamarack swamps, shores, marshy or swampy ground, and meadows.

Including here *C. laricina* Bright, distinguished previously by Mackenzie and by Hermann.

40. C. sterilis Willd.

Map 341. Marshy shores, bogs, meadows, and cedar thickets, often in marly ground; particularly characteristic of sandy-marly flats and interdunal swales along the northern shores of Lakes Michigan and Huron.

337. Carex howei

338. Carex atlantica

339. Carex angustior

12. DEWEYANAE

1. Perigynia ca. 0,8–1.2 mm wide and ca. 4–5 times as long as wide, conspicuously nerved on dorsal face, weakly to strongly nerved on ventral face . . . . 41. **C. bromoides**
1. Perigynia ca. 1.3–1.6 mm wide and usually ca. 3–3.5 times as long as wide, faintly nerved or nerveless on both faces . . . . 42. **C. deweyana**

41. **C. bromoides** Willd.   Fig. 182

Map 342. Wet or less commonly upland woods, borders of ponds in woods, damp thickets and banks, low meadows and swales. The Keweenaw County record (*Farwell 702*, Sept. 20, 1888, BLH) is somewhat dubious, as several species attributed to that date seem unlikely for the county (see note under *C. jamesii*, no. 69). However, in view of recent collections from nearby Baraga, Houghton, and Ontonagon counties, it is tentatively accepted here.

The pistillate scales are frequently tinged with orange and the leaves tend to be somewhat narrower than in the next species.

42. **C. deweyana** Schw.   Fig. 183

Map 343. In almost all kinds of dense to open, moist to dry woods and thickets, particularly characteristic of beech—maple stands; not in bogs.

The pistillate scales, except for the green along the midvein, are very pale and translucent, giving a characteristic silvery appearance to the inflorescence.

13. OVALES

This is a notoriously difficult group, and there is a strong temptation to recognize fewer species. The species as recognized by Mackenzie, Fernald, and Hermann are mostly maintained here, but with the realization that some individuals will not run satisfactorily in the key. Great care must be exercised in examining perigynia, with good lighting and magnification of at least 15× frequently necessary. Length and width (and location of widest part) of perigynia and their ratio must be carefully measured, not estimated.

340. Carex cephalantha

341. Carex sterilis

342. Carex bromoides

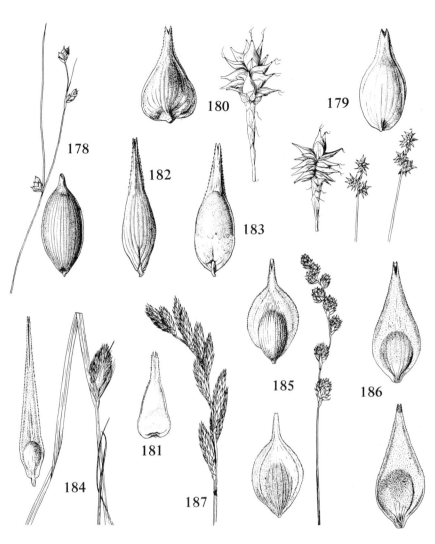

178. *Carex trisperma* ×⅔;
     perigynium (dorsal) ×8
179. *C. interior* ×⅔; terminal
     spikelet ×3; perigynium
     (dorsal) ×8
180. *C. atlantica,* terminal spikelet
     ×3; perigynium (dorsal) ×8
181. *C. angustior,* perigynium
     (ventral) ×6
182. *C. bromoides,* perigynium
     (dorsal) ×6

183. *C. deweyana,* perigynium
     (dorsal) ×6
184. *C. sychnocephala* ×⅔;
     perigynium (dorsal) ×8
185. *C. argyrantha* ×⅔; perigynia
     (dorsal above, ventral below)
     ×6
186. *C. aenea,* perigynia (dorsal
     above, ventral below) ×6
187. *C. muskingumensis* ×⅔

1. Bracts leaf-like, the broadest 2–3.5 mm wide, many times exceeding the spikelets (which are crowded in a dense head, fig. 184); perigynia very narrowly lanceolate, not over 1 mm wide . . . . 43. C. sychnocephala
1. Bracts not resembling the leaves, narrower than 2 mm most or all their length and not over twice as long as the inflorescence; perigynia various
  2. Pistillate scales about or fully as long as the perigynia *and* nearly the same width as the beaked portion (not necessarily the body), so that the apical portion of each perigynium is largely concealed; anthers ca. 1.5–3 mm long
    3. Perigynia lanceolate, ca. 1.5 mm broad, the beak prominently slender and at the very tip nearly terete, hyaline, and smooth-margined; body ± membranous and nerveless ventrally . . . . 44. C. praticola
    3. Perigynia broader, ± ovate, the beak flattened and serrulate to the tip; body usually firmer and often at least sparsely nerved ventrally
      4. Inflorescence stiff, the spikelets close together, mostly overlapping; pistillate scales nearly as wide as the bodies of the perigynia, almost concealing them . .
      . . . . 45. C. adusta
      4. Inflorescence ± lax or flexuous, the lowermost spikelets usually remote; pistillate scales distinctly narrower than bodies of perigynia (the wings of which clearly protrude at maturity)
        5. Perigynia whitish green throughout, even in maturity, with several very distinct nerves on ventral face (fig. 185) and with the wing often broadened and ± undulate or erose at the junction of beak and body . . . . . 46. C. argyrantha
        5. Perigynia at maturity brownish, at least in basal half, with at most a few faint nerves (or nerved toward base only) on ventral face (fig. 186), the wing without a broadened undulate or erose area . . . . 47. C. aenea
  2. Pistillate scales (or most of them) both shorter and narrower than beaks of perigynia, so the mature perigynia are largely exposed apically; anthers various
    6. Perigynia (6.5) 7–9 mm long; larger spikelets on each culm ca. 1.5–2.5 cm long and a third to a fifth as thick in the middle, tapered to both ends (fig. 187) . . .
    . . . . 48. C. muskingumensis
    6. Perigynia shorter; spikelets usually shorter and a third or more as thick as long
      7. Mature perigynia more than 2 mm broad at widest part
        8. Pistillate scales (especially upper ones) mostly narrowed into slender awn-tips (fig. 188); anthers 1.5–2 mm long
          9. Beak of perigynium about half as long as the body; perigynia 2.2–2.5 (3) mm wide, the broadest part about or below the middle (of total length from base of perigynium to tip of beak); spikelets with distinctly clavate bases (formed by staminate scales), in an elongate, flexuous inflorescence, the lowermost widely separated . . . . 49. C. straminea
          9. Beak of perigynium less than half as long as body; perigynia (2.6) 2.8–3.5 mm wide, the broadest part at or slightly above the middle; spikelets with inconspicuous slightly clavate bases or rounded, the lowermost nearly or quite overlapping in a rather stiff, ± erect inflorescence
          . . . . 50. C. alata
        8. Pistillate scales merely acute or acuminate, not prolonged into awn-tips; anthers various
          10. Leaf sheaths green-nerved on ventral surface almost to their summit, the hyaline area a very short belt or rapidly tapering to a point; perigynia up to 2.8 mm wide, often broadest at about middle (of *total* length)
            11. Edges of perigynia rather broadly tapered (not curved) to the base, the body thus ± diamond-shaped in general outline (fig. 189); nerves on ventral face (over achene) absent or 1–2 and very faint; perigynia

2.1–2.6 (2.8) mm wide and (3.6) 4–5 mm long; leaf blades usually less
than 2.5 mm wide, occasionally 3 (3.5) mm . . . . 51. **C. suberecta**
11. Edges of perigynia ± curved to base, the body thus ± orbicular or obovate;
nerves and size of perigynia and width of leaves various but not combined
as above
12. Perigynia 2.4–2.8 (3) mm wide, nerveless on ventral face of body;
spikelets 5–9 (most often 6–7), very crowded, often so close as to be
forced nearly at right angles to the rachis; widest leaves 3–5 mm broad
(very rarely to 8 mm) . . . . 52. **C. cumulata**
12. Perigynia 1.7–2.2 (2.5) mm wide, usually nerved on ventral face (over
the achene); spikelets 2–10 (most often 3–6), ascending, somewhat
overlapping or strung out in a cylindrical inflorescence; widest leaves up
to 2.5 (3) mm broad
13. Tips of perigynia ± widely spreading at maturity; pistillate scales
subacuminate, with very sharp tips . . . . 53. **C. albolutescens**
13. Tips of perigynia ± appressed; pistillate scales acute but with the very
tip bluntish . . . . 54. **C. longii**
10. Leaf sheaths mostly with prominent hyaline stripe on ventral face;
perigynia of various sizes, but usually broadest below the middle
14. Achenes scarcely 1.5 mm long; spikelets with rather conspicuous short
clavate bases formed by the staminate scales, ± separated toward the base
of the inflorescence; perigynia 1.7–2.3 (2.5) mm wide, 2.6–3.9 mm long,
with abruptly narrowed beak about half as long as the body . . 55. **C. festucacea**
14. Achenes mostly 1.6–2 mm long; spikelets usually ± rounded basally
(sometimes clavate in *C. brevior* or others), all crowded or the lower
separated; perigynia often broader or longer or both, with shorter and less
abruptly narrowed beak
15. Perigynia very thin and membranous, (3.7) 4–6 mm long, (2.1)
2.5–4.5 mm broad
16. Body of perigynium finely but distinctly nerved (over the achene) on
both faces (fig. 190); anthers ca. (2.7) 3–4 mm long; perigynia
4.5–6 mm long, (2.5) 3–4.5 mm broad; southern Lower Peninsula . .
. . . . 56. **C. bicknellii**
16. Body of perigynium nerveless or very faintly nerved (over achene) on
ventral face; anthers ca. 1.8–2.5 mm long; perigynia (3.7) 4–4.6
(5) mm long, (2.1) 2.5–2.8 (3.1) mm broad; Upper Peninsula and
northern Lower Peninsula south to Muskegon and (?) Bay counties. .
. . . . 57. **C. merritt-fernaldii**
15. Perigynia firmer, thin and membranous only at very margins, (3.4)
3.6–4.5 (4.8) mm long, 2.1–2.6 (3.1) mm wide
17. Body of perigynium with 3–5 (6) fine but distinct nerves (over achene)
on ventral face . . . . 58. **C. molesta**
17. Body of perigynium nerveless or very faintly nerved (over achene) on
ventral face
18. Perigynia firm, the wing itself essentially nerveless; tips of beaks often
exceeding tips of scales by 1 mm or more; perigynia 3.6–4.2 mm
long; widest leaf blades 2–3 (3.5) mm broad; southern Lower
Peninsula . . . . 59. **C. brevior**
18. Perigynia rather thin, 1–2 nerves usually apparent in the wing; tips of
beaks mostly exceeding tips of scales by less than 1 mm, usually
about 0.5 mm; perigynia (3.7) 4–4.6 mm long; widest leaf blades
2.5–4.2 (4.5) mm broad; Upper Peninsula and northern Lower
Peninsula south to Muskegon and (?) Bay counties . . 57. **C. merritt-fernaldii**

7. Mature perigynia not over 2 mm broad
19. Wing of perigynium rather distinctly broader above the middle of the body, much narrowed or obsolete basally; leaf blades 3–7 (9) mm wide; sterile culms well developed, with mostly spreading leaves; mature perigynia 2.9–5 (5.5) mm long; anthers ca. 0.7–1.3 (1.8) mm long
  20. Spikelets in a dense, stiff, crowded inflorescence
    21. Perigynia (3) 3.5–5 times as long as wide, appressed-ascending
      22. Perigynia 2.7–4 (4.5) mm long, 0.6–1.1 mm wide, the wing very narrow or obsolete; leaves up to 3.2 (rarely 5) mm wide; perigynia with nerves on ventral face usually very weak or absent . . . . 65. **C. crawfordii**
      22. Perigynia 4–5 (5.5) mm long, 1–1.3 (1.6) mm wide, the thin wing usually conspicuous above the middle of the body; leaves 3–7 mm wide; perigynia nerved on both faces . . . . 63. **C. tribuloides**
    21. Perigynia less than 3.5 times as long as wide, ascending or spreading
      23. Perigynia appressed-ascending or loosely spreading, usually winged to the base, their subtending scales acute to subacuminate and sharp at the tips . . . . 60. **C. bebbii**
      23. Perigynia strongly spreading or recurved at tips in maturity, the wings obsolete basally; scales usually acute but bluntish or often minutely notched and narrowly hyaline-bordered at the tips . . . . . . 61. **C. cristatella**
  20. Spikelets less crowded, the lowermost slightly overlapping or quite separate
    24. Perigynia 2.9–3.6 (4.2) mm long, their tips becoming spreading or recurved at maturity; spikelets ca. 4–8 (9) mm long
      25. Spikelets crowded in a ± dense, stiff inflorescence; perigynia strongly spreading or recurved at tips in maturity . . . . 61. **C. cristatella**
      25. Spikelets, at least the lowermost, usually separated, in a ± lax or flexuous inflorescence; perigynia loosely spreading-ascending at maturity . . . . 62. **C. projecta**
    24. Perigynia (3.6) 4–5 mm long, their tips appressed or loosely ascending; spikelets up to 11 (13) mm long
      26. Tips of perigynia ± appressed-ascending; spikelets appearing distinctly longer than thick and somewhat tapered toward both ends, usually rather crowded in a stiff inflorescence . . . . 63. **C. tribuloides**
      26. Tips of perigynia ± loosely spreading-ascending; spikelets subglobose, scarcely (mostly not more than 1–2 mm) longer than thick, the lowermost usually separated in a ± lax or flexuous inflorescence . . . . 62. **C. projecta**
19. Wing of perigynium not strongly narrowed below the middle and/or plants with one or more other exceptions to alternate combination: leaves usually not over 3 mm wide; leafy sterile culms poorly developed or with leaves stiff, crowded, and strongly ascending (rather than soft and spreading); mature perigynia 2.5–5.2 (5.6) mm long; anthers various
  27. Perigynia about twice as long as wide, or shorter, and broadest at about the middle (of *total* length), with broadly elliptic to suborbicular body, 1.7 mm or more wide, usually ripening in July . . . . [go to couplet 13]
  27. Perigynia of various shape and width, with narrowly elliptic body and/or broadest below the middle, ripening at various times
    28. Perigynia 4.1–5.2 (5.6) mm long (very rarely only 3.6 mm), about 2.5–3.5 times as long as broad, 1.3–2 mm wide, widest at or near the middle and often ± asymmetrical (the widest part on one side not opposite the widest on the other, or of different width) . . . . . 64. **C. scoparia**
    28. Perigynia not over 4 mm long (very rarely 4.2 or 4.4 mm), broadest below the middle, the length, width, ratio, and shape not combined as above

29. Body of perigynium at maturity suborbicular, as broad as long or broader; perigynia ripening about the latter half of June, 1.7 mm wide or wider, conspicuously winged; staminate scales forming rather conspicuous clavate bases to the spikelets . . . . 55. C. festucacea

29. Body of perigynium ovate-lanceolate or elliptic, longer than broad; perigynia ripening June–August (September), often less than 1.7 mm wide, the wing often inconspicuous or even obsolete basally; staminate scales various

    30. Perigynia about (3) 3.5–5 times as long as wide, 2.7–4 mm long, 0.6–1.1 mm wide; inflorescence stiff, of crowded, much overlapping spikelets . . . . 65. C. crawfordii

    30. Perigynia about 1.5–2.5 times as long as wide, 2.5–4 (very rarely 4.4) mm long, 1.1–1.9 mm wide; inflorescence various

        31. Spikelets crowded, much overlapping, in a stiff compact head, ± rounded at bases (the staminate scales inconspicuous, not forming narrowed clavate bases); anthers 0.8–1.2 (rarely 1.5) mm long; perigynia 1.1–1.6 mm wide, 2.5–3.5 mm long . . . . 60. C. bebbii

        31. Spikelets usually not crowded, at least the lowermost overlapping only a little or quite separate, the staminate scales often forming conspicuous clavate bases; anthers ca. (1) 1.2–2 mm long; perigynia (1.4) 1.5–1.8 (1.9) mm wide, (2.8) 3–4 (very rarely 4.4) mm long

            32. Broadest leaves (very rarely 2.2) 2.5–6 mm wide; lowermost spikelets nearly or quite overlapping in a stiffish or somewhat lax inflorescence in which the rachis immediately below the second spikelet (from bottom) is usually 0.5–0.7 (1) mm wide; fertile culms often with 4 or more leaves . . . . 66. C. normalis

            32. Broadest leaves 1.4–2.3 (very rarely 2.7 or 3.5) mm wide; inflorescence characteristically very lax and elongate, the lower spikelets well separated (fig. 193), the rachis immediately below the second one often less than 0.5 mm wide; fertile culms often with only 3–4 leaves . . . . 67. C. tenera

43. C. sychnocephala Carey   Fig. 184

Map 344. Very rare, moist sandy-mucky river bottoms and drying lake shores.

44. C. praticola Rydb.

Map 345. Collected once in Michigan — if the label be trusted, in woods near the Montreal River adjacent to Mt. Houghton, Keweenaw County (*Farwell*

343. Carex deweyana

344. Carex sychnocephala

345. Carex praticola

1614½, Aug. 22, 1898, BLH). A northern and western species, but reported by Butters and Abbe (Rhodora 55: 131. 1953) from several places in Cook County, Minnesota. Reports from Isle Royale all appear to be based on misidentifications, chiefly of *C. aenea*.

45. **C. adusta** Boott

Map 346. In dry open ground or sometimes on moist shores.

46. **C. argyrantha** Tuckerman   Fig. 185

Map 347. Sandy aspen woods, especially along roads and clearings; thickets; wooded dunes and less often in richer deciduous woods — quite local.

47. **C. aenea** Fern.   Fig. 186

Map 348. Dry open ground, cut- and burned-over woods, sandy ridges, roadsides, less often in moist ground or forested areas.

Some Isle Royale specimens have in the past been referred to *C. praticola*, but the perigynia are too ovate, with some ventral nerves, and darkened on the basal half, to be that species.

48. **C. muskingumensis** Schw.   Fig. 187

Map 349. Swamp forests (deciduous), floodplains, and swales.

346. Carex adusta

347. Carex argyrantha

348. Carex aenea

349. Carex
muskingumensis

350. Carex straminea

351. Carex alata

49. **C. straminea** Willd.

Map 350. Known in Michigan only from Kalamazoo County, where it has been collected in marshes and swamp borders; often forms large clumps.

In some manuals, this is called *C. richii* Mack.

50. **C. alata** Torrey   Fig. 188

Map 351. Swampy ground, bogs, marshes, shores, meadows.

51. **C. suberecta** (Olney) Britton   Fig. 189

Map 352. Marshy shores, ditches, swamps, and low ground.

52. **C. cumulata** (Bailey) Fern.

Map 353. Moist often sandy ground and burned-over polytrichum bogs.

53. **C. albolutescens** Schw.

Map 354. Wet ground, apparently very local.

54. **C. longii** Mack.

Map 355. Moist, often sandy open ground; borders of marshes; and open woodland.

55. **C. festucacea** Willd.

Map 356. Marshes and low ground, very rare in Michigan.

352. Carex suberecta

353. Carex cumulata

354. Carex albolutescens

355. Carex longii

356. Carex festucacea

357. Carex bicknellii

56. **C. bicknellii** Britton   Fig. 190

Map 357. Dry or sometimes moist sandy open ground, especially prairie-like habitats and along railroads.

The culms grow in small tufts from a stout rhizome, which tends to be more distinct in this species than in its relatives. (TL: "Michigan")

57. **C. merritt-fernaldii** Mack.

Map 358. Dry sandy or rocky soil, sometimes under aspens or jack pine.

The thinner perigynia will usually distinguish this species from the next two, although they are sometimes not so membranous as in *C. bicknellii*. A very immature collection from Bay County is probably this species (*Dreisbach 4908*, MICH, PH).

58. **C. molesta** Mack.

Map 359. Low ground; open moist fields and swales as well as swampy woods and thickets along rivers.

By some authors united with the following species, as var. *molesta* (Mack.) Gates. The pistillate scales in *C. molesta* reach only to the base of the beak of the perigynium; in *C. brevior* the scales reach at least to the middle of the beak. Hermann also stresses the more clavate bases of the spikelets in *C. brevior*, which usually has a less congested inflorescence than *C. molesta*, in which the spikelets have broadly rounded bases and are ± crowded in a stiff inflorescence.

59. **C. brevior** (Dewey) Mack.

Map 360. Dry open or sandy prairie-like ground, bluffs, or rarely in moist places. The Alpena County collection (*Wheeler* in 1895, MICH) has the sheaths green ventrally and does not appear like fully typical *C. brevior*. See also comments under the previous species.

60. **C. bebbii** (Bailey) Fern.   Fig. 191

Map 361. One of our commonest species. Wet sandy shores, meadows,

358. Carex
     merritt-fernaldii

359. Carex molesta

360. Carex brevior

swales, ditches, stream banks, cedar and tamarack swamps, damp woods, usually in ± open ground or clearings.

Some specimens approach *C. cristatella*, and hybridization between the two has been postulated. Superficially the species tends to resemble *C. crawfordii*. The perigynia usually ripen July—August.

61. C. cristatella Britton

Map 362. Wet woods (deciduous or coniferous) and thickets, swales, marshes, shores, ditches, and meadows; rarely in oak upland woods or rich hardwoods.

The spikelets are more separated than usual in f. *catelliformis* (Farw.) Fern. (TL: South Rockwood [Monroe County]), which thus resembles *C. projecta*.

62. C. projecta Mack.

Map 363. In wet meadows and ditches or, more often, swampy woods and thickets or depressions in upland woods.

A variable species, sometimes resembling the preceding and sometimes the following; perhaps suspect as to its origin. The dorsal surface of the leaf sheaths is very weak, whitish or pale green and hyaline between the green nerves; however, this condition is sometimes also found in the preceding species and the next. The typically lax or flexuous inflorescence of this species with remote spikelets is very distinctive when well developed, but is often obscure in young, depauperate, and occasional other individuals. *C. cristatella* is sometimes distinguished from *C. projecta* and *C. tribuloides* by the obvious distention of the perigynium over the included achene; this condition also occurs sometimes in *C. projecta* and (less commonly) in *C. tribuloides*.

63. C. tribuloides Wahl.

Map 364. Damp to wet ground generally: marshes, swales, ditches, and shores; swampy woods, alder thickets, shaded borders — very rarely on uplands.

Sometimes confused with the next species, especially when immature. The pistillate scales are usually ± blunt and hyaline-tipped, while in *C. scoparia* they tend to be narrowly acuminate to very sharp tips. The leaves are generally narrower in *C. scoparia*.

361. Carex bebbii 362. Carex cristatella 363. Carex projecta

64. **C. scoparia** Willd.

Map 365. Moist open ground: sandy (often marly) shores, stream borders, meadows and wet fields, borders of swamps, thickets, ditches, marshes, temporary ponds in woods; rarely in dry sand or peaty ground.

The perigynia ripen from late June (southern Michigan) to August, most commonly in July, and are generally broader than in the preceding species.

65. **C. crawfordii** Fern.   Fig. 192

Map 366. Usually on wet sandy shores or in meadows, ditches, and marshy ground; occasionally in woods, especially in clearings and along roads; and on dry sandy ridges (especially near Lake Superior).

The perigynia ripen July–September, usually in August or late July.

66. **C. normalis** Mack.

Map 367. Usually in moist ground, damp fields, thickets, woods; but also sometimes in dry open ground. A collection labeled as from Keweenaw County (*Farwell 683*, Sept. 6, 1888, BLH, MICH, MSC) is not mapped; many Farwell records for that month appear to be confused (see discussion under *C. jamesii*, no. 69), and this species is not otherwise known in Michigan north of the middle of the Lower Peninsula.

The perigynia ripen in middle or late June or sometimes early July. They are not noticeably asymmetrical compared with the next species, and are generally widest at about a third or more of the total length from base to tip of beak.

67. **C. tenera** Dewey   Fig. 193

Map 368. Dry or usually moist open ground or woodland: meadows, swales, and fields; grassy clearings and trails in swampy woods; thickets and borders of woods and rivers. Quite local northward.

The perigynia ripen from May or early June to mid-July. They are often asymmetrical, at least on one side the widest part being from a fourth to a third of the total length. One collection has been seen in which the largest perigynia exceed 2 mm in breadth.

364. Carex tribuloides

365. Carex scoparia

366. Carex crawfordii

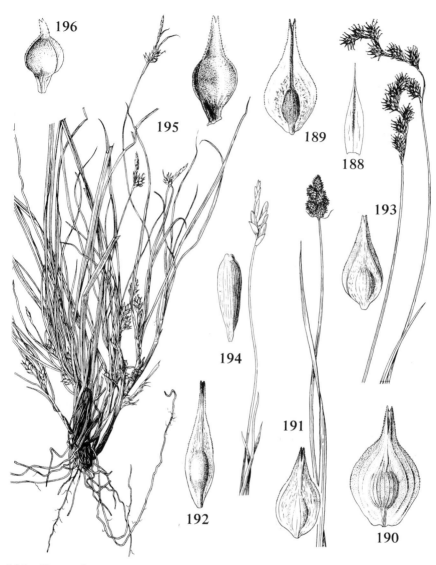

188. *Carex alata*, pistillate scale ×8
189. *C. suberecta*, perigynium (dorsal) ×6
190. *C. bicknellii*, perigynium (dorsal) ×6
191. *C. bebbii* ×⅔; perigynium (dorsal) ×6
192. *C. crawfordii*, perigynium (dorsal) ×8
193. *C. tenera* ×⅔; perigynium (dorsal) ×6
194. *C. leptalea* ×1⅓; perigynium (dorsal) ×6
195. *C. rugosperma* ×⅔; perigynium (var. *tonsa*) ×6
196. *C. umbellata*, perigynium ×6

14. POLYTRICHOIDEAE

68. C. leptalea Wahl. Fig. 194

Map 369. An easily recognized species of damp open woods and peaty ground, often abundant in bogs and coniferous swamps (or mixed hardwood–conifer stands), including marly situations, and particularly in open areas and along trails; also in wet grassy fields and clearings.

Plants with the perigynia more overlapping and longer (ca. 3.5−5 mm compared with the typical form in which they are not over 3.5 mm) have been segregated as var. *harperi* (Fern.) Weathb. & Grisc. − a variety of chiefly coastal distribution but known in Michigan from several counties from the Straits of Mackinac southward: Antrim, Cass, Cheboygan, Emmet, Ingham, Kalamazoo, Kent, Leelanau (mainland), Mackinac (mainland and Bois Blanc Island), Macomb, and St. Clair.

15. PHYLLOSTACHYAE

1. Pistillate scales mostly narrower than the perigynia, not concealing them; staminate scales about 4 or more; wider leaf blades 2−2.6 (3.5) mm broad; anthers ca. 0.5−1.1 mm long . . . . 69. C. jamesii
1. Pistillate scales much wider than the perigynia, embracing and nearly concealing them; staminate scales about 3 (usually hidden by upper perigynia); wider leaf blades (3) 3.5−5.5 mm broad; anthers ca. 1.3−1.6 mm long . . . . 70. C. backii

367. Carex normalis

368. Carex tenera

369. Carex leptalea

370. Carex jamesii

371. Carex backii

372. Carex rossii

69. **C. jamesii** Schw.

Map 370. Rich deciduous woods and ravines, especially beech–maple and floodplain woods.

Note: Specimens of *C. davisii, C. jamesii, C. lurida, C. squarrosa,* and *C. virescens* labeled by Farwell with the date September 20, 1888, and purporting to be from Keweenaw County, have been seen. Even if one should assume that the date is actually a date of accessioning or mounting in his herbarium (later interpreted by him in an unpublished manuscript as a collecting date), these species would be so far from the range otherwise known for them that the records are not considered reliable enough to map. A sheet of the same number (*698*) of *C. jamesii* in the Gray Herbarium is not dated and bears a note by Farwell: "Probably introduced in imported hay as also CC. Davisii, virescens, squarrosa, & lurida, as I found them only in the season of 1888." If it were not for Farwell's apparent labelling of his early specimens, at least, long after collecting and for his "known unmethodical habits in preserving and labelling his specimens" (Hermann, 1951, p. 484), one might be more inclined to accept these specimens – obviously not collected as late in the season as September – as one-time waifs. A Gray Herbarium sheet of the *C. virescens* number (*697*) is dated "Aug. 1888" without further comment. Old Farwell specimens of *C. blanda, C. bushii, C. normalis, C. prairea,* and *C. torreyi,* attributed to Keweenaw County, and of *C. angustior* attributed to Oakland County, although bearing different dates, are likewise rejected on phenological and phytogeographical grounds (see Hermann, 1951). A few other records somewhat more plausible in the light of current information on the distribution of the species (e. g., *C. bromoides* and *C. sprengelii*) have been at least tentatively accepted despite their understandable rejection by Hermann.

70. **C. backii** Boott   Fig. 197

Map 371. Dry rocky and sandy ground, both open and in second-growth woods, rather local.

16. MONTANAE

1. Pistillate spikelets on culms of varying length, at least some of the culms short (up to ca. 5 cm long) and partly hidden among the tufted leaf bases (fig. 195) [be sure to sample a population adequately to obtain some plants bearing elongate culms] ; some culms bearing only staminate or only pistillate spikelets; anthers ca. 1.5–2 mm long
2. Bract of the lowest non-basal pistillate spikelet leaf-like, equalling or exceeding the tip of the staminate spikelet; remnants of old leaves only slightly breaking into fibrous shreds at the base
3. Perigynia ca. 3–3.5 mm long, including a beak ca. 1 mm or a little longer (or only 0.5 mm in some basal spikelets); staminate spikelets ca. 5–12 mm long; rhizomes very stout . . . . 71. **C. rossii**
3. Perigynia ca. 2–2.7 (3) mm long, the beak ca. 0.4–0.6 mm; staminate spikelets 2–2.5 mm long, often very inconspicuous; rhizomes slender . . . . 72. **C. deflexa**
2. Bract of the lowest non-basal pistillate spikelet scale-like or bristle-like, not exceeding the staminate spikelet (or *all* spikelets often on short basal culms, but foliage and culms stiffer and much more scabrous than in nos. 71 & 72, which nearly always have some elongate culms); remnants of old leaves breaking into copious fibrous shreds at the base

4. Perigynia (3) 3.2–4 mm long, the beak 1.2–1.6 (2) mm, about half as long as
the body or longer (fig. 195) . . . . 73. **C. rugosperma**
4. Perigynia 2.5–2.9 mm long, the beak 0.4–0.9 mm, about a fourth to a third
(rarely half) as long as the body (fig. 196) . . . . 74. **C. umbellata**
1. Pistillate spikelets all on elongate culms (no short crowded basal ones); the
staminate spikelets on the same culms (in very rare cases the plants dioecious or
the terminal spikelet only partly staminate); anthers various
  5. Main body of perigynium, not including spongy-tapered base or beak, orbicular
to short-obovoid, about the same diameter as length; anthers 2.1–3.7 mm long;
plants either with the widest leaves 3–8 mm broad or with elongate shallow
rhizomes
    6. Widest leaves (at least the oldest dry ones) (3) 3.3–5 (8) mm broad; cauline
leaves above base of plant (when present on culm) usually with the ligule longer
than the width of the leaf; bract subtending the middle (and sometimes the
lowest) pistillate spikelet(s) ± scarious-lobed at base, the awn-like to leaf-like
usually green blade arising from between the lobes; staminate spikelet ca. 1–2
(2.5) mm thick; plants without elongate rhizomes . . . . 75. **C. communis**
    6. Widest leaves 1.5–2.9 (very rarely 3.5) mm broad; cauline leaves with ligule no
longer than the width; bracts subtending middle pistillate spikelets tapered to
apex, without an elongate awn-like or leaf-like blade (the lowermost bract often
green but seldom lobed); staminate spikelet ca. 2–3.5 (5) mm thick; plants with
stout, shallow elongate rhizomes with fibrous sheaths
      7. Beak of perigynium long (1–1.6 mm), half or more as long as the body . . . .
. . . . 76. **C. lucorum**
      7. Beak of perigynium very short (less than 1 mm, usually no more than 0.5 mm),
much less than half as long as the body . . . . 77. **C. pensylvanica**
  5. Main body of perigynium ± elliptic (to slightly obovoid or oblong), definitely
longer than thick; anthers ca. 1.3–2.2 (2.5) mm long; plants with mostly narrow
leaves and no elongate rhizomes   (and otherwise not fitting either lead of
couplet 6)
    8. Bodies of mature perigynia mostly distinctly exceeding their scales; beak of
perigynium ca. 0.4–0.7 mm long
      9. Perigynia ca. 2–2.7 (3) mm long, minutely puberulent to short-hairy; culms
very slender (seldom over 0.4 mm thick) and mostly surpassed by the leaves .
. . . . 72. **C. deflexa**
      9. Perigynia ca. 3–3.8 mm long, definitely short-hairy; culms usually 0.5 mm or
more in thickness and surpassing the leaves . . . . 78. **C. peckii**
    8. Bodies of mature perigynia about as long as their scales or even slightly shorter;
beak of perigynium ca. 0.5–1.4 mm long
      10. Widest leaves (at least the oldest dry. ones) (3) 3.3–5 (8) mm broad; bract
subtending the middle (and sometimes also the lowest) pistillate spikelet(s) ±
scarious-lobed at base, the blade awn-like to leaf-like, usually green, arising
from between the lobes . . . . 75. **C. communis**
      10. Widest leaves not over 3 mm broad; bracts tapering to apex, rarely with a
green awn
        11. Culms weak, usually loosely spreading or arching and usually shorter than
the leaves; at least the middle and upper staminate scales strongly narrowed
to a sharp, often slightly scabrous tip [use 15X lens!] , the midrib prominent
to the end or even slightly excurrent . . . . 79. **C. emmonsii**
        11. Culms firm, ± erect, usually surpassing the leaves; staminate scales ± obtuse
or broadly acute, the midrib usually weak or absent just below the tip. . . .
. . . . 80. **C. artitecta**

71. **C. rossii** Boott

Map 372. Rocky bluffs, summits, and Lake Superior shore in Keweenaw County.

72. **C. deflexa** Hornem.

Map 373. Moist, often sandy soil (or thin soil over rock) in open ground, on creek borders, or more often on knolls or at borders of cedar swamps and trails and roads in mixed or coniferous woods.

73. **C. rugosperma** Mack.   Fig. 195

Map 374. Generally in dry sandy (sometimes rocky) ground, including dunes and oak, aspen, or pine woodlands. Both varieties apparently occur more or less throughout the state in the same habitats and sometimes together, although var. *tonsa* has not yet been recorded from Isle Royale.

Some authors have considered this to be the true *C. umbellata* Schk. ex Willd. (see note under the next species). Plants with thick rigid leaves and perigynia glabrous (or nearly so) may be referred to var. *tonsa* (Fern.) E. Voss [*C. tonsa* (Fern.) Bickn.]. Those who unite this species with the next would call these plants *C. umbellata* var. *tonsa* Fern. They are often growing with pubescent-fruited plants, as evidenced by the mixed collections in herbaria.

74. **C. umbellata** Willd.   Fig. 196

Map 375. Sandy or gravelly ground, rock crevices, moist banks and sandy excavations; gravelly shores and coniferous thickets near the Great Lakes.

Those who consider the epithet *umbellata* to belong to the preceding species call this one *C. abdita* Bickn. Some authors have combined the two as *C. umbellata*, but this species in a narrower sense seems fairly distinctive in its small, short-beaked fruit.

75. **C. communis** Bailey

Map 376. Deciduous woods of all kinds except the wettest, especially on sandy soils, in disturbed areas, along roads, and in clearings; in thin, rocky mixed

373. Carex deflexa

374. Carex rugosperma

375. Carex umbellata

woods in the northern part of the state, toward Lake Superior; sometimes under hemlock or red pine, but mostly not associated with conifers.

Plants very rarely occur with the terminal spikelet entirely pistillate or with only a few staminate flowers; these may be called f. *gynandra* (Farw.) E. Voss (TL: Livonia [Wayne County]). Others occur with only staminate spikelets, the plants thus (and very rarely) dioecious. Specimens with the lowest bract leaf-like and tending to have wider leaves and a less conspicuous staminate spikelet have been called var. *wheeleri* Bailey (TL: Ionia County).

76. **C. lucorum** Link

Map 377. Usually in dry sandy aspen, pine, or oak woods (sometimes with *C. pensylvanica*) and roadsides; recorded once from a tamarack swamp (Crawford County).

Sometimes treated as a variety [var. *distans* Peck] of the next species, but distinctive in its long beak.

77. **C. pensylvanica** Lam.

Map 378. Particularly common in dry, usually sandy open ground or under jack pine, oaks, or aspen; dune ridges and thin rocky soil; also in deciduous woods, even on somewhat low ground.

Plants with the terminal spikelet almost wholly pistillate may be called f. *androgyna* Hermann (TL: East Lansing, Ingham County). The pistillate spikelets are often small (few-flowered) and inconspicuous in this species, in comparison with the staminate spikelets (or with *C. communis*). Some very immature specimens placed here by their thick staminate spikelets, rhizomes, narrow leaves, and scale-like bracts may actually be *C. lucorum*.

A single Michigan collection has been seen of var. *digyna* Boeckl. [*C. heliophila* Mack.] : Kalamazoo County, Sunset Lake, east bank (*Rapp 6957*, May 26, 1941, WMU). The perigynia of this variety have broad (ca. 1.8 mm) suborbicular bodies, while in typical var. *pensylvanica* they are smaller (ca. 1.5 mm wide) and often obscurely angled.

376. Carex communis

377. Carex lucorum

378. Carex pensylvanica

78. **C. peckii** Howe

Map 379. Usually in beech—maple—hemlock woods, but also in ± open sandy or rocky woods and cedar swamps.

79. **C. emmonsii** Dewey

Map 380. Moist to dry usually sandy open ground and wooded dunes, often near lakes or marshes.

This species and the next are often distinguished only with great difficulty. The perigynia in our specimens of *C. emmonsii* tend to be a little shorter (2–2.8 mm long) than in *C. artitecta* (2.5–3.4 mm), but this may not be a valid distinction, as the measurements are not in accord with figures from elsewhere in the range of the species.

80. **C. artitecta** Mack.

Map 381. Woods and clearings, especially on sand dunes; less often in low ground. Probably overlooked by collectors, as well as being local.

17. SCIRPINAE

81. **C. scirpoidea** Michaux   Fig. 198

Map 382. Crevices and thin soil over rock and gravelly calcareous shores, in areas at least seasonally damp: Thunder Bay Island, Alpena County; Detour and Drummond Island, Chippewa County; and near Eagle River, Keweenaw County.

The scales of the pistillate spikelets are dark brown to purplish; those of the staminate spikelets may be nearly white. The spikes are normally solitary, but pistillate ones may sometimes be branched. Plants with slender incurved leaf blades (in contrast to the usual flat ones) have been called var. *convoluta* Kuek. (TL: Thunder Bay Island), but plants from the type locality vary a great deal in leaf width and flatness (e. g., *Voss 13289*, MICH, UMBS, MSC, BLH, GH, NY, CAN, WIS).

379. Carex peckii

380. Carex emmonsii

381. Carex artitecta

18. DIGITATAE

1. Terminal spikelet pistillate at base; basal spikelets usually present, on long capillary peduncles; pistillate scales abruptly truncate and awned (fig. 199); anthers (1.6) 2–3 mm long . . . . 82. C. pedunculata
1. Terminal spikelet usually entirely staminate; basal spikelets not present; pistillate scales obtuse or acuminate, not awned; anthers various
    2. Staminate spikelet ca. 4–6 mm long; pistillate spikelets less than 10 mm long; pistillate scales obtuse, minutely ciliate, distinctly shorter than the perigynia; anthers ca. 1–1.5 mm long . . . . 83. C. concinna
    2. Staminate spikelet ca. 10–22 mm long; pistillate spikelets (often staminate at their tips) ca. (8–) 10 mm long; pistillate scales mostly acute to acuminate, glabrous, and equalling or exceeding the perigynia; anthers ca. (1.7) 2–3.5 mm long . . . . 84. C. richardsonii

82. C. pedunculata Willd.   Fig. 199

Map 383. Most characteristic and common in beech–maple woods, easily recognized by its early fruiting (May), red bases, narrow leaves, and lack of strong rhizome; also in mixed woods and under conifers in both dryish and moist sites (usually on hummocks in swampy woods) and in open rocky ground.

83. C. concinna R. Br.   Fig. 200

Map 384. Very local, edges of cedar and balsam thickets near gravelly calcareous shores of the northern ends of Lakes Michigan and Huron, associated with such plants as *C. eburnea*, *C. capillaris*, *Linnaea borealis*, and *Arctostaphylos uva-ursi*. Locally abundant on Summer Island, Delta County.

84. C. richardsonii R. Br.

Map 385. Very local, flourishing for only a few days in spring or early summer. Sandy open ground, bluffs, borders of oak woods in southern Michigan; northward, known from Huron Bay, Drummond Island (*Hagenah 6996* in 1970), moist mossy openings among cedars on shore of Big Shoal Bay, Drummond Island (*Hiltunen & Hayes 3371* in 1961, WUD), and the gravelly summit (bare, grassy, or with *C. pensylvanica*) of West Bluff, Keweenaw Co. (several collectors).

382. Carex scirpoidea   383. Carex pedunculata   384. Carex concinna

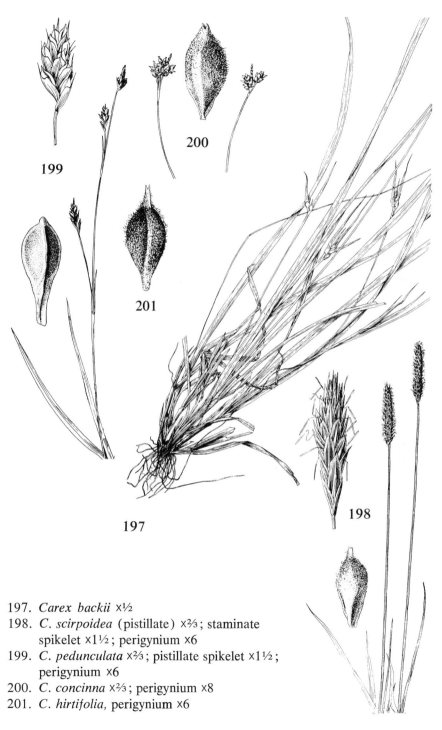

197. *Carex backii* ×½
198. *C. scirpoidea* (pistillate) ×⅔; staminate
     spikelet ×1½; perigynium ×6
199. *C. pedunculata* ×⅔; pistillate spikelet ×1½;
     perigynium ×6
200. *C. concinna* ×⅔; perigynium ×8
201. *C. hirtifolia,* perigynium ×6

85. **C. hirtifolia** Mack.   Fig. 201

Map 386. Rich deciduous woods (beech—maple, floodplains, moist ravines) and thickets.

20. ALBAE

86. **C. eburnea** Boott   Fig. 202

Map 387. Usually under cedar (also fir and white spruce) on damp sandy or calcareous gravelly soil near the shores of Lakes Michigan and Huron (very rarely Superior), forming dense carpets; also inland in boggy cedar thickets, on logs in wet conifer swamps, and marly, clay, or rocky bluffs and banks.

21. BICOLORES

1. Mature perigynia golden-orange when fresh (drying dark brown or, especially if immature, ± white-pulverulent); terminal spikelets mostly all staminate (occasionally with a very few perigynia); pistillate scales ± loosely spreading, distinctly shorter than the mature perigynia (usually averaging 3/4 or less as long), most of them acute to cuspidate . . . . 87. **C. aurea**
1. Mature perigynia white-pulverulent when fresh; terminal spikelets usually staminate at base only, with several to numerous perigynia apically; pistillate scales ± appressed, nearly (averaging about 3/4) to quite as long as the perigynia, most of them blunt to acute . . . . 88. **C. garberi**

87. **C. aurea** Nutt.

Map 388. In almost all kinds of moist open ground, as well as bogs, swamps of cedar or tamarack, and low woods and thickets.

Striking in its golden, rather fleshy perigynia when ripe — readily stripped for an unusual, if sparse, nutty nibble. The type locality is "on the shores of Lake Michigan" — perhaps the northern end of the lake, in Michigan, though possibly in Wisconsin, where Nuttall also traveled in 1810.

385. Carex richardsonii

386. Carex hirtifolia

387. Carex eburnea

88. **C. garberi** Fern.   Fig. 203

Map 389. Wet sandy, gravelly, or marly shores, limestone pavements, inter-dunal flats, and edges of cedar thickets, chiefly near the Great Lakes although at some inland lakes; in rock crevices along Lake Superior.

Dry or immature material is sometimes difficult to distinguish from *C. aurea* and a few collections are quite intermediate. In characteristic *C. garberi* the perigynia are more granular, densely crowded, and strongly overlapping than in *C. aurea*.

22. PANICEAE

1. Perigynium with a beak ca. 1 mm long . . . . 89. **C. vaginata**
1. Perigynium beakless or with a short apiculus
  2. Leaves very glaucous, becoming channeled or 3-angled . . . . 90. **C. livida**
  2. Leaves slightly if at all glaucous, with blades flat (except sometimes at the very end)
    3. Base of plant with strongly reddened, essentially bladeless sheaths; rhizomes rather stout, shallow or very superficial, also reddish; perigynia ± 2-ranked; plants of rich woods . . . . 91. **C. woodii**
    3. Base of plant usually only brown or pale, occasionally reddish, the sheaths with well developed blades; rhizomes deep, slender, and paler; perigynia 3–6-ranked; plants of low ground . . . . 92. **C. tetanica**

89. **C. vaginata** Tausch   Fig. 204

Map 390. Mossy swamps of cedar and mixed conifers, quite local here at the southern edge of its range; a circumpolar species.

The American plant is sometimes distinguished from the Eurasian one as *C. saltuensis* Bailey.

90. **C. livida** (Wahl.) Willd.

Map 391. In bogs and wet peaty ground with sphagnum.

Plants of our region (the southern part of the range) were once considered a distinct species, *C. grayana* Dewey. If now considered a named variety of the circumpolar *C. livida*, the correct name is var. *radicaulis* Paine.

388. Carex aurea   389. Carex garberi   390. Carex vaginata

91. C. woodii Dewey

Map 392. Rich, often low, deciduous woods, especially beech–maple; an early species, generally fruiting about mid-May or even earlier.

See note under *C. ormostachya* (no. 100).

92. C. tetanica Schk.

Map 393. Low marshy or boggy ground, meadows, shores, wet prairies, damp woodland, etc., often in marly places. Very local northward.

Some plants in the southernmost Lower Peninsula have been referred to var. *meadii* (Dewey) Bailey [*C. meadii* Dewey], but the distinction is not clear and there are a number of transitional collections. The perigynia of var. *meadii* tend to be larger (3–4.2 mm long) in thicker (4.5–7 mm) and often more densely flowered spikelets. In var. *tetanica* the larger perigynia are 2.5–3.5 mm long, in spikelets ca. 3–5.5 mm thick. At least some perigynia in var. *meadii* have a vague and very short strongly outcurved beak (the orifice around the style thus bent to one side), while in var. *tetanica* the perigynia are ± evenly and broadly tapered or rounded at the apex (in *C. woodii*, they are narrowed).

23. LAXIFLORAE

1. Larger leaf blades (especially on sterile shoots) mostly 1–4 cm broad (or if largest blades as narrow as 8 mm, the bases of plants and the staminate scales strongly reddish); anthers mostly 2.5–4.5 mm long (shorter in *C. albursina*)
   2. Bases of plants and staminate scales strongly tinged with reddish; perigynia 4–6.5 mm long
      3. Sheaths of cauline bracts and leaves bladeless or nearly so; perigynia (3.5) 4–5 mm long . . . . 93. **C. plantaginea**
      3. Sheaths of cauline bracts and leaves with flat green blades; perigynia (3.7) 5–6.5 mm long . . . . 94. **C. careyana**
   2. Bases of plants and staminate scales pale or brownish, not reddish; perigynia 2.6–4.2 mm long
      4. Angles of culm very narrowly winged; pistillate scales broadly obtuse or truncate, scarcely toothed at apex; staminate spikelet sessile or nearly so; edges of sheaths usually ± minutely serrulate or roughened; anthers ca. 1.5–2.2 (very rarely 2.8) mm long . . . . 95. **C. albursina**
      4. Angles of culm not (or only rarely) winged; pistillate scales usually narrowly acuminate, short-awned, or cuspidate at apex; staminate spikelets short- or long-peduncled (rarely nearly sessile); edges of sheaths smooth or serrulate; anthers various
         5. Lower pistillate spikelets on peduncles scarcely exserted beyond the summit of the sheath; perigynia ± sharply triangular with flattish sides; edges of sheaths and both surfaces of leaf blades (not margins) perfectly smooth . . 96. **C. platyphylla**
         5. Lower pistillate spikelets on peduncles long-exserted beyond the summit of the sheath; perigynia ± rounded-triangular with swollen sides; edges of sheaths and surfaces of leaf blades smooth or roughened . . . . [go to couplet 8]
1. Larger leaf blades less than 1 cm broad (less than 8 mm if strongly reddish at the base); anthers mostly (1.5) 1.8–3.5 (4.5) mm long
   6. Perigynia ± sharply triangular with flattish sides, short-tapering at the base; lower pistillate spikelets on elongate filiform spreading (or even drooping) peduncles

7. Pistillate spikelets (except sometimes the upper ones) with 1–2 staminate
flowers at the base; leaf blades rather light glaucous green, the wider ones 5–9
(very rarely one 12) mm broad . . . . 97. **C. laxiculmis**
7. Pistillate spikelets without staminate flowers at the base; leaf blades deep or
bright green, the wider ones 2.5–4 (5) mm broad . . . . 98. **C. digitalis**
6. Perigynia ± rounded-triangular with swollen sides, long-tapering (or nearly
stipitate) at the base; lower pistillate spikelets usually on erect or ascending
peduncles
8. Sides of perigynia with at most 1 main nerve, otherwise nerveless or each with
up to 6 obscure nerves (fig. 205); perigynium with a straightish or slightly bent
short beak . . . . 99. **C. leptonervia**
8. Sides of perigynia each with 7 or more conspicuous nerves; perigynium with
straightish or strongly bent beak
9. Angles of bract sheaths smooth or nearly so (granular in *C. ormostachya*); beak
of perigynium usually straight or slightly bent
10. Perigynia mostly twice as long as wide, or shorter, abruptly contracted to a
very short bent tip . . . . 100. **C. ormostachya**
10. Perigynia mostly more than twice as long as wide, tapered to the straightish
beak . . . . 101. **C. laxiflora**
9. Angles of bract sheaths minutely ciliate-serrulate; beak of perigynium strongly
bent
11. Widest leaves 8 mm or more broad; pistillate scales broadly obtuse or
truncate, at most scarcely toothed at apex; staminate spikelet sessile or
nearly so . . . . 95. **C. albursina**
11. Widest leaves often less than 8 mm broad; pistillate scales acuminate, awned,
or cuspidate; staminate spikelet sessile or peduncled
12. Staminate spikelet peduncled, elevated above pistillate spikelets and
(usually) ends of the bracts; pistillate spikelets scattered; staminate and
sometimes pistillate scales often strongly flushed with orange-brown; bases
of plants usually red-tinged . . . . 102. **C. gracilescens**
12. Staminate spikelet sessile or at most short-peduncled, exceeded by one or
more of the bracts (fig. 210); upper pistillate spikelets ± crowded; scales
not (or only slightly) flushed with orange-brown; bases of plants not
red-tinged . . . . 103. **C. blanda**

93. **C. plantaginea** Lam.   Fig. 208

Map 394. Usually in rich deciduous woods, especially beech–maple–
hemlock stands, but sometimes in lower ground, as under cedar. Collected

391. Carex livida   392. Carex woodii   393. Carex tetanica

"among the pines" in Clare County by the U. S. Government Surveyor in 1849 (MSC).

Easily recognized by the broad evergreen leaves, red bases, and naked culms.

94. **C. careyana** Dewey

Map 395. Rich deciduous woods.

95. **C. albursina** Sheldon   Fig. 211

Map 396. Very characteristic of rich, often moist beech–maple and mixed deciduous woods; occasionally under oaks or oak–hickory.

Sometimes treated as a wide-leaved variety of *C. laxiflora*. Wide-leaved (blades to 2.5 cm broad) plants of the latter as here treated occasionally occur, but may generally be distinguished from *C. albursina* by some or all of the following tendencies: longer-awned or cuspidate scales; longer anthers; longer and straighter beak; smooth bract sheaths (except in the usually narrow-leaved var. *serrulata*); and more conspicuous peduncled staminate spikelet.

96. **C. platyphylla** Carey

Map 397. The only Michigan collections known are from wooded slopes of the Black River valley west of Jeddo, St. Clair County (*Hermann 7372* in 1936, MICH, NY; *Hiltunen 4335* in 1970, WUD, MICH).

Grows with *C. plantaginea* and *C. albursina*, both also broad-leaved but

394. Carex plantaginea

395. Carex careyana

396. Carex albursina

397. Carex platyphylla

398. Carex laxiculmis

399. Carex digitalis

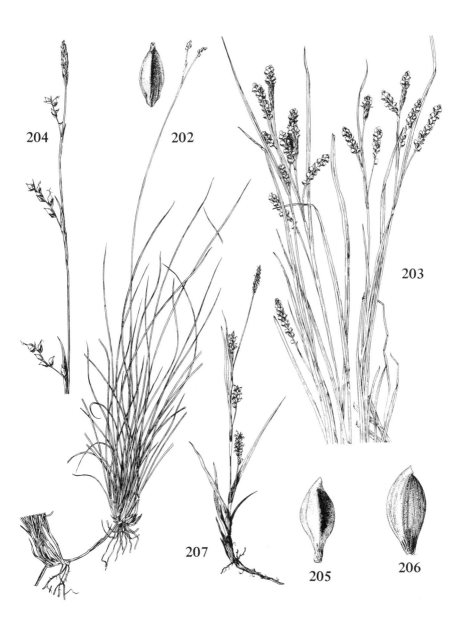

202. *Carex eburnea* x⅔; perigynium x8
203. *C. garberi* x⅔
204. *C. vaginata* x⅔
205. *C. leptonervia*, perigynium x6
206. *C. blanda*, perigynium x6
207. *C. crawei* x⅔

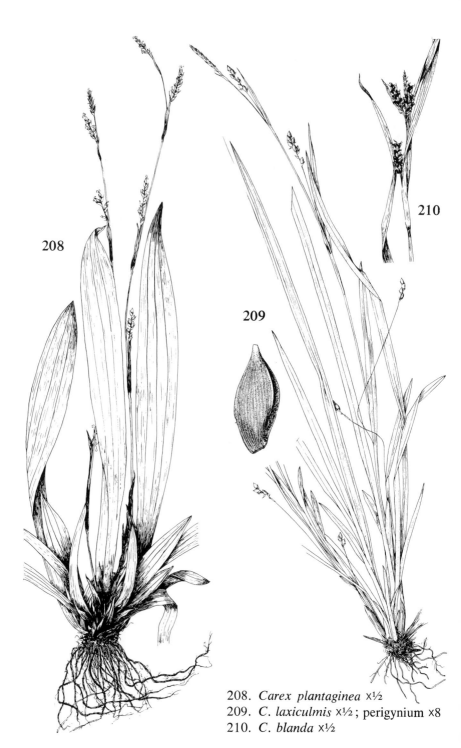

208. *Carex planta ginea* x½
209. *C. laxiculmis* x½; perigynium x8
210. *C. blanda* x½

neither of these having the strongly glaucous appearance of *C. platyphylla*, which also differs from *C. albursina* in its evergreen leaves, the old ones shriveling at the tips, and from *C. plantaginea* in the absence of red color at the base. The angles of the perigynia are relatively sharp and the sides flat in *C. platyphylla*.

97. C. laxiculmis Schw. Fig. 209
Map 398. Usually in rich deciduous woods, often near streams.

Plants with the bract sheaths and culms more scabrous and the leaves less pale have been referred to var. *copulata* (Bailey) Fern. (TL: Lansing [Ingham County]), sometimes considered to be a hybrid with the next species.

98. C. digitalis Willd.
Map 399. Dry (oak, etc.) or more often rich moist deciduous woods.

99. C. leptonervia Fern. Fig. 205
Map 400. Usually in rich deciduous or mixed woods, but occasionally in cedar swamps (sometimes marly) and bogs; apparently continues to thrive in disturbed areas and clearings.

The edges of the bract sheaths are usually ± strongly but minutely serrulate. The pistillate scales are occasionally obtuse, although usually at least some of them are short-awned.

100. C. ormostachya Wieg.
Map 401. In beech—maple and hemlock—hardwoods, sandy aspen and red pine woodland, roadside clearings and open rocky woods, rarely in swampy ground.

C. woodii and *C. tetanica* might possibly run here or to *C. laxiflora* in the keys if they were not recognized as belonging to the Paniceae. Even if the prominent superficial rhizomes of *C. woodii* are not collected, that species differs in its very strongly red-colored basal sheaths. *C. tetanica* has the sides of the rachis of the pistillate spikelets minutely papillose-granular, while in the somewhat similar Laxiflorae the rachis is smooth, at least on the sides.

400. Carex leptonervia

401. Carex ormostachya

402. Carex laxiflora

101. **C. laxiflora** Lam.

Map 402. Generally in rich deciduous woods and ravines, occasionally in more open sandy woodland or thickets, especially where cut-over.

The bract sheaths are somewhat serrulate in the very uncommon var. *serrulata* Hermann.

102. **C. gracilescens** Steudel

Map 403. Usually in rich beech—maple woods, moist oak or oak—hickory woods, or other low woods and thickets; rarely in open meadows.

Intermediates with the next species are not uncommon. The perigynia of *C. gracilescens* are often somewhat shorter but with longer beak than those of *C. blanda*, but there is considerable overlap in any measurement.

103. **C. blanda** Dewey   Figs. 206, 210

Map 404. Low or upland woods, especially rich beech—maple stands and ravines, less common under oaks or in meadows and thickets. A collection (*Farwell 738* in 1890, GH) purporting to come from Keweenaw County, without further locality data, is not mapped, it being suspicious as the only Upper Peninsula record (see also note under *C. jamesii*, no. 69).

Plants intermediate between this and *C. laxiflora* have been collected in Kent County (*Bazuin 1121*, MSC) and may represent a hybrid, as both of the species were apparently growing with it.

24. GRANULARES

1. Staminate spikelet long-peduncled, elevated above summit of uppermost pistillate spikelets; lowest pistillate spikelet usually on a separate basal peduncle; culms mostly solitary from elongate rhizomes; widest leaves 1.5–4 mm broad . . 104. **C. crawei**
1. Staminate spikelet sessile or nearly so; lowest pistillate spikelet not on a basal peduncle; culms clumped, without elongate rhizomes; widest leaves 4.5–10 (13) mm broad . . . . 105. **C. granularis**

104. **C. crawei** Dewey   Fig. 207

Map 405. Sandy or gravelly hummocks and ridges (often at the edge of

403. Carex gracilescens   404. Carex blanda   405. Carex crawei

cedar), wet marly sands and beach pools, and limestone pavements near the shores of the Great Lakes; inland very locally in marly bogs, calcareous cedar thickets, and prairie-like marshes. Often associated with *C. garberi* and *Eleocharis pauciflora*.

105. C. granularis Willd.   Fig. 212

Map 406. Low open ground of all kinds (moist fields, shores, borders of swamps, trails and clearings in wet woods); also in moist woods, coniferous swamps, and crevices of rock shores along Lake Superior.

Certain plants have been segregated as var. *haleana* (Olney) Porter or *C. haleana* Olney, supposedly distinguished by slightly less inflated, more ellipsoid and narrow perigynia subtended by more prominently awned scales. The distinctions seem not at all decisive nor correlated with each other, and two species are not here maintained. Typical *C. granularis* is considered to be restricted to the southern half of the Lower Peninsula, while var. *haleana* occurs throughout the range indicated by the map.

Rare specimens of *C. gracillima* with the terminal spikelet entirely staminate might be keyed here, but can be distinguished from *C. granularis* by the long peduncle of the terminal spikelet, the more slender spikelets drooping on elongate peduncles at maturity, and the strongly reddened bases of the plant.

25. OLIGOCARPAE

1. Leaf sheaths strongly hispidulous; perigynia ca. 4–5.5 mm long; plants brownish at
  base . . . . 106. C. hitchcockiana
1. Leaf sheaths glabrous; perigynia ca. 3.5–4 mm long; plants reddish at base . . . .
  . . . . 107. C. oligocarpa

106. C. hitchcockiana Dewey   Fig. 213

Map 407. Rich deciduous woods (beech–maple and alluvial woods), often in slightly disturbed areas.

107. C. oligocarpa Willd.

Map 408. Rich deciduous woods.

406. Carex granularis   407. Carex hitchcockiana   408. Carex oligocarpa

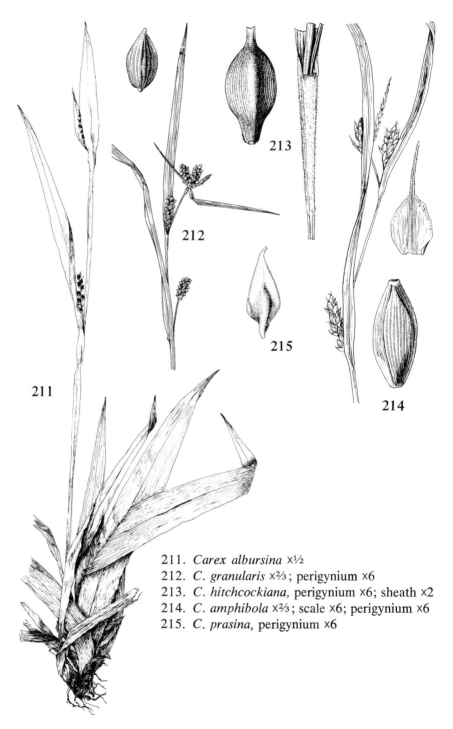

211. *Carex albursina* ×½
212. *C. granularis* ×⅔; perigynium ×6
213. *C. hitchcockiana,* perigynium ×6; sheath ×2
214. *C. amphibola* ×⅔; scale ×6; perigynium ×6
215. *C. prasina,* perigynium ×6

26. GRISEAE

1. Larger perigynia ca. 2.5–3.5 mm long and 1.5–1.7 mm thick; staminate spikelet long-peduncled, generally elevated well above the summit of the uppermost pistillate spikelet; peduncles and axis of inflorescence scabrous . . . . . 108. **C. conoidea**
1. Larger perigynia ca. 4–5 mm long and 2–2.5 mm thick; staminate spikelet sessile or nearly so; peduncles and axis of inflorescence essentially smooth . . 109. **C. amphibola**

108. **C. conoidea** Willd.

Map 409. Low ground, moist grassy places, etc. The collection mapped from Keweenaw County (Cedar Creek, in meadows, *Farwell 12407* in 1940, BLH) is so far out of range as to be suspicious, but there is no evidence in Farwell's field notes to question it.

An occasional (especially immature) specimen of *C. tetanica* might be keyed here if the perigynia are impressed-nerved and not clearly tapered at the base; *C. conoidea* has strongly rough-awned pistillate scales and a very scabrous axis in the inflorescence.

109. **C. amphibola** Steudel   Fig. 214

Map 410. Wooded riverbanks and floodplains, swampy or upland woods, less often in wet open ground.

Specimens from our part of the country belong to var. *turgida* Fern. (long known as *C. grisea*).

27. GRACILLIMAE

1. Sheaths and leaf blades (at least below) ± softly pubescent (sometimes sparsely so)
   2. Upper pistillate scales with a distinct prolonged awn, nearly equalling or exceeding the perigynia; only the terminal spikelet staminate at base (the lateral entirely pistillate) . . . . 110. **C. davisii**
   2. Upper pistillate scales merely acute (or at most with a tip less than 0.5 mm long), distinctly shorter than the perigynia; lateral as well as terminal spikelets usually with a few staminate flowers at base . . . . 111. **C. formosa**
1. Sheaths and blades glabrous
   3. Perigynia strongly angled, gradually tapering into a beak often nearly as long as body (fig. 215); bract of lowest pistillate spikelet sheathless or with sheath up to 1.2 cm long; terminal spikelet entirely staminate or at most with a few perigynia . . . . 112. **C. prasina**
   3. Perigynia obscurely angled or nearly terete, essentially beakless; bract of lowest spikelet with sheath 1.5–8 cm or more in length; terminal spikelet staminate at base, pistillate toward apex . . . . 113. **C. gracillima**

110. **C. davisii** Schw. & Torrey

Map 411. Rich floodplain woods and banks; also collected "along roadside" in Monroe County, presumably near a stream.

111. **C. formosa** Dewey   Fig. 216

Map 412. Rich beech–maple woods and ravines; also meadows.

112. **C. prasina** Wahl.   Fig. 215

Map 413. Rich deciduous woods, especially in low places and along streams.

113. **C. gracillima** Schw.

Map 414. A common sedge of woods, especially beech—maple, moist oak or oak—hickory stands, edges of woodland ponds or streams, and mixed swampy woods; northward, also in coniferous swamps (cedar, fir, spruce, tamarack), particularly along trails and clearings, and at borders of bogs.

28. SYLVATICAE

1. Leaf sheaths and blades (at least toward the base) ± hairy; pistillate spikelets 1–2.5 cm long . . . . 114. **C. castanea**
1. Leaf sheaths and blades glabrous (at most the lowermost bladeless sheaths minutely hispidulous); pistillate spikelets mostly (2) 2.5–6.5 cm long
  2. Perigynia short-stipitate, the achene within sessile or nearly so; broadest leaves (5) 6–10 (12) mm wide; pistillate scales mostly awned or cuspidate . . . . . 115. **C. arctata**
  2. Perigynia sessile but the achene within on a definite short stipe ca. 0.5–1 mm long; broadest leaves 2.5–4.5 (5.5) mm wide; pistillate scales mostly not awned . . . . 116. **C. debilis**

114. **C. castanea** Wahl.

Map 415. Cedar and cedar—fir woods and borders, bogs and wet meadows, rich moist deciduous or mixed woods, jack pine plains, rock pools along Lake

409. Carex conoidea

410. Carex amphibola

411. Carex davisii

412. Carex formosa

413. Carex prasina

414. Carex gracillima

Superior. Especially characteristic of margins of mixed coniferous woods and here often found with *C. capillaris* which it resembles in general habit (although larger and pubescent).

115. **C. arctata** Boott   Fig. 217

Map 416. Especially characteristic of rich deciduous and beech–maple–hemlock woods, including wooded dunes; also in rocky open woods, mixed woods, and under conifers such as cedar, pine, or hemlock, particularly in moist ground.

The perigynia tend to be more strongly nerved in this species than in the next. Occasionally the lowermost sheaths are minutely hispidulous or at least roughened, but not pubescent as in *C. castanea*. Occasional specimens with the slender long spikelets of *C. arctata* and sheaths nearly as pubescent as in *C. castanea* have generally been assumed to be a hybrid of the two, called *C. × knieskernii* Dewey. Specimens of this have been collected in Michigan in Emmet, Keweenaw, and Ontonagon counties. *C. arctata* appears to hybridize occasionally with other species as well.

116. **C. debilis** Michaux

Map 417. In moist ground of all sorts, open or (usually) more or less wooded (hardwoods rather than conifers).

Specimens from our region are var. *rudgei* Bailey.

29. CAPILLARES

117. **C. capillaris** L.   Fig. 218

Map 418. Moist, often calcareous and/or sandy ground at edges of conifer (especially cedar and fir) woods and thickets along the northern shores of Lakes Michigan and Huron; also along roads, trails, clearings, and similar openings – even on mossy logs in rivers and streams – through such woods and boggy ground; on rock shores of Lake Superior.

415. Carex castanea

416. Carex arctata

417. Carex debilis

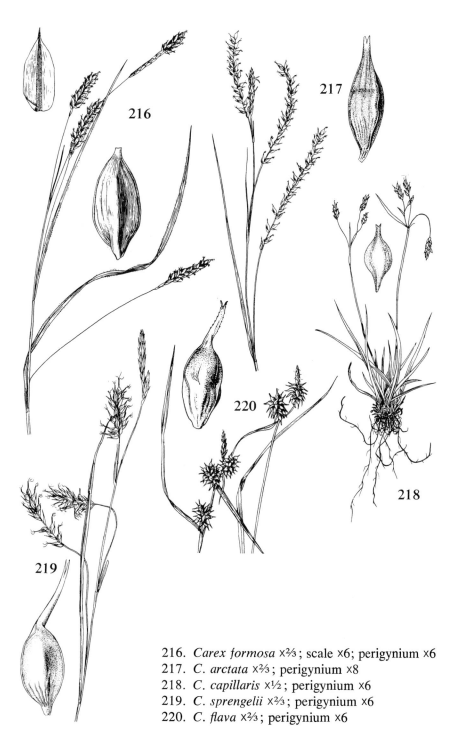

216. *Carex formosa* ×⅔; scale ×6; perigynium ×6
217. *C. arctata* ×⅔; perigynium ×8
218. *C. capillaris* ×½; perigynium ×6
219. *C. sprengelii* ×⅔; perigynium ×6
220. *C. flava* ×⅔; perigynium ×6

118. **C. sprengelii** Sprengel   Fig. 219

Map 419. Moist to dry woods and thickets, often on riverbanks; occasionally in open meadows or marshy ground.

31. EXTENSAE

1. Larger perigynia ca. 2–3 mm long, horizontally spreading, the beak about a fourth to nearly half as long as the body . . . . 119. **C. viridula**
1. Larger perigynia ca. (3) 3.5–6.2 mm long, at least the beaks becoming conspicuously deflexed on lower half of spikelet, the beak nearly or fully half as long as the body
  2. Perigynia (3) 3.5–4.5 (5) mm long, the beak smooth; pistillate scales mostly not conspicuously colored; wider leaves 1.5–3.5 mm broad . . . . 120. **C. cryptolepis**
  2. Perigynia (4.5) 5–6.2 mm long, the beak (at least on many of them) very minutely and sparsely serrulate toward the tip (fig. 220); pistillate scales at maturity strongly flushed with shiny brown or reddish color, hence conspicuous in the spikelet; wider leaves (2.5) 3–5 mm broad . . . . 121. **C. flava**

119. **C. viridula** Michaux

Map 420. A very common species of sandy to rocky, often marly, open or marshy shores, beach pools, and interdunal swales; occasionally in bogs, swampy clearings, and marshes elsewhere.

Usually a short plant, but sometimes half a meter or more in height. A late-flowering phase (late July–October) generally with more crowded and greener spikelets is f. *intermedia* (Dudley) Hermann.

120. **C. cryptolepis** Mack.

Map 421. Sandy (sometimes marly or mucky) shores, ditches, swales, and marshy ground generally; occasionally in bogs.

121. **C. flava** L.   Fig. 220

Map 422. Wet shores, marshes, meadows, and ditches; frequently in sphagnum bogs and coniferous swamps (cedar, tamarack), often in marly situations.

Very rarely all the perigynia on a plant may have smooth beaks (only two

418. Carex capillaris

419. Carex sprengelii

420. Carex viridula

such collections have been seen from Michigan), but the large perigynia in rounded spikelets should help to place such abnormal individuals. Some collections from Cheboygan County have been determined by Hermann as *C. flava* × *C. viridula*; Fernald & Pease collections from Schoolcraft County are similar. In these, the large scales and serrulate beaks are as in *C. flava*, but the perigynia are distinctly smaller.

32. VIRESCENTES

1. Perigynia pubescent; terminal spikelet pistillate at apex, staminate at base
  2. Pistillate spikelets linear-cylindric, the lower ones 1.8–3.5 cm long; anthers 1.5–2.5 mm long; culms usually surpassing the leaves . . . . 122. **C. virescens**
  2. Pistillate spikelets ellipsoid or thick-cylindric, the lower ones (0.5) 0.7–1.5 (2) cm long; anthers 0.7–1.3 (1.6) mm long; culms often shorter than the leaves
  . . . . 123. **C. swanii**
1. Perigynia glabrous; terminal spikelet staminate or pistillate at apex
  3. Terminal spikelet entirely staminate; perigynia ellipsoid-cylindric, rather faintly nerved; anthers 1.7–2.8 mm long. 124. **C. pallescens**
  3. Terminal spikelet pistillate at apex; perigynia obovoid-orbicular, ± compressed, strongly nerved (especially on dorsal face); anthers 1.3–1.8 (2.2) mm long . . . .
  . . . . 125. **C. hirsutella**

122. **C. virescens** Willd.
  Map 423. Scarce in dry, often sandy woods or rarely in moist open or shaded ground.

123. **C. swanii** (Fern.) Mack.   Fig. 221
  Map 424. Dry open fields and woods (oak, etc.) or, more often, rich deciduous woods or thickets bordering low ground.
  By some considered as a variety [var. *minima* Bailey] of the preceding species. The two are often easily confused. The perigynia of *C. virescens* are generally ellipsoid, while those of *C. swanii* tend to be short-obovoid.

124. **C. pallescens** L.
  Map 425. Our few collections are from a diversity of moist sites: peaty meadows, grassy border of beech–maple woods, fields and clearings, cedar woods.

421. Carex cryptolepis   422. Carex flava   423. Carex virescens

American plants have been segregated from those of Eurasia as var. *neogaea* Fern.

125. C. hirsutella Mack.
Map 426. Upland oak woods; low open ground; and shady borders of ponds, marshes, and swamps.

33. CAREX

This is the group long known as the Hirtae, but under the present Code of Nomenclature it would be called Carex, if treated as a section, since it includes the type species of the genus, *C. hirta* (see Mich. Bot. 11: 31–32. 1972). Although the groups are not here formally recognized as sections in the nomenclatural sense, in the spirit of the Code the group name Carex is adopted.

1. Leaf sheaths and blades and staminate scales soft-hairy; perigynium strongly ribbed, the beak with strong spreading teeth ca. 1 mm or more long . . . . . 126. C. hirta
1. Leaves and scales glabrous; perigynium ribbed or not, the beak with teeth rarely as long as 1 mm
  2. Perigynia ca. 4.5–6.5 mm long, strongly ribbed, the beak nearly half as long as the body, including teeth ca. 0.5–0.7 (1) mm long (fig. 222); plants usually of dry and sandy habitats . . . . 127. C. houghtoniana
  2. Perigynia ca. 3–4.5 mm long, with obscure ribs (at least apically) largely hidden by the dense pubescence, the short beak less than half as long as the body and with teeth not over 0.5 (0.7) mm long (fig. 223); plants usually of wet habitats
    3. Leaf blades flat or with revolute margins, the broadest ones (2.2) 2.5–4.5 (5) mm wide . . . . 128. C. lanuginosa
    3. Leaf blades involute-filiform (or triangular-channeled), not over 2 mm wide . . . . 129. C. lasiocarpa

126. C. hirta L.
Map 427. A European species locally naturalized in North America. Collected in 1923 (*Farwell 6584*, MICH, BLH, WUD, GH) on a vacant lot at Shelbyville, Allegan County.

In the several sheets of the Michigan collection, the style is ± contorted toward its base and appears indurated and continuous with the achene, although this is not normally the case in the Carex (Hirtae) group.

424. Carex swanii   425. Carex pallescens   426. Carex hirsutella

127. C. houghtoniana Dewey   Fig. 222

Map 428. Sandy, gravelly, or rocky open ground, ranging from moist (shores, swampy woods borders and clearings) to very dry (jack pine plains, sandy blowouts, dune ridges) — often in somewhat disturbed sites such as along roads and railroads, spreading by strong rhizomes.

The type material of this sedge was collected on a lucky Friday the 13th of July, 1832, by Douglass Houghton on dry sandy jack pine ridges at Lake Itasca, Minnesota, a few minutes before he and Henry Rowe Schoolcraft first arrived at that lake, the source of the Mississippi (see Mich. Bot. 1: 66. 1962).

The epithet would be in accord with better usage if it were spelled *houghtonii* and it was early "corrected" to this ending; however, there is no sanction in the International Code of Botanical Nomenclature for making such a change from the spelling as originally published.

128. C. lanuginosa Michaux

Map 429. Sandy or marly shores (often in shallow water), marshes, meadows, swales, riverbanks, and rarely in dry sandy woodland (jack pine or oak–aspen), spreading by strong rhizomes.

The culms tend to be more sharply 3-angled and rougher in this species than in the next, of which it is sometimes treated as var. *latifolia* (Boeckl.) Gilly.

129. C. lasiocarpa Ehrh.   Fig. 223

Map 430. In the same wet habitats as the preceding (if anything, in even wetter sites and deeper water) and occasionally growing with it; but also, and most characteristically, in sphagnum bogs where the well developed rhizome system makes it one of the major components of floating mats; remaining as a relic under black spruce, tamarack, and cedar.

The species also grows in Eurasia, and American plants have been segregated as var. *americana* Fern.

34. ANOMALAE

130. C. scabrata Schw.   Fig. 224

Map 431. Usually in springy places, creek borders, ravine bottoms, seepy

427. Carex hirta   428. Carex houghtoniana   429. Carex lanuginosa

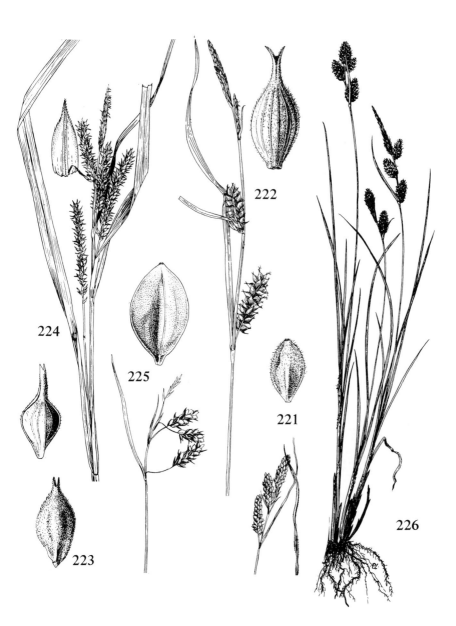

221. *Carex swanii* ×⅔; perigynium ×8
222. *C. houghtoniana* ×⅔; perigynium ×6
223. *C. lasiocarpa,* perigynium ×6
224. *C. scabrata* ×⅔; perigynium ×8; scale ×8
225. *C. paupercula* ×⅔; perigynium ×8
226. *C. buxbaumii* ×⅖

hillsides, or other wet spots in rich deciduous woods; less often in coniferous swamps or wet open clearings and hollows.

35. PENDULINAE

131. C. flacca Schreber

Map 432. A native of Europe and locally naturalized in the eastern United States. Collected by Farwell (*1509* in 1896, BLH; *1808a* in 1903, BLH), doubtless a waif, on the shores of Belle Isle in the Detroit River.

36. LIMOSAE

1. Pistillate scales nearly or quite as broad as the perigynia and often only slightly if at all longer; staminate spikelet (12) 15–30 (50) mm long; plants strongly rhizomatous . . . . 132. C. limosa
1. Pistillate scales distinctly narrower than perigynia, generally with narrowly acuminate tips much exceeding them; staminate spikelet 5–12 (15) mm long; plants loosely clumped . . . . 133. C. paupercula

132. C. limosa L.

Map 433. Chiefly in bogs, including marly ones, often on the open mat at the edge of water; also on marshy open shores.

430. Carex lasiocarpa

431. Carex scabrata

432. Carex flacca

433. Carex limosa

434. Carex paupercula

435. Carex buxbaumii

133. **C. paupercula** Michaux   Fig. 225
    Map 434. In sphagnum bogs, coniferous swamps (tamarack, spruce, cedar), boggy hollows, and occasionally wet open ground; rock crevices along Lake Superior.
    The terminal spikelet rarely has a few perigynia, but the plants can be distinguished from *C. atratiformis* by their longer pistillate scales and pubescent roots. The leaves of *C. paupercula* tend to be flatter and broader than those of *C. limosa*, which has very slender, often ± involute blades.

37. ATRATAE

1. Pistillate scales mostly awned or narrowly acuminate, exceeding the perigynia; sheaths at base of plant becoming filamentose; throughout the state . . . .
. . . . 134. **C. buxbaumii**
1. Pistillate scales obtuse or acute, equalling or shorter than the perigynia; sheaths not filamentose; rare plants of Lake Superior region
  2. Pistillate spikelets ± sessile and erect, rather crowded; anthers ca. 1 (1.2) mm or less long . . . . 135. **C. media**
  2. Pistillate spikelets peduncled, usually spreading or drooping; anthers ca. 1.5–2.5 mm long . . . . 136. **C. atratiformis**

134. **C. buxbaumii** Wahl.   Fig. 226
    Map 435. Shores, meadows, marshes, and old bogs, often forming large stands; sometimes in marly bogs and marshes; in crevices and hollows of the rock shore of Lake Superior.

135. **C. media** R. Br.
    Map 436. Crevices of rock shores of Lake Superior and open woods on thin rocky soil, Isle Royale and mainland of Keweenaw County. A boreal species frequent on the north shore of Lake Superior and Isle Royale.

136. **C. atratiformis** Britton
    Map 437. Several collections have been seen from Isle Royale (including Passage Island), where this species grows at the borders of woods and in rock crevices along the shore. It is found in similar situations on rock outcrops of the

436. Carex media   437. Carex atratiformis   438. Carex haydenii

north shore of Lake Superior in Ontario, and is very local on the south shore (Keweenaw Point, *J. W. Robbins* in 1863, MO; Presque Isle, *Leila J. Laughlin* in 1914, OS). A boreal species which ranges south to northern New England and Lake Superior.

38. ACUTAE

The species of this group are often very difficult to distinguish from each other without basal parts.

1. Fertile culms of current year with conspicuous bladeless sheaths at base, not surrounded by dried-up bases of the previous year's leaves but arising laterally
   2. Perigynia suborbicular to obovoid, 2–2.3 mm long at maturity, broadest at or slightly above middle, rather abruptly contracted to a minute apiculus, at least the lower ones in a spikelet much exceeded by the spreading scales; lower leaf sheaths very slightly or not at all filamentose, the intact sheaths smooth ventrally; ligule longer than width of leaf blade; plants with short, ascending rhizomes . . . . 137. **C. haydenii**
   2. Perigynia elliptic to ovate, mostly 2.2–2.7 (3.3) mm long at maturity, broadest at or slightly below the middle, ± tapered to apex, as long as or longer than the scales (rarely exceeded by scales); lower leaf sheaths ± strongly filamentose or if not, then the ligule shorter than width of leaf blade; plants with long horizontal rhizomes
      3. Ligule shorter than width of leaf blade (often nearly horizontal); lower sheaths not filamentose; ventral surface of sheaths smooth . . . . 138. **C. emoryi**
      3. Ligule longer than width of leaf blade (deeply inverted V-shaped); lower sheaths ± strongly filamentose; membranous ventral surface of sheaths usually minutely scabrous . . . . 139. **C. stricta**
1. Fertile culms of current year mostly lacking bladeless sheaths at base, arising centrally from tufts of dried-up bases of previous years's leaves
   4. Perigynia essentially nerveless, except sometimes at the base only; staminate spikelets usually 2 or more . . . . 140. **C. aquatilis**
   4. Perigynia conspicuously few-ribbed on both sides; staminate spikelet 1
      5. Plants tufted, without long rhizomes; scales with central green portion about as wide as the darker margins; leaves mostly overtopping spikelets . . . 141. **C. lenticularis**
      5. Plants with long rhizomes; scales with very slender green portion scarcely if at all broader than midrib; leaves mostly shorter than summit of culm . . . 142. **C. nigra**

137. **C. haydenii** Dewey
Map 438. The only Michigan collection with habitat data is from Midland County: woods along Salt River (*Dreisbach 8056* in 1934, MICH). A collection from the herbarium of William Boott (MSC) from "Grass Lake" in 1858 is presumably from Jackson County. A collection from Kalamazoo County previously referred to this species has been reidentified by Hermann as *C. aquatilis* var. *altior*.

138. **C. emoryi** Dewey
Map 439. Wet ground along rivers and around lakes and ponds.

139. **C. stricta** Lam.

Map 440. Wet, usually open ground: marshes, ditches, meadows (often springy), edges of lakes and streams, bogs (marly or not), occasionally in cedar or tamarack swamps (especially in clearings); often in large clumps and may form extensive meadows of "wild hay."

Most of our material of this common species is referred to var. *strictior* (Dewey) Carey, distinguished by the minutely scabrous ventral surface of the sheaths and usually paler or glaucous foliage. Typical *C. stricta* with smooth sheaths is known from several counties scattered the length of the state from Isle Royale to St. Joseph County.

140. **C. aquatilis** Wahl.   Fig. 227

Map 441. Wet meadows, shores, lakes, streams, bogs, marshes, and ditches, sometimes in water up to a foot deep.

Including var. *altior* (Rydb.) Fern., sometimes treated as a distinct species, *C. substricta* (Kuek.) Mack. Typical *C. aquatilis* is supposed to have more narrow, elliptic perigynia (1–1.5 mm wide), broadest well below the apex, while in var. *altior* the perigynia are strongly obovate (1.5–2.5 mm wide), broadest near the apex. Typical var. *aquatilis* as identified by Hermann is apparently restricted to the region from Marquette County to Isle Royale, while the common Michigan plant is var. *altior*, which is usually a taller and coarser plant than var. *aquatilis*.

141. **C. lenticularis** Michaux   Fig. 228

Map 442. Wet sandy shores of Lake Superior and inland lakes; pools in rock crevices of Isle Royale and the Pictured Rocks; in southern Michigan collected in springs at Plaster Creek, Kent County (*Cole* in 1897, AQC), and meadow at Marl Lake, Oakland County (*Farwell 4933* in 1918, BLH).

142. **C. nigra** (L.) Reichard

Map 443. Collected once (*Farwell 1820a* in 1904, BLH) on the shores of Belle Isle in the Detroit River. Assumed to be a waif here, the range normally from Massachusetts and Rhode Island northward and eastward to Greenland (also Europe).

439. Carex emoryi

440. Carex stricta

441. Carex aquatilis

39. CRYPTOCARPAE

1. Sheaths scabrous-hispidulous; bodies of most or all pistillate scales on lower part of spikelet truncate or tapered at summit (fig. 230) . . . . 143. **C. gynandra**
1. Sheaths smooth; bodies of most if not all pistillate scales shallowly lobed at summit (on each side of base of the awn, fig. 229) . . . . 144. **C. crinita**

143. **C. gynandra** Schw.   Fig. 230

Map 444. Low wet openings, clearings, stream borders, and roadsides in swampy (chiefly coniferous) woods; moist shores and thickets.

Sometimes treated as a variety of the next species, as var. *gynandra* (Schw.) Schw. & Torrey, but quite consistently distinct.

144. **C. crinita** Lam.   Fig. 229

Map 445. Swamp forests (especially wet borders and clearings), ponds, ditches, and wet hollows in deciduous woods, river borders, and marshes.

40. ORTHOCERATES

145. **C. pauciflora** Lightf.   Fig. 232

Map 446. One of the commonest and most distinctive species of this genus in sphagnum bogs in the northern part of the state, more local southward; usually on open mats but also under black spruce, tamarack, and less often cedar.

442. Carex lenticularis

443. Carex nigra

444. Carex gynandra

445. Carex crinita

446. Carex pauciflora

447. Carex michauxiana

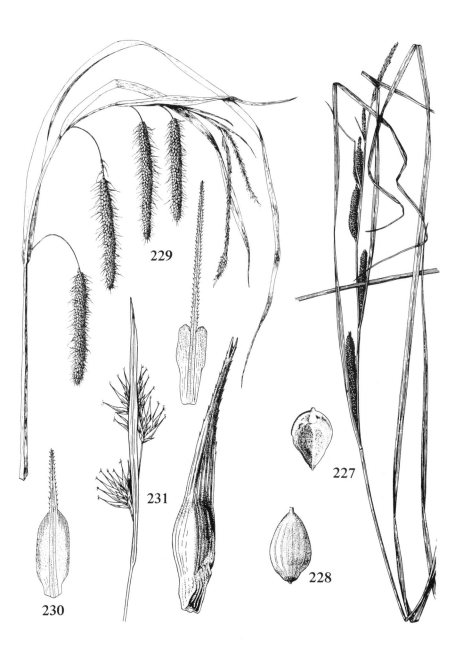

227. *Carex aquatilis* (var. *altior*) ×½; perigynium ×6
228. *C. lenticularis,* perigynium ×6
229. *C. crinita* ×½; scale ×8
230. *C. gynandra,* scale ×8
231. *C. michauxiana* ×⅔; perigynium ×6

41. FOLLICULATAE

1. Broadest leaf blades 1.5–3.5 mm wide; sheaths of bracts concave at mouth; staminate spikelet sessile or very short-peduncled, scarcely if at all projecting above the pistillate spikelets .... 146. **C. michauxiana**
1. Broadest leaf blades 5–17 mm wide; sheaths of bracts with a ± prolonged lobe at mouth; staminate spikelet usually peduncled, its tip projecting well above the pistillate spikelets .... 147. **C. folliculata**

146. **C. michauxiana** Boeckl. Fig. 231

Map 447. Bogs and wet shores, usually near Lake Superior; particularly characteristic of peaty-sandy ditches, interdunal boggy pools, and ponds with fluctuating water levels.

147. **C. folliculata** L.

Map 448. Wet woods and cedar swamps, apparently very uncommon and local.

42. PSEUDO-CYPEREAE

1. Perigynia ± reflexed at maturity, rather asymmetrical or flattened-triangular in cross section, of rather tough texture with close and very strong ribs; teeth at end of beak 0.5–2.2 mm long
    2. Teeth at end of beak (1) 1.2–2.2 mm long and strongly divergent-curved (fig. 233); perigynia 5.5–7 mm long .... 148. **C. comosa**
    2. Teeth at end of beak 0.5–1 (1.3) mm long and ± straight, parallel or slightly divergent (fig. 234); perigynia 4–5.5 (6.5) mm long ...... 149. **C. pseudo-cyperus**
1. Perigynia spreading-ascending, ± round in cross section, thin-textured and much inflated, with few to many ribs or nerves; teeth at end of beak not over 0.7 mm long (or rarely 1 mm in *C. hystericina*)
    3. Staminate scales (except sometimes the lowermost) smooth or nearly so at tip, merely acute or acuminate; perigynia with about 8 nerves (fig. 235); culms solitary or close together, arising from elongate rhizomes ...... 150. **C. schweinitzii**
    3. Staminate scales (some or all of them) with a rough awn (rarely all awnless); perigynia about 8–20-nerved; culms clumped
        4. Achene with rough-papillate surface; perigynia about 8–10-nerved .... 151. **C. lurida**
        4. Achene smooth; perigynia about 15–20-nerved (fig. 236) ...... 152. **C. hystericina**

448. Carex folliculata

449. Carex comosa

450. Carex pseudo-cyperus

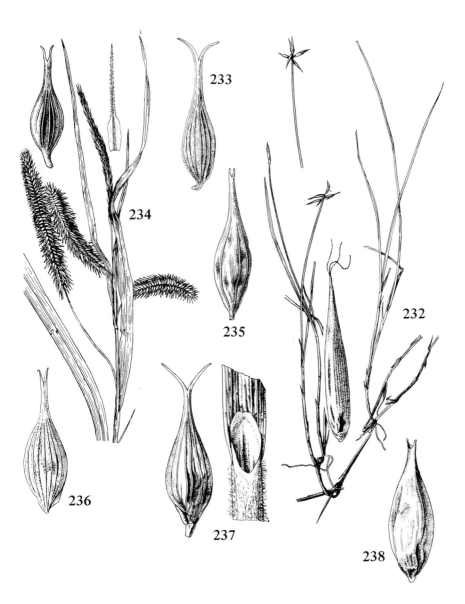

232. *Carex pauciflora* ×⅔; perigynium ×6
233. *C. comosa,* perigynium ×6
234. *C. pseudo-cyperus* ×⅔; perigynium ×6; scale ×6
235. *C. schweinitzii,* perigynium ×6
236. *C. hystericina,* perigynium ×6
237. *C. atherodes,* perigynium ×6; sheath ×1
238. *C. hyalinolepis,* perigynium ×6

148. **C. comosa** Boott   Plate 3-C; fig. 233
Map 449. Marshes and wet shores along lakes, ponds, and rivers; bogs and open spots in swamps.

149. **C. pseudo-cyperus** L.   Fig. 234
Map 450. Marshes, ditches, lake shores (often in shallow water), edges of pools and streams, cedar swamps, and bogs.
Occasional specimens are very close to the preceding species.

150. **C. schweinitzii** Schw.   Fig. 235
Map 451. Evidently rather local; in my experience, a species of wet ground along cold spring-fed rivers and brooks, in mixed or coniferous cover.

151. **C. lurida** Wahl.
Map 452. Marshes, swamps, wet woods and clearings, ditches, margins of lakes and streams.
The nerves of the perigynium are ± distinct in the beak of this species and the preceding, while in the next they become fully confluent. *C. lurida* is placed in the Vesicariae by some authors.

152. **C. hystericina** Willd.   Fig. 236
Map 453. One of our commonest species, in wet ground everywhere (except

451. Carex schweinitzii

452. Carex lurida

453. Carex hystericina

454. Carex subimpressa

455. Carex trichocarpa

456. Carex atherodes

sphagnum bogs), especially on shores, along ditches and edges of streams, in meadows and marshes.

The terminal spikelet frequently has some perigynia at the base or apex. Occasional plants intermediate with *C. pseudo-cyperus* occur and may be hybrids with that species (these have been seen from Lapeer, Roscommon, and Washtenaw counties). The original spelling of the specific epithet is often "corrected" from the original form to *hystricina*.

43. PALUDOSAE

1. Perigynia pubescent
  2. Perigynia not over 6 mm long, the ribs of the body usually obscure; main part of beak and its teeth each about 0.5 mm long; style usually ± sinuous; leaf sheaths (when intact) copiously red-dotted ventrally . . . . 153. **C. subimpressa**
  2. Perigynia ca. 7–9 mm long, the ribs of the body very strong and conspicuous; main part of beak and its teeth each ca. 1–2 mm long; style straight; leaf sheaths not copiously red-dotted ventrally (though often reddened toward summit) . . .
  . . . . 154. **C. trichocarpa**
1. Perigynia glabrous
  3. Leaf sheaths and usually at least the lower blades beneath pubescent; perigynia ca. 8–10 mm long, including apical teeth ca. 1.5–3 mm long (fig. 237) . . . . . .
  . . . . 155. **C. atherodes**
  3. Leaf sheaths and blades glabrous; perigynia 5–7.5 mm long, including apical teeth not over 0.8 mm long
    4. Lower leaf sheaths scarcely if at all pinkish, seldom filamentose; ligule shorter than width of leaf blade or slightly longer; mature perigynia nerveless or very delicately nerved (fig. 238) . . . . 156. **C. hyalinolepis**
    4. Lower leaf sheaths strongly reddened, becoming filamentose on ventral side (fig. 239); ligule usually twice or more as long as width of leaf blade; mature perigynia often rather strongly nerved . . . . 157. **C. lacustris**

153. **C. subimpressa** Clokey

Map 454. Very local: low wet meadow at Cedar Lake [Montcalm County] (*Davis* in 1897, MICH); Big Portage marsh, Waterloo Tp., Jackson County (*Churchill* in 1956, MICH); S. Rockwood, Monroe County (*Hebert* in 1953, ND).

The Jackson County collection is atypical in having rather prominent ribs on the perigynia and straight styles; however, the size of the perigynia, beaks, and teeth are appropriate for *C. subimpressa* and Hermann has confirmed this determination. This is a species of uncertain status, often suspected of being a hybrid and originally described as *C. hyalinolepis* × *C. lanuginosa*.

154. **C. trichocarpa** Schk.

Map 455. Rare, in low deciduous woods, river banks, floodplain marshes, ditches.

Rarely (as in Grand Rapids specimens collected by Emma Cole) the perigynia are nearly glabrous, with only a few short stiff hairs on the beak.

155. **C. atherodes** Sprengel Fig. 237

Map 456. Marshes and boggy hollows, often in shallow water at the edges of ponds and lakes.

Sometimes specimens from shallow water will have pubescence evident only at the summit of some of the leaf sheaths.

156. **C. hyalinolepis** Steudel   Fig. 238

Map 457. Wet woods, roadside ditches, low open ground – depressions which are not permanently flooded.

157. **C. lacustris** Willd.   Fig. 239

Map 458. Marshes, swales, ditches, bog borders, river and creek banks, alder or tamarack swamps, and similar wet places, often forming extensive colonies.

Some collections from Livingston, Oakland, and Washtenaw counties have the sheaths unusually hispidulous.

44. SQUARROSAE

1. Pistillate scales with a long awn exceeding the beak of the perigynium; achenes ca.
   1.5 mm long. . . . . 158. **C. frankii**
1. Pistillate scales awnless or short-awned, much shorter than the beak, usually ± hidden among the dense perigynia; achenes ca. 2.2–3 mm long
   2. Achenes slenderly ellipsoid, slightly more than twice as long as wide, terminated by a strongly sinuous style (fig. 240); pistillate scales very sharp-tipped or short-awned . . . . 159. **C. squarrosa**
   2. Achenes broadly ellipsoid, about or slightly less than twice as long as wide, terminated by a ± straight style (fig. 241); pistillate scales ± acute in outline but blunt at the very tip . . . . 160. **C. typhina**

158. **C. frankii** Kunth

Map 459. Recorded from low open woodland in Kalamazoo County (*C. R. & F. N. Hanes*, several collections 1939–1946, WMU, MICH); and from margins of moist fields and thickets in Monroe County (*Hiltunen 3792* in 1961, WUD; *Voss 13547 & 13548* in 1970, MICH); also found by Hebert at Maple Beach, Wayne County (in 1931, ND), and at Hubbard Lake, Alcona County (in 1950, ND).

159. **C. squarrosa** L.   Fig. 240

Map 460. A species of swampy woods and thickets; our specimens almost all lack habitat data.

457. Carex hyalinolepis   458. Carex lacustris   459. Carex frankii

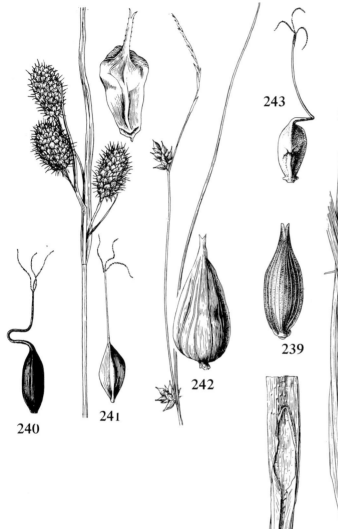

239. *Carex lacustris* ×½; perigynium ×6;
    sheath with summit breaking into
    fibers ×1
240. *C. squarrosa,* achene ×6
241. *C. typhina* ×⅔; perigynium ×6;
    achene ×6
242. *C.· oligosperma* ×⅔; perigynium ×6
243. *C. tuckermanii,* achene ×4

160. **C. typhina** Michaux   Fig. 241

Map 461. Very rare in Michigan, in swamps (e. g., dense maple–ash floodplain forest).

45. VESICARIAE

1. Leaf blades and bracts involute-filiform, wiry; pistillate spikelets 3–15(18)-flowered, subglobose or short-oblong (not over 2 cm long); staminate spikelet solitary (rarely a second small one present) . . . . 161. **C. oligosperma**
1. Leaf blades and bracts flat; pistillate spikelets oblong to long-cylindric; staminate spikelets normally 2 or more (or 1 in *C. retrorsa*)
  2. Perigynia (4) 4.5–7 mm thick; achenes with a deep notch or constriction on one angle (fig. 243) . . . . 162. **C. tuckermanii**
  2. Perigynia 2.5–3.5 (4) mm thick; achenes not notched
    3. Lowest pistillate bract 3 or more times as long as the entire inflorescence; mature perigynia 7–12 mm long, at least the lower reflexed or widely spreading (fig. 244); staminate spikelet often 1, its base (or base of lowest staminate spikelet if there are more than 1) slightly if at all elevated above summit of the crowded pistillate spikelets (rarely all spikelets remote) . . . . 163. **C. retrorsa**
    3. Lowest pistillate bract less than 3 times as long as inflorescence; perigynia (4) 4.5–7.5 (very rarely 8.5) mm long, ascending or spreading; staminate spikelets mostly 2–4, generally well elevated above the pistillate spikelets
      4. Culms with sharp and scabrous angles immediately below the lowest pistillate bract, more strongly scabrous between the spikelets; widest leaf blades 3–5 (6) mm broad; ligules about as long as wide or often distinctly longer (fig. 245); perigynia ascending . . . . 164. **C. vesicaria**
      4. Culms with blunt angles, smooth or nearly so below the lowest bract, smooth or scabrous between spikelets; widest leaf blades 4.5–10.5 mm broad; ligules about as long as wide or usually shorter (fig. 246); perigynia (at least those on lower portion of fully mature spikelet) ± widely spreading (ascending when immature) . . . . 165. **C. rostrata**

161. **C. oligosperma** Michaux   Fig. 242

Map 462. Usually in bogs, where its rhizomatous habit makes it a mat-former rivaling in importance *C. lasiocarpa* – with which it is rather frequently confused when sterile; relic in older coniferous swamps and also found on wet lake shores and in marshes.

460. Carex squarrosa

461. Carex typhina

462. Carex oligosperma

162. **C. tuckermanii** Dewey   Fig. 243

Map 463. Swampy woods and thickets, less often in wet open ground such as swales and ditches; particularly characteristic of margins of ponds, low depressions, streams, and hollows in deciduous or mixed woods (seldom if ever in bogs or coniferous sphagnum swamps).

163. **C. retrorsa** Schw.   Fig. 244

Map 464. Usually in damp muddy somewhat shaded ground, especially along swampy borders, ditches, or low places in woods, seepage areas, alder thickets, swamp forests, marshy borders of lakes, streams, and rivers; occasionally in bogs and cedar thickets.

The perigynia in this species tend to be ± asymmetrical, the body more convex above than below, thus emphasizing the reflexed aspect. See also comments under *C. lupulina* (no. 169).

164. **C. vesicaria** L.   Fig. 245

Map 465. Swamps and wet thickets, swales and ditches, marshy or boggy shores and ponds, interdunal pools, often forming large tussocks.

A more slender plant than most specimens of the next species, not producing elongate rhizomes, the leaf blades weakly if at all septate-nodulose.

165. **C. rostrata** Stokes   Fig. 246

Map 466. In very wet muddy or sandy places: marshy stream margins, lake shores, riverbanks; swamps, marshes, ponds, ditches, and wet bogs; wet trails and clearings in cedar swamps.

Typically a more robust plant than the preceding, strongly rhizomatous, with culms large and ± spongy at base, the broader leaves rather strongly septate-nodulose (at least when dry). The perigynia tend to be more abruptly beaked. Smaller plants are often rather difficult to distinguish from the preceding, an exception being likely to occur to any one of the characteristics given in the key, and it is possible that the two species occasionally hybridize.

463. Carex tuckermanii   464. Carex retrorsa   465. Carex vesicaria

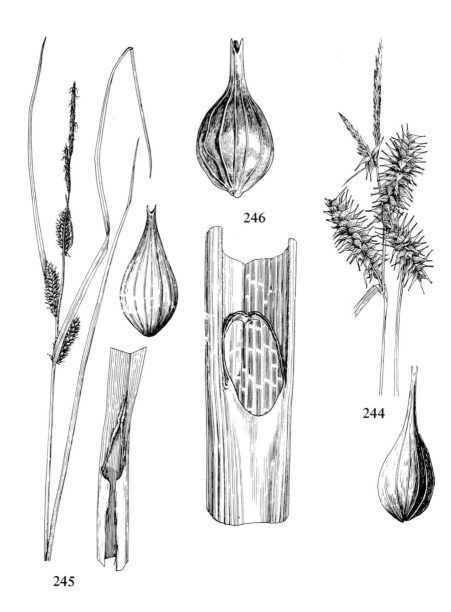

244. *Carex retrorsa* ×⅔; perigynium ×5
245. *C. vesicaria* ×⅖; perigynium ×5; summit of sheath with ligule ×2
246. *C. rostrata,* perigynium ×5; summit of sheath with ligule ×3

1. Pistillate spikelets globose or nearly so, scarcely if at all longer than wide; style straight or sinuous or contorted (especially in *C. intumescens*) just below or at the middle; beak of perigynium much shorter than the body
   2. Perigynia (7) 10–31 per spikelet, radiating in all directions, narrowed at the base to a ± broad cuneate stipe, sometimes hispidulous basally (fig. 247) . . . . 166. **C. grayi**
   2. Perigynia 2–8 (12) per spikelet, mostly spreading-ascending, rounded at the base (fig. 248), glabrous (and often very shiny) . . . . 167. **C. intumescens**
1. Pistillate spikelets cylindrical or short-oblong, usually definitely longer than broad; style strongly bent and contorted immediately above the body of the achene; beak of perigynium nearly or quite as long as the body
   3. Body of achene with broadly diamond-shaped sides, at most 0.5 mm longer than wide, the angles each with a prominent swollen knob (fig. 249) . . 168. **C. lupuliformis**
   3. Body of achene with somewhat diamond-shaped to ± elliptic or ovate sides, usually 1 mm or more longer than wide, the angles obscurely if at all knobbed (fig. 250) . . . . 169. **C. lupulina**

166. **C. grayi** Carey   Fig. 247

Map 467. Usually in rich deciduous woods, often on floodplains in woods and marshes. The Iron County collection (*F. B. Bevis 435* in 1959, MICH) is from a hemlock–white spruce–yellow birch–sugar maple stand near the University of Michigan Forestry Camp.

Occasionally the perigynia are hispidulous, especially toward the base; such plants may be called var. *hispidula* Bailey.

167. **C. intumescens** Rudge   Fig. 248

Map 468. A common species of low wet woods and thickets, including swamp forests, depressions and stream banks in oak and beech–maple woods (as well as upland sites); less often in coniferous swamps (cedar, tamarack, spruce, fir), especially along borders, roads, and clearings.

Plants with very plump ovoid perigynia (more than (4) 5 mm thick) are the typical variety; most of our plants have more elongate perigynia (up to 5 mm thick) and have been called var. *fernaldii* Bailey. Both varieties occur throughout the state, and separation of them does not appear to be significant. Immature specimens with crowded spikelets (appearing as if one) might be confused with

466. Carex rostrata   467. Carex grayi   468. Carex intumescens

C. lupulina, from which they differ in having the inner surfaces of the teeth at the end of the perigynium usually hispid; in *C. lupulina*, the teeth are ± smooth within.

168. **C. lupuliformis** Dewey   Fig. 249

Map 469. Swales, marshes, swamps, and depressions in woods — very local.

169. **C. lupulina** Willd.   Fig. 250

Map 470. Swamp forests (usually deciduous, rarely cedar or tamarack); floodplains, marshes and thickets along rivers, lakes, and ponds; mucky hollows and depressions in woods; occasionally in moist fields and ditches adjacent to wet woods; very rarely in bogs.

Plants of this species sometimes bear a very close superficial resemblance to those of *C. retrorsa* (no. 163), from which they differ in their slightly longer and thicker, strongly ascending perigynia, with more nerves. It is possible that certain apparently intermediate (but immature) specimens are hybrids with that species as well as with others, although *C. lupulina* is an extremely variable species with the spikelets sometimes much more separated than usual. The achenes of *C. lupuliformis* when slightly immature may be distinctly longer than wide, but such plants can generally be distinguished from *C. lupulina* by the very prominent knobs of the angles of the achene, as well as by the often long ± loosely flowered spikelets.

2. **Scleria**   Nut-rush

Two Coastal Plain disjunct species are known from northern Indiana and might be discovered in southwestern Michigan, *S. muhlenbergii* Steudel and *S. reticularis* Michaux. Both have glabrous foliage and a reticulate pattern on the achenes, which have a 3-lobed basal disc; the achene is slightly pubescent in *S. muhlenbergii* and glabrous in *S. reticularis*.

KEY TO THE SPECIES

1. Achenes smooth, ca. 3 mm long, including the whitish, foam-like basal disc; larger leaves (3.5) 5–7.5 mm wide; mature anthers ca. 2.5–4 mm long . . . . 1. **S. triglomerata**
1. Achenes papillose-roughened or wrinkled, ca. 1–2 mm long, including basal disc (not a foam-like crust); larger leaves not over 2 mm wide; anthers ca. 1–2.5 mm long
   2. Culms and leaves ± densely pubescent; anthers ca. 2–2.5 mm long; flower clusters 1 or 2 (rarely 3) on a culm, each subtended by an elongate leaf or leaf-like bract; disc at base of achene ornamented with several rounded tubercles . . . . 2. **S. pauciflora**
   2. Culms and leaves glabrous or nearly so; anthers ca. 1 mm long; flower clusters (1) 3–5, at most the lowest subtended by an elongate blade; disc at base of achene not tubercled, inconspicuous or apparently absent . . . . 3. **S. verticillata**

1. **S. triglomerata** Michaux

Map 471. Dry or moist open or shaded sandy ground, prairies, borders of marshes; very local.

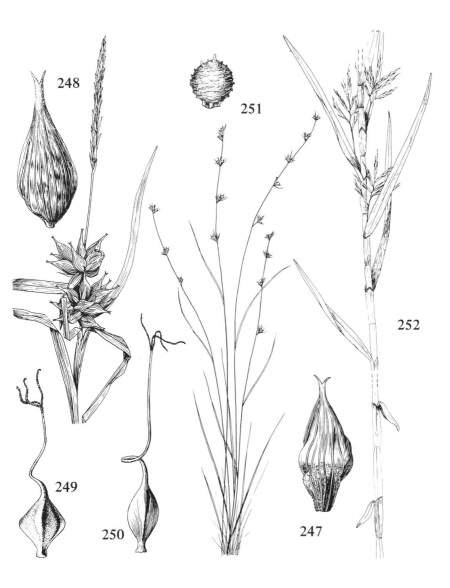

247. *Carex grayi* (var. *hispidula*), perigynium ×2½
248. *C. intumescens* ×⅔; perigynium ×3
249. *C. lupuliformis,* achene ×6
250. *C. lupulina,* achene ×6
251. *Scleria verticillata* ×½; achene ×10
252. *Dulichium arundinaceum* ×½

2. **S. pauciflora** Willd.

Map 472. Dry sandy or gravelly open ground; barely ranging as far north as Michigan.

2. **S. verticillata** Willd.   Fig. 251

Map 473. Sandy or often marly shores, interdunal flats, meadows, and marshes. Our commonest species of *Scleria*, but easily overlooked.

3. **Dulichium**

1. **D. arundinaceum** (L.) Britton   Fig. 252   Three-way Sedge

Map 474. Marshes and marshy shores, hollows, ponds, swales, ditches, and river margins, seldom in water over 1 or 2 feet deep; also in bogs, on open mats and relic in tamarack swamps; often abundant where found, forming large beds. On any two adjacent culms from the same rhizome, one will have the leaves spiraled clockwise and the other, counterclockwise.

A very distinctive plant; the numerous leaves are perfectly 3-ranked, although the culms are terete or nearly so (and hollow).

4. **Cyperus**   Nut-grass; Umbrella Sedge

REFERENCE

Mohlenbrock, Robert H. 1960. The Cyperaceae of Illinois I. Cyperus. Am. Midl. Nat. 63: 270–306.

KEY TO THE SPECIES

1. Styles 2-cleft; achenes lenticular or biconvex; anthers 0.3–0.5 (0.6) mm long
  2. Fully mature achene black with whitish incrustation in transverse lines, the surface cells (under considerable magnification) elongate; scales not over 2 (2.2) mm long, with very little or no reddish color . . . . 1. **C. flavescens**
  2. Fully mature achenes drab or brown, the whitish incrustation, if present, ± evenly reticulate, and the surface cells not elongate; scales mostly 2–2.5 (3) mm long, prominently marked with reddish (occasionally some entirely pale)

469. Carex lupuliformis   470. Carex lupulina   471. Scleria triglomerata

3. Reddish color of scales concentrated near the apex, usually prolonged toward base of scale in narrow bands bordering edge and midrib; style ± persistent, conspicuous at maturity of spikelet . . . . 2. **C. diandrus**

3. Reddish color of scales concentrated near the base and toward the edge, sometimes extending to apex and throughout the sides; style deciduous, often not conspicuous at maturity of spikelet . . . . 3. **C. rivularis**

1. Styles 3-cleft; achenes 3-sided; anthers various

  4. Tips of scales definitely spreading outward

    5. Scales with a narrowly acuminate or awn-like spreading-recurving tip ca. 0.5 mm long; anthers ca. 0.2−0.3 mm long; sides of scales distinctly ribbed or nerved. .

. . . . 4. **C. aristatus**

    5. Scales merely acute and spreading at the tip; anthers ca. 0.5−0.7 mm long; sides of scales distinctly cellular, at most with one obscure nerve . . . . . . 5. **C. acuminatus**

  4. Tips of scales not spreading, the backs of the scales ± straight (or slightly incurved) from base to apex

    6. Spikelets at maturity mostly 1.5−5 mm wide, forming a globose or hemispherical head, or ascending (or ± loosely disposed) in a cluster (see fig. 254) with central axis not over 1 cm long or rarely 1.5 cm; rachilla at most with very narrow wing; anthers 0.5−1.1 mm long; perennials from a hard knotty rhizome, of dry usually sandy habitats

      7. Culms 1−2.5 mm thick below the inflorescence, ± antrorsely scabrous on the sides and sharp angles; margins of involucral and cauline leaves scabrous; blades of broadest cauline leaves 3−5 mm wide; achenes (2) 2.2−2.5 mm long; larger scales 3−4 mm long, including (at least on upper scales) an awn-like tip ca. 0.5−1 mm long . . . . 6. **C. schweinitzii**

      7. Culms 0.4−1 (1.2) mm thick below the inflorescence, smooth, rather obtusely angled; margins of leaves various; broadest cauline blades 1−3 mm wide (or rarely to 5 mm in *C. houghtonii*); achenes 1.6−2 mm long (not including tiny style base); scales 2−3 mm long, including a tiny tip less than 0.5 mm long (or this absent)

        8. Involucral leaves strongly ascending, slightly scabrous or smoothish toward base; cauline leaves smooth or nearly so; inflorescence hemispherical or (usually) narrower, with loosely ascending spikelets (often 1−7 similar inflorescences on separate rays) . . . . 7. **C. houghtonii**

        8. Involucral leaves mostly widely spreading or recurved at maturity, the margins of these and of the cauline leaves scabrous; inflorescence a dense hemispherical head (occasionally 1 or 2 similar inflorescences on separate rays) . . . . 8. **C. filiculmis**

472. Scleria pauciflora   473. Scleria verticillata   474. Dulichium arundinaceum

6. Spikelets linear, mostly up to 1.5 mm wide (occasionally to 2.3 mm in *C. strigosus*), pinnately arranged ± at right angles to a distinct central axis (often over 1 cm long, see figs. 256, 257) (lowermost spikelets sometimes reflexed, upper and immature ones sometimes ascending); rachilla rather distinctly winged; anthers 0.2–0.5 mm long (0.6–1.7 in *C. esculentus* only); annuals or perennials (a hard knotty rhizome only in *C. strigosus*) of moist habitats

    9. Scales not overlapping on the same side of spikelet (the tip of one scale not quite reaching the base of the next above it, fig. 255) . . . . 9. **C. engelmannii**

    9. Scales overlapping (the tip of each overlapping the base of the next one above it on the same side of spikelet, fig. 256)

       10. Scales (3) 3.5–4.5 (5) mm long, with yellowish-golden sides and greenish midrib; plants usually with hard knotty rhizomes . . . . 10. **C. strigosus**

       10. Scales 1.1–2.7 mm long, often marked with reddish or dark brown; plants fibrous-rooted (but often large) annuals or (*C. esculentus*) with numerous slender scaly rhizomes

          11. Wings of rachilla chaff-like, readily deciduous at maturity; achenes ca. 0.7–0.8 mm long; scales 1.1–1.5 (1.6) mm long . . . . 11. **C. erythrorhizos**

          11. Wings of rachilla not separating; achenes ca. 1.1–1.4 mm long; scales (1.5) 1.7–2.7 mm long

             12. Anthers 0.2–0.6 mm long; plants fibrous-rooted, without scaly rhizomes; rachilla wings embracing much of the basal half of the achenes; rachilla disarticulating between the achenes . . . . 12. **C. odoratus**

             12. Anthers (0.6) 0.8–1.5 (1.7) mm long at maturity; plants with numerous slender scaly rhizomes (these ultimately producing tubers); achenes (often poorly developed) only slightly covered by the wings of the rachilla; rachilla not disarticulating into segments . . . . 13. **C. esculentus**

1. C. flavescens L.

Map 475. Very local, on moist sandy-mucky shores and low grassy areas.

The achenes are very characteristic when fully mature, but are brown and lack the transverse white wrinkles when young. The dark achene visible through the translucent sides of the essentially colorless scales should not be confused with the colored scales of related species. North American plants of this widespread species have been distinguished as var. *poaeformis* (Pursh) Fern.

2. C. diandrus Torrey

Map 476. Sandy to muddy shores, banks, and wet hollows, occasionally at margins of bogs.

475. Cyperus flavescens   476. Cyperus diandrus   477. Cyperus rivularis

3. **C. rivularis** Kunth   Fig. 253

Map 477. Wet sandy or muddy, often marshy shores and banks of creeks, ponds, and lakes; wet marshy hollows, low fields, and meadows; ditches and swales.

4. **C. aristatus** Rottb.

Map 478. Sandy or muddy banks, shores, and margins of rivers, creeks, and lakes.

Our plants have often been distinguished from those of the Old World and tropics as *C. inflexus* Muhl. The dried plants frequently have a pleasant odor, like that of sweet-clover (*Melilotus*).

5. **C. acuminatus** Torrey & Hooker

Map 479. A single Michigan collection has been seen, from Union Lake, Oakland County (*D. J. Ameel* in 1928, MICH).

Occasional specimens of *C. erythrorhizos* have the tips of the lower scales very slightly spreading and might run here in the key. They are readily distinguished from *C. acuminatus* by their shorter anthers, ribbed scales, and abundant pinnate inflorescences.

6. **C. schweinitzii** Torrey

Map 480. Sandy shores, dunes, dry prairies, fields, and roadsides.

Occasional specimens with smoothish culm or smaller achenes and scales than usual appear to approach the next species, but in each case all other characters point to *C. schweinitzii*. Specimens from the vicinity of Caseville, Huron County, in particular come close to *C. houghtonii* although one of them has been confirmed by O'Neill as *C. schweinitzii*. Quite probably a hybrid complex is involved here, for other Huron County specimens appear somewhat intermediate between *C. filiculmis* and *C. houghtonii*.

7. **C. houghtonii** Torrey

Map 481. Sandy, usually disturbed places such as dunes, shores, trails and roads in jack pine, oak, or aspen woodland.

478. Cyperus aristatus

479. Cyperus acuminatus

480. Cyperus schweinitzii

Resembles the preceding species rather closely, but generally smaller. The type material of this species, which is named for Michigan's first State Geologist, Douglass Houghton, was collected by him on August 4, 1831, at Lac des Isles, Wisconsin [Sawyer County, on portage to Lac Court Oreilles, ca. 3 mi. from Namekagon River], while he was on the first Schoolcraft expedition to the source of the Mississippi River.

8. C. filiculmis Vahl   Fig. 254

Map 482. Usually in dry sandy open ground, including oak or jack pine woodlands, roadsides, old fields, dunes, and dry prairies; occasionally in moister places on banks of lakes and ponds.

A fairly distinctive species, and one of our commonest. The achenes and scales tend to be longer than in the preceding two species.

9. C. engelmannii Steudel   Fig. 255

Map 483. Sandy or muddy shores of lakes and rivers, borders of marshes, sand bars, etc. Very often growing with *C. odoratus* and almost suggestive of an abnormal form of that species.

10. C. strigosus L.   Fig. 256

Map 484. Sandy to muddy shores, creek and river banks, exposed flats, low marshy ground and ditches, meadows, moist thickets; rarely in waste places such as roadsides, railroads, and vacant ground.

481. Cyperus houghtonii

482. Cyperus filiculmis

483. Cyperus engelmannii

484. Cyperus strigosus

485. Cyperus erythrorhizos

486. Cyperus odoratus

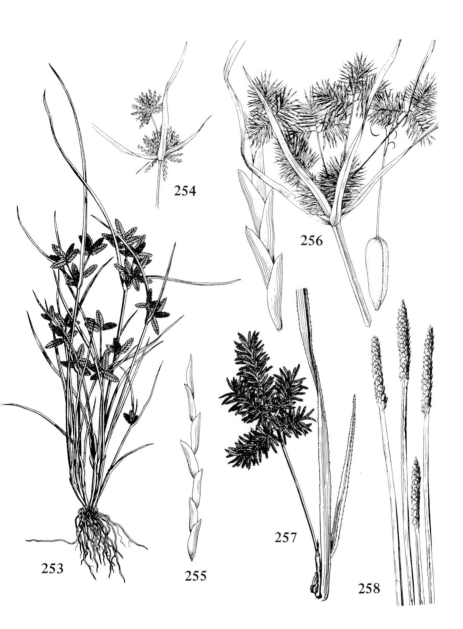

253. *Cyperus rivularis* ×⅔
254. *C. filiculmis* ×½
255. *C. engelmannii,* spikelet ×5
256. *C. strigosus* ×½; spikelet ×5; achene ×10
257. *C. erythrorhizos* ×⅘
258. *Eleocharis quadrangulata* ×½

11. C. erythrorhizos Muhl.   Fig. 257

Map 485. In the same habitats as *C. odoratus* and often growing with it. Often distinguishable superficially by a richer chocolate-brown color to the mature spikelets, together with the very dense short scales.

12. C. odoratus L.

Map 486. Sandy or muddy shores, riverbanks, borders of marshes, exposed flats, ditches, etc.; sometimes spreading into waste ground and low fields.

A widespread species in warmer climates around the world. Plants of the central and western United States have been segregated by some authors as *C. ferruginescens* Boeckl. If retained in the inclusive *C. odoratus*, however, our plants may be called var. *squarrosus* (Britton) Gilly.

13. C. esculentus L.

Map 487. Moist shores, ditches, riverbanks, fields, and marshes; gardens, cultivated fields, and waste places — sometimes a bad weed; in sand (even dunes), mud, or clay.

Seldom collected with tubers, but the slender scaly rhizomes resembling roots are very characteristic; together with the pale scales and large anthers, they make this one of the most easily recognized species.

5. Eleocharis   Spike-rush

Many of our specimens were examined for this project by H. K. Svenson, who also kindly checked an early draft of the key.

REFERENCE

Svenson, Henry Knute. 1957. Scirpeae (continuatio). N. Am. Flora 18: 505–556 [*Eleocharis*, pp. 509–540].

KEY TO THE SPECIES

1. Mature spikelet scarcely if at all thicker than main portion of culm; scales persistent; culms quadrangular, triangular, or terete and cross-partitioned
  2. Culms terete, cross-partitioned, appearing as if jointed (with narrow, light-colored bands on surface) . . . . 1. E. equisetoides
  2. Culms angled, not clearly partitioned nor appearing jointed
    3. Culms sharply 4-angled, stout (3–5 mm thick); spikelet (1.2) 2–5 cm long . . .
      . . . . 2. E. quadrangulata
    3. Culms 3-angled, not over 2 mm thick; spikelet 1–2 (2.5) cm long . . . 3. E. robbinsii
1. Mature spikelet decidedly thicker than culm, with scales usually deciduous; culms terete (or sometimes flattened or many-ridged), not cross-partitioned
  4. Tubercle appearing as if a slender or minute conical continuation of the body of the achene, slightly differentiated in texture or color, not separated by a constriction nor appearing as a distinct apical cap; stigmas 3; summit of leaf sheath without a prominent tooth
    5. Fertile culms 20–70 cm tall, flattened, stout; sterile culms often as long or

longer and rooting at their tips; spikelets (7) 9–17 mm long . . . . . . 4. **E. rostellata**
5. Fertile culms up to 35 cm tall, but some if not all in a tuft less than 20 cm tall,
very slender; culms not rooting at tips; spikelets (2) 3–7 mm long
  6. Plants not over 5 cm tall; achenes ca. 1–1.3 mm long, including tiny tubercle;
spikelets 2–3 mm long . . . . 5. **E. parvula**
  6. Plants (at least most culms) usually over 5 cm tall; achenes 2–2.5 mm long;
spikelets 4–7 mm long . . . . 6. **E. pauciflora**
4. Tubercle differentiated in shape as well as texture, and usually separated from
body of achene by a narrow constriction, forming a distinct apical cap; stigmas 2
or 3; leaf sheaths in several species (especially those likely to be confused on
tubercle characters) with a prominent tooth at summit
  7. Achenes 3-sided (the angles sharp, or obscure and the achene plumply rounded);
styles 3-cleft; surface of achene normally ridged, reticulate, roughened, or in a
few species only minutely cellular or punctate
    8. Achenes white or pearly, with prominent longitudinal ridges connected by
numerous minute cross-bars (fig. 259); basal scales of spikelet fertile
      9. Culms (conspicuously spongy) and scales rather pale green; anthers not over
0.6 mm long; bristles overtopping tubercle . . . . 7. **E. radicans**
      9. Culms (not spongy) and scales deep green, the culms often with reddish basal
sheaths and the scales with a reddish brown band on each side; anthers 0.6–1
(1.2) mm long; bristles equalling achene, shorter, or absent . . . . . 8. **E. acicularis**
    8. Achenes colored, greenish, yellow, golden, brown, or black, and reticulate,
smoothish, or roughened, but not as described above; basal scales of spikelets
sterile
      10. Plants tufted to densely cespitose, without rhizomes; achenes greenish, olive,
or black, smoothish to finely reticulate
        11. Culms capillary, very unequal in length, the longest up to 18 (or
occasionally 35) cm tall, with many short ones crowded at base; achenes
minutely reticulate or punctate, light greenish yellow or olive, the awl-like
tubercle with basal flange not over a fourth the width of the achene;
anthers less than 0.5 mm long; bristles equalling or overtopping the
tubercle . . . . 9. **E. intermedia**
        11. Culms stout, flattened, mostly over 20 cm long; achenes smoothish or
minutely punctate, very dark glossy brown to black, the pale depressed
tubercle completely covering the summit of the body (fig. 260); anthers ca.
1.3–2 mm long; bristles scarcely reaching summit of achene . 10. **E. melanocarpa**
      10. Plants with very stout rhizomes; achenes yellow, golden, or brown, the
surface strongly papillate-roughened or honeycombed
        12. Achene yellowish to dark brown, the 3 angles narrowly keeled or at least
strongly ribbed . . . . 11. **E. tricostata**
        12. Achene yellow to orange or golden-brown, bluntly angled or nearly terete
          13. Culms very strongly flattened and often ± twisted, with obscure ridges;
scales at middle of spikelet reddish brown with narrow, deeply bifid
scarious whitish tips mostly 0.6–1 mm long . . . . 12. **E. compressa**
          13. Culms slightly or not at all flattened, prominently ridged; scales at middle
of spikelet deep reddish brown to nearly black, with short, entire,
lacerate, or bifid tips mostly not over 0.6 mm long . . . . 13. **E. elliptica**
  7. Achenes 2-sided (lenticular or biconvex); styles 2-cleft (or often 3-cleft in nos.
14 & 18–20); surface of achene smooth (sometimes minutely cellular), usually
± shiny
    14. Summit of leaf sheaths thin and membranous, cleft on one side, usually
whitish; achene olive green to brown, ca. 1–1.5 mm long, including the green
tubercle; anthers ca. 0.6–1 mm long . . . . 14. **E. olivacea**

14. Summit of leaf sheaths thin to firm, obliquely to ± squarely truncate, not split (sometimes with a tooth); achenes and anthers various

    15. Achenes shiny dark purplish to black, not over 1 mm long, including the tiny tubercle (which is scarcely 0.3 mm broad and 0.2 mm tall); anthers ca. 0.3–0.4 mm long; plant delicate, with culms ca. 0.5 mm thick or less . . . .
. . . . 15. **E. caribaea**

    15. Achenes yellow to brown, 1.1–2.8 mm long, including the tubercle (larger than in the preceding); anthers and stature various

        16. Plants perennial, with stiff culms and rhizomes; scales acute to acuminate at apex (or somewhat obtuse); achenes 1.5–2.8 mm long, including tubercle; anthers ca. (1) 1.2–2.7 mm long

            17. Basal sterile scale of spikelet solitary, completely or nearly encircling spikelet at the base; fertile scales ovate, acute to somewhat obtuse at apex; culms slender, usually not over 0.7 (rarely as much as 1.1) mm thick . . . . 16. **E. erythropoda**

            17. Basal sterile scales of spikelet 2 or 3, the lower one not encircling the spikelet; fertile scales ovate to ± lanceolate, often with more acuminate (and sometimes recurved) tips than in the preceding species; culms often stouter, 0.5–4 mm thick . . . . 17. **E. smallii**

        16. Plants annual, with soft, easily compressed, densely tufted culms; scales broadly rounded at apex; achenes 1.1–1.5 (1.7) mm long, including the strongly flattened tubercle; anthers (0.3) 0.4–0.7 mm long

            18. Base of tubercle slightly less than two-thirds as wide as the broadest part of the mature achene; scales purplish brown . . . . 18. **E. ovata**

            18. Base of tubercle at least two-thirds as wide as broadest part of mature achene; scales brown or reddish brown (rarely flushed with purple)

                19. Tubercle very depressed, not over a fourth of the total length of the achene, nearly or quite as wide as the truncate body, on which it appears as a flattish cap (fig. 261) . . . . 19. **E. engelmannii**

                19. Tubercle ± deltoid, more than a fourth the total length of the achene and about three-fourths to nearly as wide as the broadest part of the body (fig. 262) . . . . 20. **E. obtusa**

1. **E. equisetoides** (Ell.) Torrey
Map 488. Local on wet shores of lakes and ponds or in shallow water on sandy, marly, or peaty bottoms.

2. **E. quadrangulata** (Michaux) R. & S.   Fig. 258
Map 489. Along margins of lakes and ponds and in shallow water (up to about 2 feet) on sandy, marly, or peaty bottoms.

The roots sometimes bear distinct pedicelled tubers. Both this species and the preceding are distinctive large robust plants.

3. **E. robbinsii** Oakes
Map 490. Sandy to peaty shores, lakes and ponds, wet marshy bogs, and thriving (fruiting) on recently exposed shores.

The three angles of the culms may not be clearly seen in pressed specimens, which may nevertheless be identified by the combination of spikes scarcely thicker than the culm and slender culm (compared with the two preceding species). Numerous limp, capillary culms are often produced when the plant is growing in water; these are contrasted with the similar structures of *Juncus militaris* and *Scirpus subterminalis* — with both of which this species may grow — in couplets 22–24 of Key A, p. 54.

4. E. rostellata Torrey

Map 491. Marshes, shores, bogs, exposed mud flats, boggy meadows, often in calcareous sites.

The culms are flattened, and often arching and rooting at the tips. The resulting loops may catch the foot of the collector, calling his attention to a plant he might otherwise overlook. See also comments under *E. melanocarpa* and *E. tricostata*.

5. E. parvula (R. & S.) Link

Map 492. Known in Michigan only from a "salt lick" near Hubbardston, Ionia County (*Wheeler* in 1887, MSC, NY) and brackish banks of the Maple River in Clinton County (*E. F. Smith* in 1875, MICH). A species of both Atlantic and Pacific coasts, extremely local inland in North America.

A small, delicate plant. The scales are acute to obtuse or rounded, greenish to straw-colored (sometimes with a little brownish on the sides), the lowermost sterile.

487. Cyperus esculentus

488. Eleocharis equisetoides

489. Eleocharis quadrangulata

490. Eleocharis robbinsii

491. Eleocharis rostellata

492. Eleocharis parvula

6. **E. pauciflora** (Lightf.) Link

Map 493. Almost always on wet sandy, gravelly, usually marly shores and flats, sometimes in marshy places or bogs; also reported from a cedar bog and a dry clay bank.

American plants may be differentiated from the European as var. *fernaldii* Svenson. The scales, especially the middle ones, are acute to acuminate or prolonged, usually dark brown with thin broad scarious margins, the lowermost fertile. See note under *E. acicularis*.

7. **E. radicans** (Poiret) Kunth

Map 494. A species of more southern range, collected only twice in Michigan: Olivet [Eaton County] (*H. L. Clark* in 1905, GH); Mud Lake bog (mixed with *E. olivacea*), Washtenaw County (*L. Wehmeyer & A. H. Smith* in 1930, MICH, BLH). Ranging from the West Indies and South America northward to southern California, Mexico, Texas, and Oklahoma, *E. radicans* is quite isolated at stations in Michigan and southern Virginia.

8. **E. acicularis** (L.) R. & S.   Fig. 259

Map 495. Wet sandy or mucky shores and hollows, mud flats, riverbanks.

Small plants, with capillary culms, often about 5 cm tall (sometimes 10 cm or even more if submersed); the tufts (connected by slender rhizomes) may form dense carpets, both on wet shores and submersed. Two other common small species (though usually with slightly thicker culms) are sometimes confused with this one, at least when mature achenes are not present: *E. intermedia* and *E. olivacea*. In *E. intermedia* the anthers are less than 0.5 mm long; in the other two species, slightly over 0.5 mm. The basal scales of the spikelets in *E. acicularis* are fertile; in *E. intermedia* and *E. olivacea*, they are sterile. *E. olivacea* and to some extent *E. acicularis* both have the sheaths thin, whitish, and loose at the summit. *E. pauciflora* when very young may be readily distinguished by its much larger anthers (ca. 1.5−2.5 mm) and rather prominent blunt often slightly swollen tips on the sterile culms.

9. **E. intermedia** Schultes

Map 496. Sandy-mucky shores, stream margins, marshy ground; especially characteristic of exposed mud flats, drying lakes, and beach pools.

10. **E. melanocarpa** Torrey   Fig. 260

Map 497. Moist sandy, mucky, or boggy shores of a large number of small lakes in the southwestern Lower Peninsula, especially by receding lakes. A species of disjunct distribution, not found between Michigan and northern Indiana and the Coastal Plain (and inland Virginia).

The culms tend to arch and root at the tip. Plants lacking achenes may thus resemble *E. rostellata*, from which they may be easily distinguished by the presence of a definite narrow tooth (up to 1 mm long) at the summit of the leaf sheath in *E. melanocarpa*; in *E. rostellata* the sheaths are truncate with at most a callous swelling.

11. E. tricostata Torrey

Map 498. Another Coastal Plain disjunct species, found in Michigan only at a site in Ottawa County later destroyed by construction of a motel; it was a sandy pool in oak woods about a mile southeast of Fruitport (*Bazuin 3230* in 1941, MICH, MSC, BLH; and *5027* in 1942, MSC, GH). The species may well occur elsewhere in the southwestern part of the state.

The summit of the leaf sheath has a distinct tooth which, along with the very stout rhizome and absence of perianth bristles, distinguishes this species from *E. rostellata*, to which it might be keyed since the tubercle is very small and scarcely constricted.

12. E. compressa Sulliv.

Map 499. The best material of this species in the state is found in shallow soil and crevices of dolomite on Drummond Island; a few other collections appear also to be *E. compressa*: cranberry-sedge zone of tamarack bog in Pennfield Tp., Calhoun County (*G. E. Crow 95* in 1967, MSC); 3–4 mi. east of Mackinaw City on the bay, Cheboygan County (*Bazuin 7959* in 1948, BLH); borders of a ditch, Alto, Kent County (*F. P. Daniels* in 1898, MSC).

See discussion under the next species.

493. Eleocharis pauciflora

494. Eleocharis radicans

495. Eleocharis acicularis

496. Eleocharis intermedia

497. Eleocharis melanocarpa

498. Eleocharis tricostata

13. E. elliptica Kunth

Map 500. Low often marly ground: beaches, stony and often marshy shores, interdunal pools and flats, sandy swales and ditches; occasionally in bogs and along trails in conifer swamps; sometimes in shallow water, with culms over 5 dm long, although normally they are much shorter.

Included here are plants referred by Svenson to *E. compressa* var. *atrata* Svenson, which seems to be more closely related to *E. elliptica* — from which, in fact, I find it indistinguishable. The two taxa intergrade considerably in characteristics of culm (slightly to not at all flattened), achene and tubercle size and shape, scale shape (ovate-elliptic to ovate-lanceolate) and color (reddish brown to black). (See also Drapalik & Mohlenbrock in Am. Midl. Nat. 64: 501–502. 1960.) Typical *E. compressa* as here narrowly defined differs slightly more sharply, as indicated in the key, but there might be good grounds for considering all of these plants as belonging to a single variable taxon, in which case the previous species would be called *E. elliptica* var. *compressa* (Sulliv.) Drapalik & Mohlenb. The very flattened, ± twisted culms of *E. compressa* are striking in the field; some plants from Thunder Bay Island, Alpena County, in similar situations, do not have this aspect although they approach *E. compressa*.

14. E. olivacea Torrey

Map 501. Moist sandy to muddy shores, exposed mud flats, bog mats; often in marly places and sometimes in several inches of water.

The styles are occasionally 3-cleft.

15. E. caribaea (Rottb.) S. F. Blake

Map 502. This is a widespread tropical species, occurring on the Coastal Plain in the southern United States and rarely found in the Great Lakes region. Wet sandy, marly borders of ponds in northwestern Washtenaw County and northeastern Jackson County.

In all of our specimens, the brownish bristles are shorter than the achene, although they are usually longer in this species. Inland plants of *E. caribaea* have been called var. *dispar* (E. J. Hill) S. F. Blake; also known as *E. geniculata* (L.) R. & S.

499. Eleocharis compressa   500. Eleocharis elliptica   501. Eleocharis olivacea

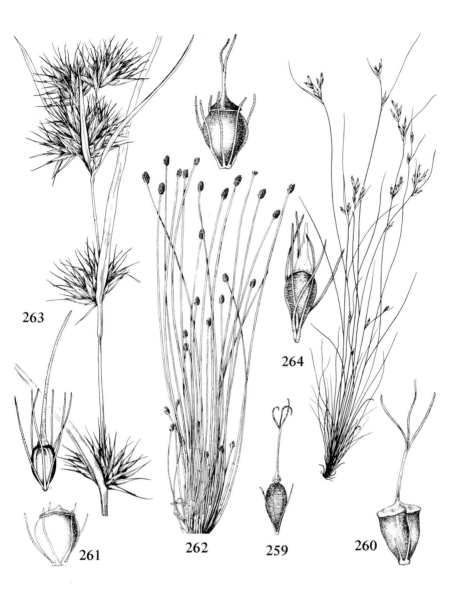

259. *Eleocharis acicularis,* achene ×20
260. *E. melanocarpa,* achene ×10
261. *E. engelmannii,* achene ×10
262. *E. obtusa* ×½; achene ×20
263. *Rhynchospora macrostachya* ×½; achene ×2
264. *R. capillacea* ×½; achene (f. *capillacea*) ×10

16. **E. erythropoda** Steudel

Map 503. Low ground of all kinds: sandy, gravelly, or muddy shores, stream margins, and riverbanks, often in shallow water; wet meadows; bogs and conifer swamps (cedar, tamarack).

The spikelet is frequently infested with an ergot-like fungus (*Claviceps*). The achenes (including tubercle) are (1.5) 1.6–1.8 (2) mm long. See comments under the next species. Long known as *E. calva* Torrey, an invalid name.

17. **E. smallii** Britton

Map 504. In almost all kinds of wet places; especially common in shallow water of marshes and along marshy shores and river margins; also in bogs, wet meadows, swamp borders, etc.

The achenes range from 1.5 to 2.8 mm long. Plants with slender culms and small achenes are rather easily confused with the preceding species, but large robust plants are easily recognized. The summits of the leaf sheaths in *E. smallii* tend to be blacker than the merely darkened ones of *E. erythropoda*. The tubercle ranges from dome-shaped (no longer than broad) to elongate in both species, though tending to be shorter and more conical in the preceding one. Included in the European *E. palustris* (L.) R. & S. by many eastern American authors.

502. Eleocharis caribaea

503. Eleocharis erythropoda

504. Eleocharis smallii

505. Eleocharis ovata

506. Eleocharis engelmannii

507. Eleocharis obtusa

18. E. ovata (Roth) R. & S.

Map 505. Very local in muddy or other moist places.

Occasional tall specimens of *E. olivacea* with soft culms might superficially resemble this species, but differ in lacking firm summits on the leaf sheaths and in having both the body and the tubercle (not flattened) of the achene greenish.

19. E. engelmannii Steudel   Fig. 261

Map 506. Very local in low ground.

20. E. obtusa (Willd.) Schultes   Fig. 262

Map 507. Moist sandy to muddy shores and fields, marshes, waterholes; especially characteristic of exposed mud flats and drying shores of receding lakes, where the large clumps of soft culms are easily recognized.

Intermediate specimens with the two preceding species are occasionally found. The three have sometimes been considered as representing a single variable species.

6. Rhynchospora   Beak-rush

A magnification of 15× or more is recommended to see the nature of the barbs on the perianth bristles; look toward the end of the bristle to see the barbs to best advantage.

KEY TO THE SPECIES

1. Body of achene ca. 4.5–5 mm long, with a stout beak about 3 or more times as long exserted beyond the scale; anthers ca. 3.5–4 mm long . . . . . . 1. **R. macrostachya**
1. Body of achene less than 2.5 mm long, the beak or tubercle shorter; anthers ca. 1–2.5 mm long
  2. Achenes prominently transversely ridged or wrinkled; tubercle ± deltoid, ca. 0.5–0.7 mm wide and nearly as high; bristles shorter than body of achene, minutely antrorsely barbed . . . . 2. **R. globularis**
  2. Achenes smooth or nearly so; tubercle longer; at least the longer bristles about equalling or surpassing tubercle, antrorsely or retrorsely barbed or smooth
    3. Perianth bristles minutely antrorsely barbed; tubercle usually green, basally serrulate; scales deep brown; filaments broadly flattened; plants with slender rhizomes . . . . 3. **R. fusca**
    3. Perianth bristles minutely retrorsely barbed or smooth (in *R. alba,* ascending-villous at base); tubercle and scales various; filaments narrower; plants tufted, not rhizomatous
      4. Bristles ca. 9–12, ± antrorsely villous at base; scales pale brown or (especially when fresh) whitish . . . . 4. **R. alba**
      4. Bristles 6 (very rarely 9 in *R. capillacea*), not hairy at base; scales brown
        5. Spikelets mostly several and spreading in fan-shaped clusters; achenes broadly obovoid, 1.1–1.7 mm long (not including tubercle), with prominent whitish margins; leaves flat . . . . 5. **R. capitellata**
        5. Spikelets mostly 1–4 (7) in each ± narrow, ascending, ellipsoid cluster; achenes oblong-ellipsoid to narrowly obovate, 1.8–2.4 mm long (including a prominent stipe-like base, but not the tubercle), often with a less prominent pale margin; leaves involute-setaceous . . . . 6. **R. capillacea**

1. **R. macrostachya** Gray   Fig. 263

Map 508. Local, but often common where found, on sandy-mucky, often marly, shores and in marshy hollows; in a marly bog adjacent to Muskegon State Park. As with its most interesting associates (*Fuirena squarrosa, Hemicarpha micrantha, Psilocarya scirpoides*, etc.), seems to thrive best on recently exposed shores and dried lake beds. This is a plant of primarily Coastal Plain distribution, extending northward in the Mississippi valley and recurring in the western Great Lakes area. A large, conspicuous species, it serves in the field as a good indicator of the sort of habitat where one should drop to his hands and knees in search of the smaller Coastal Plain disjuncts.

Sometimes classified as *R. corniculata* (Lam.) Gray var. *macrostachya* (Gray) Britton.

2. **R. globularis** (Chapman) Small

Map 509. The only Michigan collection of this species was made in "prairie bogs" at Sturgis (*F. P. Daniels* in 1898, MSC), where it was noted as rare.

The typical variety is primarily a Coastal Plain plant; in the interior our specimens are referred to var. *recognita* Gale.

3. **R. fusca** (L.) Ait. f.

Map 510. Sandy-peaty shores, interdunal hollows, sandy excavations, boggy

508. Rhynchospora
macrostachya

509. Rhynchospora
globularis

510. Rhynchospora fusca

511. Rhynchospora alba

512. Rhynchospora
capitellata

513. Rhynchospora
capillacea

meadows; occasionally on bog mats but more often at remnants of bogs and often in somewhat marly places.

The very slender involute-filiform leaves of this species will generally distinguish it quickly from the flat-bladed *R. capitellata*, if good achenes are not available for examination of the barbs on the bristles.

4. **R. alba** (L.) Vahl

Map 511. Most often in bogs (sometimes marly) and open conifer swamps (tamarack, black spruce) and on boggy shores, but also in wet meadows.

5. **R. capitellata** (Michaux) Vahl

Map 512. Low open ground: moist sandy or peaty shores, wet meadows and marshy places, ditches and swales, marly bogs.

Sometimes considered a variety of *R. glomerata* (L.) Vahl, in which case the correct epithet is var. *minor* Britton. In f. *discutiens* (Britton) Gale, the bristles are not barbed; this form has been collected very rarely in Michigan (Allegan, Muskegon, and Newaygo counties).

6. **R. capillacea** Torrey   Fig. 264

Map 513. Wet sandy or stony shores, boggy beach pools, interdunal flats, clearings or trails in bogs, peaty marshes and meadows, usually in calcareous situations.

Plants with the bristles smooth (not barbed) are about as common as the typical barbed form, and are f. *leviseta* (Gray) Fern. The two forms sometimes grow together. The surface of the achene in this species is minutely pebbled with smoothish wrinkles, and is obscurely marked with transverse dark lines.

7. **Cladium**

1. **C. mariscoides** (Muhl.) Torrey   Plate 3-D; fig. 265   Twig-rush

Map 514. Sandy, boggy, or marshy shores, sometimes in shallow water; sphagnum bogs, where the rhizomes are an important component of floating mats; ponds and interdunal swales; often in marly places.

514. Cladium mariscoides   515. Bulbostylis capillaris   516. Psilocarya scirpoides

The achenes are very distinctive, being somewhat acorn-shaped at maturity, squarely truncate and often a little flaring at the base.

8. Bulbostylis

1. **B. capillaris** (L.) Clarke   Fig. 266
Map 515. Dry sandy open ground: shores, fields, "blow-outs," and roadsides in dry prairies.
See comments under *Fimbristylis autumnalis*.

9. Psilocarya

1. **P. scirpoides** Torrey   Fig. 267   Bald-rush
Map 516. Local on sandy-mucky shores, becoming abundant when water levels recede; also in sandy-peaty lake beds and in a marly bog at the edge of Muskegon State Park. This is one of our most striking Coastal Plain disjunct species, occurring from Massachusetts to North Carolina with no intermediate stations known before it reappears in Michigan, central Wisconsin, and northern Indiana. A smut, *Cintractia psilocaryae*, has been found infesting the spikelets at stations in all these inland states as well as along the coast.

A similar species, *P. nitens* (Vahl) Wood, was once known from the dune country of northwestern Indiana, and might be discovered in southwestern Michigan. It has a short, broad tubercle in contrast to the elongate beak of *P. scirpoides*.

10. Hemicarpha

1. **H. micrantha** (Vahl) Pax   Fig. 268
Map 517. Locally common on sandy-mucky shores of receding lakes, but easily overlooked because of its small size; its rather curly culms are quite inconspicuous when growing in a dense stand of *Fimbristylis autumnalis*, *Scirpus smithii*, and other associates. In Michigan, *Hemicarpha* is usually found in the same habitats as our Coastal Plain disjuncts, although it is actually of much wider distribution in the eastern and western United States. Although not known from the Detroit area in Michigan, it was collected long ago on the Detroit River at Amherstburg, Ontario.

11. Scirpus   Bulrush

REFERENCES

Schuyler, Alfred E. 1964. A Biosystematic Study of the Scirpus cyperinus Complex. Proc. Acad. Phila. 115: 283–311.
Schuyler, Alfred E. 1967. A Taxonomic Revision of North American Leafy Species of Scirpus. Proc. Acad. Phila. 119: 295–323.

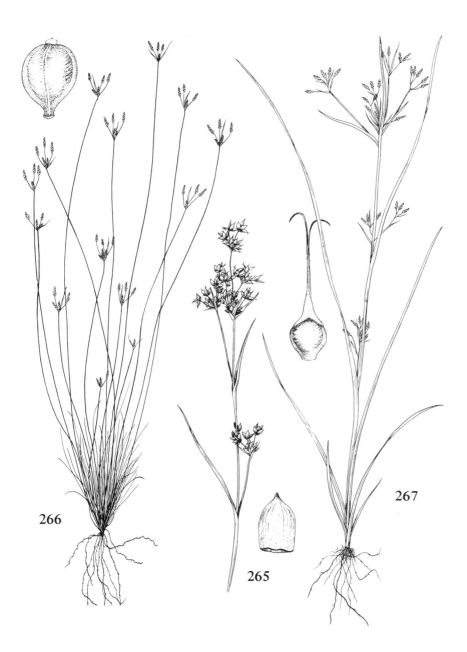

265. *Cladium mariscoides* ×½ ; achene ×5
266. *Bulbostylis capillaris* ×½ ; achene ×30
267. *Psilocarya scirpoides* ×½ ; achene ×10

KEY TO THE SPECIES

1. Spikelet single, erect, and terminal, not exceeded by an involucral bract (the lowermost scale with ± prolonged and callous tip, this sometimes exceeding spikelet by as much as 3 mm in *S. cespitosus*)*
  2. Culms terete or several-ridged, smooth; anthers 1.1–2.5 mm long (not including a short prolonged tip); body of achenes 1.5–1.7 mm long . . . . 1. **S. cespitosus**
  2. Culms 3-angled, scabrous; anthers 0.6–1.1 mm long; achenes various
    3. Perianth bristles ciliate, slightly or not at all exceeding the blunt achene; scales of spikelet not more than 7; achenes ca. 1.6–1.8 mm long . . . . 2. **S. clintonii**
    3. Perianth bristles smooth, several times as long as the apiculate achene at maturity; scales of spikelets slightly more than 7; body of achene less than 1.5 mm long . . . . 3. **S. hudsonianus**
1. Spikelets (1) 2–many, apparently lateral or terminal, often exceeded by one or more involucral bracts
  4. Inflorescence apparently lateral, one erect or somewhat divaricate 3-angled or terete involucral bract appearing to be a continuation of the culm, from the side of which the inflorescence thus appears to burst (see figs. 270, 271, 272b)
    5. Spikelets (at least several of them) distinctly pedicelled (congested in a form of *S. acutus*); culms terete, stout, 5–20 mm thick at base, often over 1 m tall
      6. Styles 3-cleft; achenes 3-sided; scales glabrous on the back; culms firm; perianth bristles 2–4 (5); mature achenes ca. 2.5 mm long, including short apiculus . . . . 4. **S. heterochaetus**
      6. Styles 2-cleft; achenes plano-convex (flat on one side and rounded on the other) or biconvex; scales puberulent on back and achenes shorter, or culms soft and easily compressed; bristles mostly 6
        7. Culms rather soft and easily compressed, pale blue-green when fresh; spikelets ovoid (about twice as long as wide, or shorter), in a ± open, lax inflorescence; scales ± shiny, rich orange-brown, often with prominent greenish midrib, the margins ciliate but the backs essentially glabrous (puberulence and swollen red flecks, if any, limited to region of midrib); mature (dark gray or lead colored) achenes ca. 1.6–2.1 (2.4) mm long, including apiculus, barely covered by the scales . . . . 5. **S. validus**
        7. Culms firm and dark olive green when fresh; spikelets ovoid to cylindrical (often 2.5 or more times as long as wide), usually in a stiffer, sometimes condensed, inflorescence; scales duller, basically pale or whitish brown, the midrib not strongly contrasting, the margins often more copiously ciliate than in the preceding species, and the backs copiously flecked with shiny red dots, often puberulent (fig. 272a); mature achenes ca. 2.2–2.7 mm long, including apiculus, completely hidden by the scales . . . . 6. **S. acutus**
    5. Spikelets 1–few, crowded, sessile or nearly so (rarely one on a short pedicel); culms 3-angled or terete (if terete, then slender, soft, and not over 1 m tall)
      8. Spikelet 1, strongly ascending, the involucral bract surpassing its tip by not more than 15 (20) mm; leaves normally many, capillary, submersed; culm seldom over 1 mm thick; anthers (2.1) 2.5–3.5 mm long; achenes 3-sided, the body ca. 2.5–3 mm long . . . . 7. **S. subterminalis**
      8. Spikelets usually more than 1 and the involucral bract surpassing them by more than 15 mm (except in smallest plants of some populations); leaves stiff and culms thicker; anthers and achenes various

*Some specimens of *S. subterminalis* with very short involucral bract might seem to run here, but may be distinguished by much longer anthers and limp, capillary, aquatic leaves.

9. Plants annual, with soft, terete or obscurely 3-angled, tufted culms; anthers
0.3–0.7 mm long
 10. Achene covered with prominent transverse ridges (fig. 269); perianth
 bristles none . . . . 8. **S. hallii**
 10. Achene smooth or obscurely pitted (fig. 270); perianth bristles present or
 absent. . . . . 9. **S. smithii**
9. Plants perennial, with elongate rhizomes; culms sharply 3-angled, at least
above; anthers 1–3.1 mm long
 11. Midrib of scale ± greenish, excurrent as a short (not over 0.5 mm) tip
 extending *beyond* the tapered (sometimes very slightly notched) apex of
 the scale; bristles slightly exceeding body of achene; rhizome soft; achene
 with apiculus 0.5 mm or more in length; styles 3-cleft and achenes 3-sided;
 leaves more than half as tall as the culms . . . . 10. **S. torreyi**
 11. Midrib of scale brown, excurrent either as a very short tip *not* exceeding
 the broad lobes at apex of scale or as a *long* (0.5–1 mm) tip equalling or
 exceeding lobes; bristles shorter than body of achene; rhizome firm and
 hard; achene with shorter apiculus; styles usually 2-cleft and achenes
 biconvex to plano-convex (occasionally some styles 3-cleft and achenes
 3-sided in a spikelet); leaves less than half as tall as the culms
  12. Scales with firm excurrent tip ca. 0.5–1 mm or more long, usually
  exceeding lobed apex of scale; culm firm, 3-angled but not winged;
  anthers (1.7) 2–3 mm long (including short scabrous tip); body of achene
  (2.3) 2.5–3 mm long; involucral bract 3.5–11 (15) cm long; plant
  common in wet ground . . . . 11. **S. americanus**
  12. Scales with excurrent tip less than 0.5 mm long, equalling or shorter than
  lobes at apex of scale; culms soft, very sharply 3-angled or -winged;
  anthers and achenes slightly smaller than in the preceding; involucral
  bract up to 5 cm long, usually shorter; plant local in salt marshes . . 12. **S. olneyi**
4. Inflorescence terminal, not appearing to be lateral, subtended by (1) 2–several
divaricate to reflexed flat leaf-like bracts (see figs. 273, 274)
 13. Perianth bristles none; spikelets pedicelled in a loose inflorescence; involucral
 bracts up to 1.5 mm wide. . . . . **Fimbristylis** (genus no. 13)
 13. Perianth bristles present; spikelets and involucres various
  14. Spikelets (10) 15–36 mm long, 5–9 mm wide; achenes 3–5 mm long,
  including apiculus; anthers ca. (2.3) 4–5 mm long; scales ± puberulent, tipped
  with a prominent outwardly curved awn; culms sharply 3-angled nearly or
  quite to the base; rhizomes with large corm-like thickenings
   15. Styles 2-cleft; achenes biconvex, 3–4 mm long; involucre with 2 well
   developed bracts; perianth bristles weak, scarcely half as long as achene,
   obscurely barbed . . . . 13. **S. paludosus**
   15. Styles 3-cleft; achenes distinctly 3-angled, 4–5 mm long; involucres mostly
   with 3–5 well developed bracts; perianth bristles strong, distinctly
   retrorse-barbed . . . . 14. **S. fluviatilis**
  14. Spikelets, achenes (0.9–1.2 mm), and anthers (0.5–1.3 mm) smaller; scales
  glabrous; culms terete or obtusely 3-angled, or sharply 3-angled toward
  summit; rhizome without corm-like enlargements
   16. Perianth bristles ± straight (or some with 1–2 kinks), slightly if at all
   exceeding achene (less than twice as long), minutely retrorsely barbed,
   sometimes rudimentary; plants mostly solitary or few together, from strong
   rhizomes; spikelets grouped into numerous hemispherical to globose heads,
   each consisting of 5–25 or more spikelets (or some on a plant as few as 3);
   base of involucre not blackened (a little reddish in nos. 16 & 17)

17. Leaf sheaths all green; bristles (often rudimentary) barbed only on apical half or third; styles 3-cleft; midrib narrow and green, excurrent on most scales as a short but distinct mucro; summit of culm smooth or slightly scabrous . . . . 15. **S. atrovirens**

17. Leaf sheaths strongly tinged with red, at least toward base of plant; bristles barbed nearly to their bases; characteristics of styles, scales, and culms not combined as above

   18. Styles 3-cleft; achenes 3-angled; bristles (3−) 6; summit of culm usually scabrous . . . . 16. **S. expansus**

   18. Styles 2-cleft; achenes 2-sided; bristles 4 (−5); summit of culm usually smooth . . . . 17. **S. microcarpus**

16. Perianth bristles much crinkled or curled at maturity, exceeding the achene, smooth (fig. 274); plants usually in dense clumps; spikelets mostly solitary or in groups of 2−5 on pedicels; base of involucre blackened or not

   19. Bristles inconspicuous, scarcely exserted beyond the scales, about twice as long as achene; spikelets cylindric, mostly solitary on pedicels of various lengths; scales orange-brown with conspicuous green midrib excurrent as a very short awn; involucre not blackened at the base (at most a little brownish) . . . . 18. **S. pendulus**

   19. Bristles conspicuously exceeding the scales and achenes at maturity; spikelets short-ovoid to nearly spherical (very rarely cylindric), solitary on pedicels or in groups of mostly 2−5; scales various but midrib not conspicuously excurrent; involucre usually blackened (sometimes reddish brown) at base . . . . 19. **S. cyperinus**

1. S. cespitosus L.

Map 518. In dense tufts: bogs, boggy shores, and cedar thickets, often in marly places; also in rock crevices and pools along Lake Superior − thence northward to the Arctic where it is a major tundra species.

American plants have been distinguished from those of western Europe as var. *callosus* Bigelow. According to Farwell, the culms of this species may arch over and root at the tip, as in *Eleocharis rostellata.*

2. S. clintonii Gray

Map 519. Dry hillsides and banks; open sandy ground such as oak and pine woodland; very local.

| 517. Hemicarpha micrantha | 518. Scirpus cespitosus | 519. Scirpus clintonii |

3. S. hudsonianus (Michaux) Fern.   Plate 3-E

Map 520. Bogs and boggy ground, including conifer swamps and beach pools; meadows and shores, especially in marly or springy places; often abundant in clearings and in sandy or peaty excavations, the mature perianth bristles looking from a distance like fresh-fallen snow.

Sometimes included in the genus *Eriophorum*, as *E. alpinum* L.

4. S. heterochaetus Chase

Map 521. Dubiously admitted to our flora on the basis of two specimens labeled as collected at Lansing (or the Agricultural College) (without collector, in 1871, MSC). These lack habitat data but presumably came from wet ground.

The spikelets on the Michigan specimens are not all solitary on pedicels, as is sometimes claimed for this species, although most of them are on very slender elongate pedicels; however, the achenes are clearly 3-sided, the culms firm, and the pale scales glabrous on the backs.

5. S. validus Vahl   Softstem Bulrush

Map 522. Wet shores and shallow water of ponds, lakes, rivers, ditches, and marshes; occasionally in bogs; on sand, marl, or peat. May form large stands by itself or with other marsh species.

'Dried specimens are especially difficult at times to distinguish from the next species. Even more than usual, one must weigh both choices in the key with the expectation that a plant will fit neither perfectly. Where the two species grow near each other, intermediate hybrids may be expected. There seems to be a tendency for the anthers of this species to be slightly shorter [1−1.6 (2.1) mm long] than in *S. acutus* [(1) 1.5−2.5 mm]. The flattened pedicels in *S. validus* are scabrous on the two edges and smooth on the sides; in *S. acutus*, they are sometimes puberulent and/or red-flecked like the scales. The common plant of the United States and Canada has been distinguished from the typical variety of the eastern American tropics as var. *creber* Fern.

6. S. acutus Bigelow   Figs. 272a, 272b   Hardstem Bulrush

Map 523. On wet shores or more often in shallow water of ponds, lakes,

520. Scirpus hudsonianus

521. Scirpus heterochaetus

522. Scirpus validus

rivers, and ditches; in bog lakes and sometimes relic in open tamarack swamps; on sand, gravel, marl, or peat. Usually in water up to about 5 feet deep, but plants with culms as long as 14 feet have been collected in deeper water. May form pure stands, as in *S. validus*, or grow with other marsh species such as *Phragmites australis*, *Typha* spp., *Scirpus americanus*, *S. validus*.

See comments under the preceding species. In the not uncommon form *congestus* (Farw.) Fern. (TL: Marl Lake [Oakland County]), the pedicels are nearly or quite suppressed, the spikelets therefore being ± densely crowded.

7. S. subterminalis Torrey   Fig. 5

Map 524. Usually submersed (except for tip of the fertile culm) in water up to 3 or 4 feet deep in lakes, ponds, bog pools, rivers, and boggy ditches, growing on sand, marl, muck, or peat.

The submersed leaves are often abundant, forming grass-like beds; they are contrasted in Key A, couplets 22–24, with similar structures of *Eleocharis robbinsii* and *Juncus militaris* which, although more local, may grow in the same places. The terrestrial form [f. *terrestris* (Paine) Fern.] has apparently been collected only very rarely in Michigan. The culms are somewhat thicker than in the aquatic form. The terrestrial form is to be expected when water levels are abnormally low.

8. S. hallii Gray   Fig. 269

Map 525. First found in Michigan by C. D. McLouth at "Five Lakes" in Muskegon County on August 22, 1900 (MSC); reported by Beal (1905). Very rare throughout its range, the species has not been attributed to Michigan in subsequent manuals. A thorough search of the Five Lakes region (all are now usually dried up) in 1959 revealed a small colony of *S. hallii* thriving amid other sedges on the sandy exposed shore of Carr Lake (*Voss 9152*, MICH, BLH, NY, GH, SMU). The plants were much more prostrate in aspect and smaller than *S. smithii*, which was common at the site. The achenes of *S. hallii* tend to be more persistent than their scales and thus are conspicuous; mature achenes are occasionally formed at the very base of the plant. Another colony was found in

523. Scirpus acutus

524. Scirpus subterminalis

525. Scirpus hallii

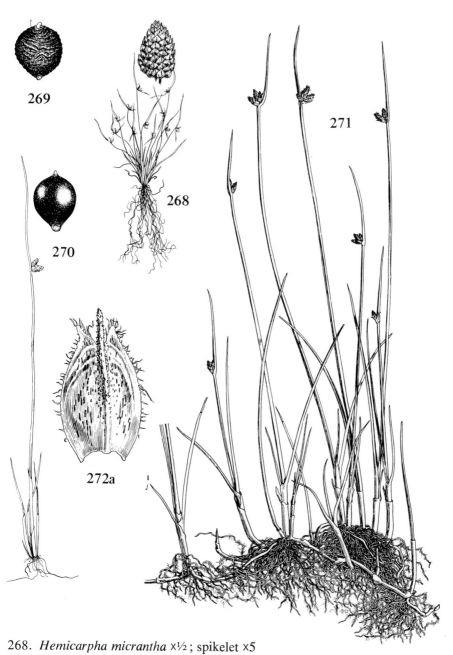

268. *Hemicarpha micrantha* ×½; spikelet ×5
269. *Scirpus hallii,* achene ×10
270. *S. smithii* ×½; achene (f. *smithii*) ×10
271. *S. americanus* ×¼
272a. *S. acutus,* scale ×8

1966 at Carr Lake by A. E. Schuyler. The locality is an excellent one for Coastal Plain disjuncts and other rare species. (See Mich. Bot. 6: 17–18. 1967.)

Sometimes recognized as *S. supinus* L. var. *hallii* (Gray) Gray.

9. **S. smithii** Gray   Fig. 270

Map 526. Sandy to mucky or boggy, sometimes marly, shores and hollows, especially where water levels have receded; frequently associated with Coastal Plain disjuncts such as *Psilocarya scirpoides* and *Fuirena squarrosa* when in the same habitats.

The perianth bristles are absent or rudimentary in the typical form; in f. *setosus* (Fern.) Fern., they are present and retrorsely barbed. In stoutly defending the distinctness of *S. purshianus* Fern. (supposedly distinguished in part by biconvex rather than plano-convex achenes), Fernald asserted that "all living botanists who accurately know living plants in the field find them abundantly distinct" (Rhodora 44: 480. 1942). However, Deam had not found them distinct in Indiana, pointing out that one can find both shapes of achenes on the same plant, and monographers differ in their interpretations. I have been unable to distinguish the species in our flora or to apply the distinctions mentioned by Koyama (Canad. Jour. Bot. 40: 918. 1962).

10. **S. torreyi** Olney

Map 527. Wet sandy or peaty shores and shallow water, very local.

11. **S. americanus** Pers.   Fig. 271   Threesquare

Map 528. One of our commonest and most easily recognized sedges. Wet sandy, marly, gravelly, or peaty shores; marshy borders of ponds, lakes, and streams; beach pools and sandy flats; often in shallow water (up to about 1 foot or even 2.5 feet).

12. **S. olneyi** Gray

Map 529. Salt marshes near the Maple and Grand rivers.

Field observations in a marsh on the north side of the Maple River in Clinton County where this species and *S. americanus* were found growing in proximity in

526. Scirpus smithii

527. Scirpus torreyi

528. Scirpus americanus

1960 indicate that there is hybridization of the two. Plants simulating *S. olneyi* in the soft, winged culms had scales approaching those of *S. americanus*. On the same day (July 23) achenes were well developed in the plants of *S. americanus*, while the fruit and anthers were much less mature in *S. olneyi* and intermediates. The two have also apparently hybridized in Gratiot County, where plants were first collected by Douglass Houghton in 1837 (Mich. Bot. 9: 241. 1970).

13. S. paludosus Nelson

Map 530. A western species, known in Michigan from a single collection from ditches at Midland (*Dreisbach 618* in 1914, MICH).

A slightly more slender plant than the next, with more congested inflorescences and paler brown scales.

14. S. fluviatilis (Torrey) Gray

Map 531. Marshes (including salt marshes), wet shores and riverbanks, swales.

Varies greatly from year to year in number of culms producing inflorescences, but the robust, sharply triangular leafy stems are easily recognized when sterile.

15. S. atrovirens Willd.   Fig. 273

Map 532. Wet ground generally: shores, marshes, meadows and low fields, ditches; moist openings, clearings, and roadsides in deciduous or coniferous woods, swamps, and thickets; creek and river banks.

Late in the season, the spikelets may produce leafy tufts [f. *proliferus* Hermann (fig. 273)]. The species is here treated in an inclusive sense. Some of our specimens (from throughout the state, from Isle Royale to Kalamazoo and Wayne counties) have been referred by Schuyler to *S. hattorianus* Mak., a species originally described from Japan. It is said to differ from *S. atrovirens*, in a restricted sense, in having more delicate perianth bristles, at most about equalling the achene, and smooth or nearly smooth leaf sheaths; whereas *S. atrovirens* has coarser bristles (the longer ones generally exceeding the achene) and usually septate-nodulose sheaths on the lower leaves (see Schuyler, Not. Nat. 398. 1967). A few specimens from southern Michigan and Keweenaw County have been annotated by Schuyler as apparent hybrids, *S. atrovirens* ×

529. Scirpus olneyi   530. Scirpus paludosus   531. Scirpus fluviatilis

hattorianus; these have abortive seeds (and, incidentally, are almost all viviparous).

16. **S. expansus** Fern.

Map 533. Marshy ground, river and ditch banks, along streams in woods.

In certain collections from Berrien, Cass, Kent, Kalamazoo, and Ottawa counties, the summit of the culm is unusually smooth; normally it is quite scabrous in this species.

17. **S. microcarpus** Presl

Map 534. Wet open or slightly shaded, often marshy, areas, including damp sandy shores of Lake Superior and nearby inland lakes, borders of thickets and swamps, riverbanks. Essentially complementary in distribution within Michigan with *S. expansus*.

In a collection from Ogemaw County, the summit of the culm is very scabrous, and it is ± so in a collection from Houghton County; normally it is smooth. Eastern American plants of this transcontinental species have been segregated as var. *rubrotinctus* (Fern.) M. E. Jones or *S. rubrotinctus* Fern.

18. **S. pendulus** Muhl.

Map 535. Borders and clearings in swampy woods, roadsides and ditches, shores, occasionally in bogs. As long ago as 1849 noted by Dr. D. Cooley as

532. Scirpus atrovirens

533. Scirpus expansus

534. Scirpus microcarpus

535. Scirpus pendulus

536. Scirpus cyperinus

537. Fuirena squarrosa

having become common during the previous 10 years in the vicinity of Washington, Macomb County; apparently a species which thrives in clearings and along damp disturbed roadsides.

Rather easily recognized at a distance by the yellow-green foliage and very lax inflorescence, the pedicels drooping at maturity. This plant was long known as *S. lineatus* Michaux, a name which applies correctly to a species of the southern states previously known as *S. fontinalis*.

19. **S. cyperinus** (L.) Kunth   Fig. 274   Wool-grass

Map 536. Widespread, often abundant, in wet places: meadows, shores, marshes (seldom in more than a few inches of water), riverbanks, ditches, bogs, deciduous or coniferous swampy woods and thickets (especially along borders, clearings, roads, etc.) – sometimes as tall as 2 meters.

In young spikelets, the bristles are not yet elongated, and the species might then be confused with others. The spikelets tend to be fewer in a group than in *S. atrovirens*, and the scales are not like those of *S. pendulus* (although the midrib may be greenish); the blackened base of the involucre in our two commonest varieties is usually a helpful character in immature material.

The species is here treated in a very broad sense. Plants vary in length of pedicels, color and length of scales, and color of involucral bracts. Our two commonest varieties are var. *cyperinus* and var. *brachypodus* (Fern.) Gilly. Plants intermediate between these two (with most characters of one of them, and one or two characters of the other) are occasional in the Upper Peninsula and northern Lower Peninsula. Although maintaining them as separate species, Schuyler (1967) states "there is abundant hybridization and introgression between them" and he treated them as a single complex in 1964.

In var. *cyperinus* [including var. *pelius* Fern.], the spikelets are mostly sessile in clusters of 3–4 (or more), with the scales brown or reddish brown and usually less than 1.5 mm long; the flowers mature ± simultaneously through the spikelet. This variety occurs throughout the state, but is less common northward than var. *brachypodus*. In the southern half of the Lower Peninsula are some plants of var. *cyperinus* with more red-brown involucral bracts, slightly longer scales, and tendency (not always expressed) for all spikelets to be pedicelled. Such plants may resemble var. *pedicellatus* but have more definitely reddened scales and bracts.

In var. *brachypodus* [= *S. atrocinctus* Fern.], the spikelets are mostly pedicelled (except in the occasional plant in which the entire inflorescence is congested, with all branches and pedicels suppressed), with the scales predominantly blackish green (sometimes with a bit of reddish, especially basally), mostly 1.5–1.6 (1.7) mm long; the flowers mature gradually from the base to the tip of the spikelet, so that spikelets are often seen with intact flowers and scales distally and a naked rachilla basally. This variety is absent from the southern half of the Lower Peninsula; it is usually a more slender plant (in culms and leaves) than var. *cyperinus*. In var. *brachypodus* and most plants of var.

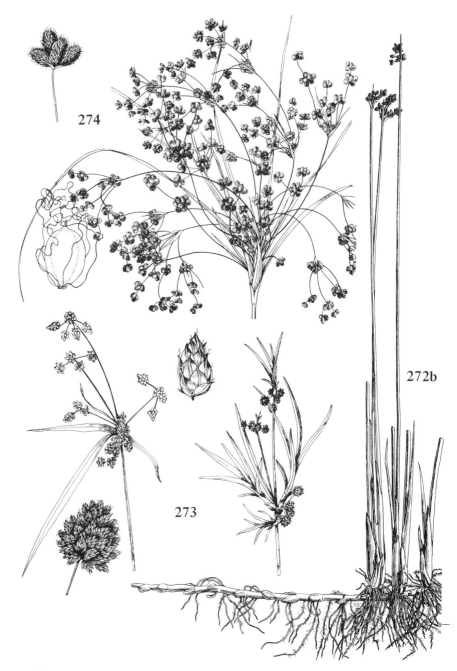

272b. *Scirpus acutus* ×⅕
273. *S. atrovirens* ×½; head ×2; spikelet ×5; proliferous inflorescence ×½
274. *S. cyperinus* ×½; head ×2; achene ×20

cyperinus the bracts of the inflorescence (in involucre and involucels) are blackish green.

Only two Michigan collections have been referred by Schuyler to var. *pedicellatus* (Fern.) Schuyler [*S. pedicellatus* Fern.] . This is a robust variety, but like var. *brachypodus* in having pedicelled spikelets maturing from the base to the apex and in having long scales; however, the scales and bracts are brown or gray-brown. It was collected in open swampy ground near L'Anse, Baraga County (*Dodge* in 1916, MICH); another 1916 Dodge collection from the Huron Mountains, Marquette County, is identified by Schuyler as intermediate between var. *cyperinus* and var. *pedicellatus*, resembling the latter but with scales less than 1.5 mm long. The other Michigan collection (*Schuyler 3108* in 1960, MICH) was made in the same Clinton County marsh along the Maple River where *S. olneyi* occurs. Some additional collections from the southwestern Lower Peninsula may also represent var. *pedicellatus*.

12. Fuirena

1. F. squarrosa Michaux   Fig. 275   Umbrella-grass

Map 537. Locally common on sandy-mucky shores, especially where the water level has receded. Our northern plants may be called var. *pumila* (Sprengel) Torrey [*F. pumila* Sprengel] . They differ from plants of the southern states in being annual and in various floral characters, but intergrades are numerous and Svenson (N. Am. Flora 18: 506. 1957) considers them a single species. The var. *pumila*, at least, is another disjunct between the Atlantic Coast and the western Great Lakes area, often found growing with other species of similar affinity.

The scales of the spikelets have long outwardly curved awns, which give the spikelets a very characteristic bristly appearance. The achenes are sharply 3-angled, elevated on short stalks.

13. Fimbristylis

REFERENCE

Kral, Robert. 1971. A Treatment of Abildgaardia, Bulbostylis and Fimbristylis (Cyperaceae) for North America. Sida 4: 57–227.

KEY TO THE SPECIES

1. Style 2-cleft, ciliate below; achenes biconvex, ca. 1.5 mm long; anthers 1.5–1.7 mm long; plant a stiff perennial . . . . 1. **F. puberula**
1. Style 3-cleft, glabrous; achenes 3-sided, ca. 0.7 mm long; anthers ca. 0.3 mm long; plant a delicate tufted annual . . . . 2. **F. autumnalis**

1. F. puberula (Michaux) Vahl

Map 538. Barely found as far north as Michigan, in prairie-like ground. Collected by the First Survey in Cass County in 1838 (MICH) and by Dodge on Harsen's Island, St. Clair County, in 1904 (MICH, BLH).

This species is here treated as defined by Kral, who assigns plants from this area to var. *puberula*. In the past, they have been variously recognized as *F. caroliniana* (Lam.) Fern., *F. drummondii* (Torrey & Hooker) Boeckl., or *F. spadicea* (L.) Vahl.

2. F. autumnalis (L.) R. & S.

Map 539. Sandy, marly, or mucky hollows and shores, especially where water levels have lowered; often abundant in a distinct zone. Very local northward.

Bulbostylis capillaris may run here in the key; the two species resemble each other closely. *Bulbostylis* differs in its capillary leaves, minute brown tubercle at the summit of the achene, and puberulent rounded scales. *Fimbristylis autumnalis* has flat leaves, no such tubercle, and glabrous acute to mucronate scales. Both have 3-sided achenes ca. 0.7–0.8 mm long and anthers ca. 0.3 mm long. *Psilocarya scirpoides* differs from both in its 2-sided achenes with long persistent style, the base of which nearly or quite covers the summit of the achene, and its larger anthers (0.6–0.9 mm). The spikelets of *P. scirpoides* are generally thicker and darker – a useful impression for field recognition. *Bulbostylis* usually does not occur on the moist shores where *Fimbristylis autumnalis* and *Psilocarya scirpoides* are often growing together.

14. Eriophorum   Cotton-grass; Bog-cotton

KEY TO THE SPECIES

1. Spikelet solitary, erect, without any involucral leaves; base of spikelet with several sterile scales . . . . 1. **E. spissum**
1. Spikelets 2 or more (very rarely 1), spreading or ± nodding (or erect in *E. virginicum*), subtended by 1 or more slender involucral leaves; base of spikelet with at most 1–2 sterile scales
  2. Cauline leaves with blades slender (up to 2 or rarely 3 mm wide), 3-angled (often channeled) their entire length; involucral leaf 1 (very rarely 2), shorter than the inflorescence; anthers 1–1.8 mm long; achenes 2.5–3.2 mm long
    3. Uppermost cauline leaf with blade about equalling or longer than its sheath; scales greenish brown or reddish brown (or occasionally becoming leaden toward edges); culms minutely scabrous above . . . . 2. **E. tenellum**
    3. Uppermost cauline leaf with blade distinctly shorter than its sheath; scales usually suffused with blackish or lead color; culms smooth . . . . 3. **E. gracile**
  2. Cauline leaves with blades flat, at least in their lower half, often wider (up to 6 mm); involucral leaves 2 or more, often exceeding inflorescence; anthers and achenes various
    4. Scales rather thick, brownish to reddish (sometimes a green band centrally), with midnerve inconspicuous (the scales either obscurely nerved or the lower ones ± equally 3–7-ribbed); anthers 0.7–1.5 mm long; mature achenes ca. 3–3.5 mm long; spikelets maturing (bristles elongating) after the middle of July (usually in August), ± congested, the pedicels short and the spikelets therefore erect or slightly spreading . . . . 4. **E. virginicum**

4. Scales thin, drab to blackish or lead colored, at least toward apex, usually with
   an evident midnerve (or if this obscure, then the anthers long); achenes often
   shorter; spikelets maturing in June or early July (though persisting into or even
   through the winter), the pedicels usually elongating, the longer ones at length ±
   nodding
   5. Midnerve of scales prominent only below the tip; anthers (2) 2.2–4.5 (5) mm
      long; summit of leaf sheaths with dark border; mature achenes mostly ca.
      2–2.3 (3) mm long . . . . 5. E. angustifolium
   5. Midnerve of scales prominent – even somewhat enlarged – to the very tip;
      anthers 0.8–1.5 mm long; summit of upper leaf sheath without dark border;
      mature achenes mostly ca. 3 (–3.4) mm long . . . . 6. E. viridi-carinatum

1. E. spissum Fern.
Map 540. Forming large tussocks in bogs and open conifer swamps (tamarack, black spruce).

Perhaps best treated as belonging to a widespread subarctic Eurasian and
North American species, as E. *vaginatum* L. ssp. *spissum* (Fern.) Hultén or E.
vaginatum var. *spissum* (Fern.) Boivin.

2. E. tenellum Nutt.
Map 541. Bogs, conifer swamps under spruce and tamarack, and occasionally
in other wet open places.

The scales somewhat resemble those of E. *virginicum*, and occasional specimens of the latter with narrow leaves may be confused. The scabrous upper
portion of the culm of E. *tenellum* is usually a helpful character (in E. *virginicum* it is scabrous at most for a few mm immediately below the inflorescence).

3. E. gracile W. D. J. Koch
Map 542. Chiefly in bogs, both on open mats and under spruce and tamarack; also in wet marshy ground and swales, and on shores.

4. E. virginicum L.   Plate 3-F   Tawny Cotton-grass
Map 543. Bogs, conifer swamps (tamarack, spruce, cedar), boggy thickets,
marshes, and open swampy or boggy ground.

538. Fimbristylis puberula   539. Fimbristylis   540. Eriophorum spissum
                                   autumnalis

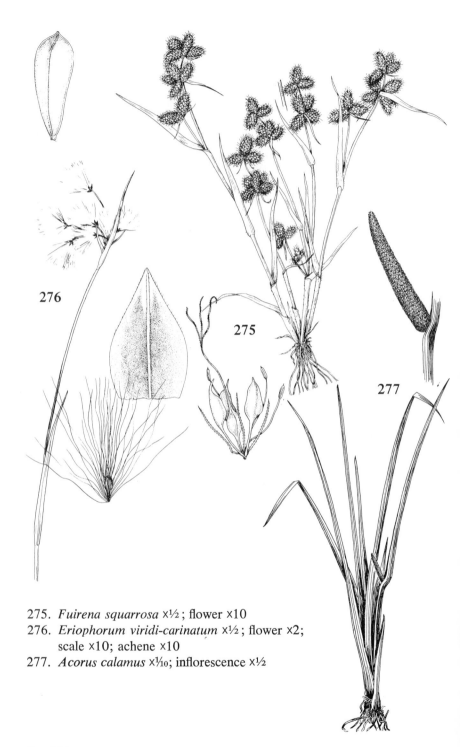

275. *Fuirena squarrosa* ×½; flower ×10
276. *Eriophorum viridi-carinatum* ×½; flower ×2;
scale ×10; achene ×10
277. *Acorus calamus* ×¹⁄₁₀; inflorescence ×½

The bristles are typically copper colored to tan, as well shown in the color plate. They are white in the rather frequent f. *album* (Gray) Wieg.

5. E. angustifolium Honck.

Map 544. Bogs and conifer swamps, especially along ditches, trails, and other openings or disturbances; marshy open ground and boggy shores.

6. E. viridi-carinatum (Engelm.) Fern.   Fig. 276

Map 545. Bogs, conifer swamps (tamarack, spruce, cedar, fir) — especially in open areas and clearings, and ± open boggy or marshy ground.

The midnerve and margins of the scales in this species tend to be minutely scabrous or ciliate.

ARACEAE   Arum Family
KEY TO THE GENERA

1. Leaves compound . . . . 1. **Arisaema**
1. Leaves simple
  2. Leaves narrow, sword-like, with ± parallel sides; spathe appearing like a
     continuation of the leaf-like peduncle (the spadix thus apparently lateral) . . 2. **Acorus**
  2. Leaves expanded; spathe clearly differentiated from peduncle
    3. Leaf blades ± sagittate, with a prominent vein extending from base of midrib

541. Eriophorum tenellum

542. Eriophorum gracile

543. Eriophorum virginicum

544. Eriophorum augustifolium

545. Eriophorum viridi-carinatum

546. Arisaema triphyllum

into each basal lobe; spathe and spadix elongate, the flowering inflorescence
usually at least 10 times as long as its diameter at the middle . . . . 3. **Peltandra**
3. Leaf blades rounded or cordate at base, with veins of basal lobes (if any) no
more prominent than others; spadix short, ellipsoid or subglobose, the
inflorescence less than 5 times as long as wide
    4. Spathe white, on peduncle 7 cm or more long; plant without strong odor . . 4. **Calla**
    4. Spathe green and/or reddish brown, the inflorescence almost sessile, barely
    emerging from the ground; plant with strong skunk-like odor . . . . 5. **Symplocarpus**

1. Arisaema

REFERENCE

Huttleston, Donald G. 1949. The Three Subspecies of Arisaema triphyllum. Bull. Torrey
Bot. Club 76: 407–413.

KEY TO THE SPECIES

1. Leaflets 3; spadix shorter than spathe . . . . 1. **A. triphyllum**
1. Leaflets 5–13; spadix very slender and long-tapering, much exceeding the spathe
at flowering time . . . . 2. **A. dracontium**

1. **A. triphyllum** (L.) Schott   Plate 4-A, 4-B   Jack-in-the-pulpit; Indian-turnip
Map 546. Perhaps most characteristic of springy spots and ravines in rich
beech–maple woods, but found in all kinds of deciduous woods, floodplains and
swamp forest, cedar swamps.

All material from Michigan is probably referable to the typical variety or
subspecies, although possibly var. *pusillum* Peck occurs in southern Michigan.
Some collections from Kalamazoo County may be var. *stewardsonii* (Britton)
Stevens [*A. stewardsonii* Britton], which has the tube of the spathe fluted or
corrugated. However, at least one of the specimens referred to this taxon by
Hanes and by Fernald has been questioned by Huttleston. Recognition of *A.
atrorubens* (Aiton) Blume as a distinct species, as is done in some manuals, is
scarcely possible in our region. Great variation exists in size, color of spathe, and
time of flowering, the latter occurring from early May to mid-July. The height of
flowering plants ranges from 12 cm to 1 m. Leaflets may be as long as 23 cm.
Aberrant individuals with 2 spathes rarely occur, and one specimen has been
seen with 4 spadices.

2. **A. dracontium** (L.) Schott   Green Dragon; Dragon-root
Map 547. Moist woods, especially along riverbanks and floodplains. (See
Mich. Bot. 1: 56–59. 1962; & 5: 114–117. 1966.)

2. Acorus

REFERENCE

Löve, Askell, & Doris Löve. 1957. Drug Content and Polyploidy in Acorus. Proc. Genet.
Soc. Canada 2: 14–17.

1. **A. calamus** L.   Fig. 277   Sweet-flag; Calamus
   Map 548. Wet open places, rather local: marshes, swales, and meadows; edges of rivers and creeks.
   Bruised parts, especially at the base of the plant, have a very pleasant aromatic fragrance, rendering the species easily identified from sterile fragments. The candied rhizome has been for generations a popular confection, with reputed medicinal properties as well, in Europe, Asia, and America. According to the Löves, the American plant is a diploid which may be distinguished, as *A. americanus* (Raf.) Raf., from the sterile triploid *A. calamus*.

3. Peltandra

1. **P. virginica** (L.) Schott & Endl.   Fig. 278   Arrow-arum; Tuckahoe
   Map 549. Shallow water (usually not over 2 feet) and muddy banks at edges of rivers and lakes, swamp forests along rivers, etc. Established but presumably introduced at margins of marshes at the Seney National Wildlife Refuge, Schoolcraft County.

4. Calla

1. **C. palustris** L.   Fig. 279   Wild Calla; Water-arum
   Map 550. Bogs, openings in coniferous swamps and thickets, shallow water at margins of boggy woods and in lakes and rivers.

5. Symplocarpus

REFERENCE

Voss, Edward G. 1964. Skunk-cabbage in Michigan. Mich. Bot. 3: 97–101.

1. **S. foetidus** (L.) Nutt.   Fig. 282   Skunk-cabbage
   Map 551. Common in low wet woods, ravines and hollows in beech—maple woods, floodplains and bottomland, stream borders, etc., in the southern half of

547. Arisaema dracontium   548. Acorus calamus   549. Peltandra virginica

the Lower Peninsula. Northward, generally very local (except on islands in the Great Lakes) and often in bogs and coniferous swamps. Common in all wet habitats of Isle Royale and also thriving on Beaver Island and the Fox Islands. This genus includes only the one species, found in eastern North America and eastern Asia. The familiar flowering spadix with its strong odor and hooded spathe is the first "wildflower" of the spring and is followed by development of the large leaves which, especially when bruised, produce a similar odor.

LEMNACEAE   Duckweed Family

Mixed collections consisting of members of any or all genera in the family are common, for the plants often grow thoroughly mixed. In some ponds, one may thrust his hand at random and gather five species.

REFERENCES

Clark, Howard L., & John W. Thieret. 1968. The Duckweeds of Minnesota. Mich. Bot. 7: 67–76.
Hartog, C. den, & F. van der Plas. 1970. A Synopsis of the Lemnaceae. Blumea 18: 355–368.

KEY TO THE GENERA

1. Plants without roots . . . . 1. **Wolffia**
1. Plants with roots
    2. Internodes (individual joints or "fronds") each with several rootlets (fig. 280), solid dark reddish purple beneath; upper surface above point of attachment of the roots often with a purple dot, from which radiate 5–7 (or more) obscure nerves; larger internodes averaging 2.5–3.5 mm in breadth, rounded to obovate . . . . 2. **Spirodela**
    2. Internodes each with a single root (very rarely forked) (fig. 281), sometimes flushed with purplish but seldom solid dark colored beneath; upper surface with at most 1–3 very obscure nerves, these never radiating from a main dark spot; internodes usually averaging smaller than 3 mm in breadth or else not rounded in general outline . . . . 3. **Lemna**

550. Calla palustris

551. Symplocarpus foetidus

552. Wolffia punctata

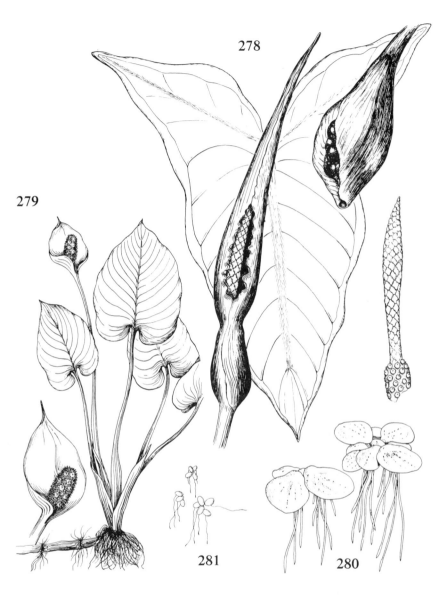

278. *Peltandra virginica* ×½ (leaf, inflorescence, fruiting inflorescence, spadix)
279. *Calla palustris* ×¼
280. *Spirodela polyrhiza* ×3
281. *Lemna minor* ×1

1. Wolffia

Water-meal

Measurements of fresh material will run somewhat larger than those given for dry. Members of this genus are the smallest known flowering plants — though they are seldom seen in flower!

REFERENCE

Dore, William G. 1957. Wolffia in Canada. Canad. Field-Nat. 71: 10–16.

KEY TO THE SPECIES

1. Key to fresh material
  2. Plants longer than broad, ± pointed at both ends, flattish on upper side (at surface of water), usually floating in a single layer (fig. 283) . . . . 1. **W. punctata**
  2. Plants ± globose, some beneath surface of water when crowded (fig. 284) . . . .
  . . . . 2. **W. columbiana**
1. Key to dried material
  3. Plants distinctly longer than broad (± oblong or shaped somewhat like segments of an orange), seldom over 0.2–0.5 mm broad by 1 mm long. . . . . . . 1. **W. punctata**
  3. Plants ± rounded in outline, the largest in a population 0.5–0.8 (1.2) mm in diameter (much shriveled and often less than 0.5 mm if not dried under pressure, but still globose, not oblong, in general outline) . . . . 2. **W. columbiana**

1. W. punctata Griseb.   Fig. 283

Map 552. In similar situations as the next species and often growing with it — very rarely growing where *W. columbiana* is absent.

2. W. columbiana Karsten   Fig. 284

Map 553. Quiet waters of ditches, cat-tail marshes, ponds, boggy pools, edges of rivers and lakes, often with other Lemnaceae.

2. Spirodela

1. S. polyrhiza (L.) Schleiden   Fig. 280

Greater Duckweed

Map 554. Floating on the surface of quiet water in ponds, lakes, ditches,

553. Wolffia columbiana   554. Spirodela polyrhiza   555. Lemna trisulca

marshes, backwaters and borders of rivers and creeks, etc., often with *Lemna minor* and less commonly with *Wolffia* spp.

3. **Lemna**   Duckweed

KEY TO THE SPECIES

1. Internodes ca. 6–15 mm long (rarely as short as 3 mm), including a conspicuous narrow stalk by which fully grown lateral internodes are attached to parents, oblong to broadly lanceolate in outline (fig. 285), often floating below surface of water in large masses . . . . 1. **L. trisulca**
1. Internodes less than 6 mm long, without stalks (or these extremely short), rounded, obovate, or narrowly oblong in outline (fig. 281), usually floating on surface of water
    2. Internodes less than 1 mm wide, narrowly oblong with ± parallel sides and rounded ends . . . . 2. **L. valdiviana**
    2. Internodes averaging more than 1 mm wide, rounded on all sides . . . . . . 3. **L. minor**

1. **L. trisulca** L.   Fig. 285   Star Duckweed
Map 555. In similar waters as the other species of Lemnaceae, but in tangled masses *beneath* the water, not floating on the surface, and hence not easily seen. Very variable in size, the largest and most luxuriant plants in cold spring-fed streams.

The margins of the internodes are ± dentate or erose, and fragmentary specimens or the very rare floating plants which have shorter segments are thus distinguishable from *L. minor*.

2. **L. valdiviana** Phil.
Map 556. Known in Michigan only from the Fox and Beers Millpond, Prairie Ronde Tp., Kalamazoo County, where collected in a few inches of stagnant water in 1943 and 1945 by Mr. and Mrs. Hanes (WMU, GH, OS).

3. **L. minor** L.   Plate 4-C; fig. 281
Map 557. Floating on the surface of standing, even stagnant water of lakes,

556. Lemna valdiviana   557. Lemna minor   558. Xyris torta

ponds, borders of streams, quiet backwaters, floodings, etc. Sometimes stranded on wet shores after lowering of water levels.

A number of collections consist in part or entirely of plants with the sheaths of the root tips tapering to a sharp point (as in *Spirodela*). These have been called *L. perpusilla* Torrey, in contrast to *L. minor* with the root tips rounded. However, following the recent disposition of Minnesota plants by Clark and Thieret, I now include all Michigan plants of this group in a single variable species, *L. minor*. Also included here are the only Michigan collections identified as *L. trinervis* (Austin) Small (*Voss 2264*, Mackinac County, MICH) and *L. obscura* (Austin) Daubs (*Grassl 3029*, Menominee County, MICH) by E. H. Daubs in 1962 but neither cited nor mapped in a report by him.

Flowers are seen very rarely on this (and other) species of the Lemnaceae. The flowering material shown in the color plate came from the Rose Lake Wildlife Experiment Station. Staminate and pistillate flowers in the Lemnaceae consist, respectively, of a single stamen and a simple pistil, which in *Lemna* are produced in a pouch or cleft in the margin of the internode.

XYRIDACEAE   Yellow-eyed-grass Family

1. Xyris   Yellow-eyed-grass

REFERENCE

Kral, Robert. 1966. Xyris (Xyridaceae) of the Continental United States and Canada. Sida 2: 177–260.

KEY TO THE SPECIES

1. Plants bulbous (swollen) and hard at the base; keel of lateral sepals ± pubescent or ciliate from apex to below the middle, with small tuft of longer hairs at apex . . . . 1. **X. torta**
1. Plants not swollen but ± flattened and soft at the base; keel of lateral sepals with pubescence absent or limited to terminal half, with or without terminal tuft
   2. Flowering bracts (not necessarily the sterile lower bracts) with central green or grayish patch obsolete or absent, or at most 2 mm long; keels of lateral sepals entire or slightly toothed toward the tip (which usually bears 2 tiny teeth but no tuft of hairs); sides of lateral sepals rather uniform opaque or translucent brownish to their edges, firm; leaves all under 2 mm in width; scape never over 1 mm in diameter; seeds ellipsoid, ca. 0.7–0.8 mm long, with distinctly contrasting darker longitudinal lines . . . . 2. **X. montana**
   2. Flowering bracts with prominent central green patch 2–3 mm long (at least the larger ones); keels of lateral sepals fringed or lacerate on terminal half, often with small tufts of hairs at apex; sides of lateral sepals becoming very pale yellow-brown or whitish, ± transparent, very thin, and membranous toward edges; largest leaves usually 2–4 mm wide; widest portions of scapes sometimes over 1 mm; seeds very plump-ellipsoid, ca. (0.4) 0.5 (0.6) mm long, with very obscure longitudinal lines . . . . 3. **X. difformis**

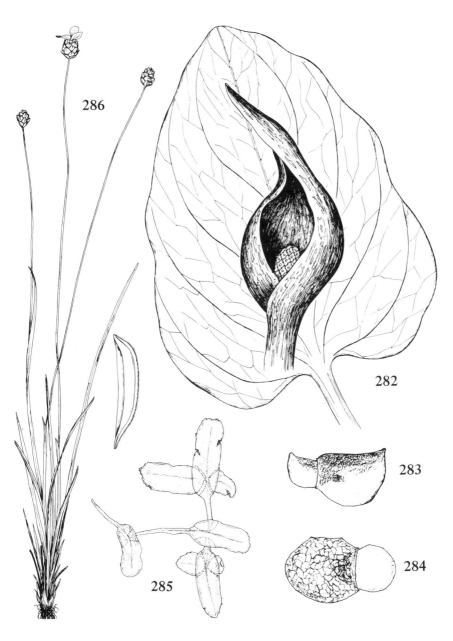

282. *Symplocarpus foetidus* ×½
283. *Wolffia punctata* ×20 (lateral view)
284. *W. columbiana* ×20 (lateral view)
285. *Lemna trisulca* ×5
286. *Xyris torta* ×½; lateral sepal ×8

1. **X. torta** Sm.   Fig. 286

Map 558. Moist to wet sandy shores and swales.

While the scapes in all our species of *Xyris* are ± twisted, they are especially so in this species.

2. **X. montana** Ries

Map 559. Wet places in sphagnum mats, sandy-mucky shores, and other boggy places.

The leaves of our plants are very frequently not papillose, as often described for this species.

3. **X. difformis** Chapman   Plate 4-E

Map 560. Sandy-peaty lake shores, sphagnum bogs, and similar acid sites.

This is the species called *X. caroliniana* Walter in most recent manuals but not by Kral, who applies the latter name to what previously had been widely known as *X. flexuosa*.

ERIOCAULACEAE   Pipewort Family

1. **Eriocaulon**

1. **E. septangulare** With.   Plate 4-D   Pipewort

Map 561. On wet sandy or boggy shores or in shallow water, the heads usually emersed (on scapes at least to 40 cm long in water, only a few cm long on land); especially characteristic of soft-water and acid lakes, where the rosettes of distinctive leaves and cross-hatched roots may form a dense turf even in deep water (3–4 feet or more). The most common and widespread rosette-former of such lakes, although locally outnumbered by *Isoëtes* spp., *Littorella uniflora*, *Juncus pelocarpus* f. *submersus*, *Lobelia dortmanna*, or other associates.

American plants have been recognized as distinct from *E. septangulare* of Great Britain; our species is then *E. pellucidum* Michaux.

559. Xyris montana   560. Xyris difformis   561. Eriocaulon septangulare

COMMELINACEAE   Spiderwort Family

KEY TO THE GENERA

1. Corolla irregular, with a white petal distinctly smaller than the 2 blue ones; inflorescence subtended by a broad folded bract or spathe less than 3 cm long . . . . 1. **Commelina**
1. Corolla regular, the petals of the same size and color; inflorescence subtended by narrow bracts 4–20 cm long, resembling the leaves . . . . 2. **Tradescantia**

1. Commelina   Day-flower

REFERENCE

Pennell, Francis W. 1938. 'Commelina communis' in the Eastern United States. Bartonia 19: 19–22.

KEY TO THE SPECIES

1. Plants decumbent or creeping, with ovate leaf blades 2–4 times as long as wide; edges of blade of spathe free for entire length . . . . 1. **C. communis**
1. Plants erect, with lanceolate or linear leaf blades 7–10 times as long as wide; edges of blade of spathe united at the base . . . . 2. **C. erecta**

1. C. communis L.   Fig. 287
Map 562. Waste ground, dumps, vacant lots, sometimes a weed in yards or even in woods; probably more widespread than the few collections would indicate. A native of Asia, locally established in North America.

Our specimens are mostly if not entirely var. *ludens* (Miq.) C. B. Clarke, with leaf sheaths ciliate at summit and the sterile lobed anthers with madder center. The margins of the sheaths are glabrous and the anthers entirely yellow in the typical variety.

2. C. erecta L.
Map 563. Apparently not collected in Michigan since August 25, 1838, when found by the botanists of the First Survey on the "banks of a small lake" in Cass County (MICH, GH) (see Mich. Bot. 9: 239. 1970).

562. Commelina communis   563. Commelina erecta   564. Tradescantia ohiensis

2. Tradescantia   Spiderwort

REFERENCES

Anderson, Edgar, & Robert E. Woodson. 1935. The Species of Tradescantia Indigenous to the United States. Contr. Arnold Arb. 9: 1–132.

Dean, Donald S. 1954 ["1953"]. A Study of Tradescantia ohiensis in Michigan. Asa Gray Bull., N. S. 2: 379–388.

KEY TO THE SPECIES

1. Pedicels and sepals glabrous or with a few hairs at base or apex of the latter . . . .
. . . . 1. **T. ohiensis**
1. Pedicels and sepals conspicuously pubescent throughout
  2. Pubescence of glandless hairs . . . . 2. **T. virginiana**
  2. Pubescence largely of glandular-viscid hairs . . . . 3. **T. bracteata**

1. T. ohiensis Raf.   Plate 4-F

Map 564. Both tetraploid and diploid plants occur in the state (Dean, 1954), the former chiefly along roadsides and railroads, in open oak woods or borders of woods, on sandy ridges, and similar dry sites; diploids are more frequent in meadows and wet ground although they grow also in dry places. Perhaps escaped from cultivation at some stations, especially northward.

White-flowered plants have been recorded from Livingston and Washtenaw counties and doubtless occur elsewhere.

2. T. virginiana L.   Fig. 288

Map 565. In situations similar to the preceding, but much less common.

Specimens with sepals sparsely pubescent their entire length are apparently hybrids with *T. ohiensis*, and have been found in Ingham, St. Clair, and Washtenaw counties.

3. T. bracteata Small

Map 566. Known only from light sand northeast of Bradley, Allegan County (*Gilbert Becker* in 1938, WMU).

565. Tradescantia   566. Tradescantia   567. Pontederia cordata
    virginiana   bracteata

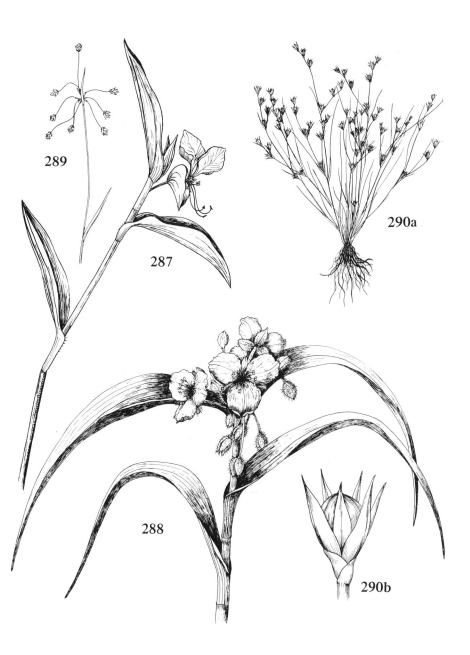

289

287

290a

288

290b

287. *Commelina communis* x¾
288. *Tradescantia virginiana* x½
289. *Luzula acuminata* x½
290a. *Juncus bufonius* x½
290b. *J. bufonius,* mature perianth, bracteoles, and fruit x5

PONTEDERIACEAE   Pickerel-weed Family

KEY TO THE GENERA

1. Flowers blue-violet, 2-lipped, in dense inflorescence; stamens 6; leaves (except for one large bract) arising from toward base of plant, mostly emersed and over 1 cm broad . . . . 1. **Pontederia**
1. Flowers yellow, regular, solitary; stamens 3; leaves less than 1 cm broad, borne along leafy stem, usually submersed . . . . 2. **Heteranthera**

1. Pontederia

1. **P. cordata** L.   Plate 5-A   Pickerel-weed
Map 567. Shallow water (seldom more than 3 feet) and marshy borders of lakes, ponds, tamarack bogs, rivers, and creeks, quite local northward in the state, often abundant where found and a striking sight in full bloom.

In Michigan we have plants with the leaf blades narrow, tapering or truncate (but not cordate) at the base as well as the more common form with blades broad, cordate (often deeply so) at the base. In addition, a form with very narrow submersed leaves (resembling *Sagittaria graminea*) has been named, as has a form with white flowers; these have not yet been reported from Michigan but probably will one day be found here.

2. Heteranthera

1. **H. dubia** (Jacq.) MacM.   Plate 5-B; fig. 291   Water Star-grass
Map 568. Shallow to deep water (recorded to 18 feet) of ponds, lakes, marshes, rivers, and creeks.

Most submersed plants are sterile or have cleistogamous flowers hidden in the bases of the submersed leaves. Flowers with bright yellow perianth occur more frequently on f. *terrestris* (Farw.) Vict. (TL: Clinton River near Utica [Macomb County]), which occurs on muddy, sandy, or peaty shores and banks, presumably stranded by a lowering of water levels (see Thieret in Mich. Bot. 11: 117–118. 1971).

568. Heteranthera dubia   569. Luzula multiflora   570. Luzula acuminata

The midvein is almost always indistinct or no more prominent than other longitudinal veins of the leaf, thus distinguishing this species vegetatively from species of *Potamogeton* (also alternate-leaved) with similar strap-shaped leaves. In addition, *Potamogeton zosteriformis* has a very flat stem (it is rounded in section in *Heteranthera*) and the stipules are entirely free from the leaf; *P. robbinsii* has partly adnate stipules, but the base of the free blade is slightly auricled, while in *Heteranthera* there are no auricles. This species is sometimes placed in the genus *Zosterella*.

JUNCACEAE   Rush Family

True rushes are easily distinguished from grasses and sedges, with which they are sometimes confused, by the presence of a true perianth of 6 tepals (see figs. 290b, 293) and a 3—several-seeded capsule rather than a 1-seeded grain or achene.

KEY TO THE GENERA

1. Widest leaf blades (usually on basal leaves) 5–10 (12) mm broad . . . . 1. **Luzula**
1. Widest leaf blades less than 5 mm broad
　2. Blades terete or strongly involute . . . . 2. **Juncus**
　2. Blades flat
　　3. Foliage ± hairy, at least toward summit of sheaths; capsules 3-seeded . . . . 1. **Luzula**
　　3. Foliage completely glabrous; capsules usually many-seeded . . . . 2. **Juncus**

1. **Luzula**   Wood Rush

KEY TO THE SPECIES

1. Flowers grouped in heads or short dense spikes; capsule usually no longer than the
　perianth . . . . 1. **L. multiflora**
1. Flowers solitary at ends of branches of the inflorescence; capsule at maturity at
　least slightly longer than perianth
　2. Inflorescence umbelliform, the bases of the primary branches close together;
　　branches unforked or at most with 1 or 2 flowers in addition to terminal one;
　　mature perianth 2.5–4 mm long; outer end of seed with conspicuous appendage
　　which may be almost as long as the body of the seed . . . . 2. **L. acuminata**
　2. Inflorescence cymose, the primary branches arising from bases distinctly spaced
　　along a short axis; branches forked at least once or twice; mature perianth
　　2–2.5 mm long; seeds without appendages . . . . 3. **L. parviflora**

1. **L. multiflora** (Retz.) Lej.

Map 569. In woods, sometimes with *L. acuminata*, but of greater ecological amplitude than the latter; also in swamp forests, boggy sites, damp grassy areas, and open ground.

This species of Europe and North America is sometimes treated as a variety or subspecies of the even wider ranging *L. campestris* (L.) DC. A few specimens

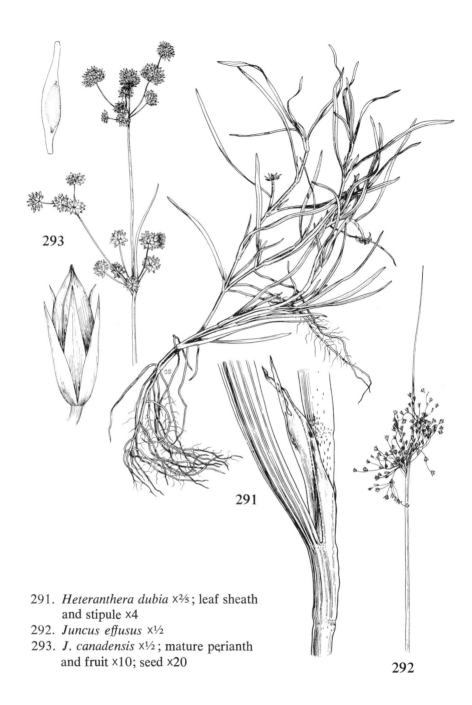

291. *Heteranthera dubia* ×⅖; leaf sheath
    and stipule ×4
292. *Juncus effusus* ×½
293. *J. canadensis* ×½; mature perianth
    and fruit ×10; seed ×20

from southern Michigan have been called var. *bulbosa* (Wood) Hermann, but they appear hardly sufficiently distinct to warrant formal recognition.

2. L. acuminata Raf.   Fig. 289

Map 570. In woods of almost all kinds except the wettest: upland oak, beech–maple, mixed conifer and hardwoods, often on thin rocky soil in the Upper Peninsula; characteristic of banks above creeks, rivers, and swamps, and in ravines, clearings, and fringes of rich deciduous woods.

Michigan specimens are typical var. *acuminata*; in some manuals called *L. saltuensis* Fern.

3. L. parviflora (Ehrh.) Desv.

Map 571. A circumpolar species of boreal range in North America, known thus far in Michigan only from Isle Royale, where found both in moist somewhat open woods and on sheltered gravelly shores.

2. Juncus   Rush

KEY TO THE SPECIES

1. Inflorescence apparently lateral, one conspicuous terete involucral bract appearing to be a straight continuation of the stem, exceeding the inflorescence, which thus looks as if bursting from the side of the stem (fig. 292); stem leaves without blades, reduced to basal sheaths
   2. Stamens 3, the anthers about equalling the length of the filaments or slightly shorter; stems in dense clumps . . . . 1. **J. effusus**
   2. Stamens 6, the anthers various; stems spaced along the rhizome singly or in small tufts (densely clumped only in the rare, glaucous *J. inflexus*)
      3. Inflorescence up to 2 cm long; sepals green or pale brown (when old); anthers about half as long as filaments or shorter; rhizome slender, less than 2 mm in diameter; involucral bract usually more than half as long as the stem below (sometimes even longer, the inflorescence then appearing to be on lower half of the plant); sheaths pale brown . . . . 2. **J. filiformis**
      3. Inflorescence over 2 cm long, or if shorter, the sepals deep brown; anthers about as long as the filaments or longer; rhizome stout (even woody), (3–) 4 mm or more in diameter; involucral bract (with rare individual exceptions in a population) less than half as long as the stem below the inflorescence; sheaths pale or more often deep brown or maroon
         4. Stems prominently ridged, glaucous, densely clumped; anthers about the length of the filaments; sepals largely green or dull brown when old . . 3. **J. inflexus**
         4. Stems smooth, neither ridged nor glaucous, arising singly or in small tufts from the rhizome; anthers at least twice as long as the filaments; sepals marked with deep brown . . . . 4. **J. balticus**
1. Inflorescence terminal, no involucral bract appearing to be a straight continuation of the stem; at least some sheaths bearing leaf blades
   5. Leaf blades flat, involute, or terete, but without hard cross-partitions
      6. Leaf blades terete, at least toward the end (sometimes firmly rounded but with a shallow or deep channel basally)

7. Inflorescence usually a fourth or more the height of the plant; adjacent flowers (single or groups of 2) separated on the ± one-sided, recurving branches of the inflorescence by a distance – at least when mature – distinctly greater than the length of the flowers; some or all flowers often reduced to sterile bulblets . . . . 14. **J. pelocarpus**
7. Inflorescence much less than a fourth the height of the plant, compact, with ± crowded flowers
   8. Inflorescence 1–4(5)-flowered; seeds over 2.5 mm long; perianth 4.5 mm or more long; largest capsules 5–7 mm long . . . . 5. **J. stygius**
   8. Inflorescence with more than 4 flowers; seeds less than 1.5 mm long; perianth less than 4.5 mm long; capsules under 5 mm long
      9. Ends of seeds with white "tails" about half as long as the slender body; sepals ca. (3.5–) 4 mm long; longest involucral bract 1–6 cm long, often under 3 cm . . . . 6. **J. vaseyi**
      9. Ends of seeds with tails absent or at most half the width of the plump body; sepals mostly smaller, 2.5–4 (averaging 3–3.5) mm long; longest involucral bract ranging to 21 cm long, usually more than 3 cm . . . . 7. **J. greenei**
6. Leaf blades flat, or mostly involute (but then clearly channeled to the end)
  10. Flowers in close heads of 2 or more; leaves obviously flat almost their entire length, 1.5 mm or more wide
    11. Inflorescence with fewer than 30 heads; stamens shorter than sepals, shriveling in maturity . . . . 8. **J. marginatus**
    11. Inflorescence with more than 30 heads; stamens about as long as the sepals (or longer), the anthers conspicuous and ± exserted in fruit . . . . . . 9. **J. biflorus**
  10. Flowers not in heads, mostly individually pedicelled or sessile on branches of the inflorescence (flowers crowded in some rare forms, but at least a few of them clearly single); leaves often involute for much or all of their length, usually less than 1.5 mm wide
    12. Blades of some foliage or involucral leaves arising near the middle of the plant
      13. Inflorescence one- to two-thirds the height of the plant; sepals acuminate, spreading, distinctly exceeding the capsules, green (or brownish in age) to the tips, with hyaline margins; plant a tufted annual with fibrous roots . . . . 10. **J. bufonius**
      13. Inflorescence much less than a third the height of the plant; sepals obtuse and incurved with deep brown areas at the tip, scarcely longer (often shorter) than the mature capsule; plant perennial from a rhizome . 11. **J. gerardii**
    12. Blades of foliage leaves all basal or nearly so, arising from lower fifth of plant; involucral leaves limited to upper third of plant
      14. Auricles at summit of leaf sheath (at least the upper ones) conspicuously produced with free portion (0.5) 1–3 (5) mm long, very thin, membranous, and fragile, pale whitish or with a smoky patch . . . . . 12. **J. tenuis**
      14. Auricles hardly prolonged, the free portion (if any) at most 0.5 mm long, thickened and coriaceous, yellow or brown at the edge . . . . 13. **J. dudleyi**
5. Leaf blades terete [may be flattened in pressing], with hard cross-partitions at regular intervals [easily felt by running one's fingernail along a leaf laid on a hard smooth surface, if not evident in dry leaf]
  15. Flowers single (or some on a plant in 2's or rarely in 3's), some or all often reduced to sterile bulblets; inflorescence usually a fourth or more the height of the plant, which is generally under 20 cm (very rarely as tall as 30 cm) . . . . 14. **J. pelocarpus**
  15. Flowers in heads of 2 or more; inflorescences usually less than a fourth the

height of the plant, which is often over 30 (nearly always over 20) cm tall (unusually small or depauperate individuals as small as 10 cm)*

16. Seeds with a long or short pale appendage or "tail," formed by prolongation of the seed coat, at both ends

    17. Majority of heads on a plant consisting of 5—many flowers each, often densely subglobose or hemispherical; perianth segments tapering to very sharp points; mature capsules often equalling or only slightly exceeding the perianth (although sometimes much longer); plants stout, over (usually much over) 30 cm tall . . . . 15. **J. canadensis**

    17. Majority of heads on a plant consisting of 2—5 (7) flowers each, narrower than hemispherical; perianth segments (tepals) sharp-pointed or blunt; mature capsules much exceeding the perianth; plants more slender, usually (15—) 20 cm or more tall

        18. Perianth (1.5) 2—2.5 mm long; tepals ± blunt or rounded at the tip, with hyaline margins often half the width of the central portion; branches of the inflorescence often divaricate or spreading . . . . 16. **J. brachycephalus**

        18. Perianth 2.5—3.5 mm long; tepals (at least the sepals) sharp-pointed, with narrower hyaline margins (usually less than half the width of the central portion); branches of the inflorescence erect or ascending . . 17. **J. brevicaudatus**

16. Seeds without pale tails, either blunt or with small *dark* points at one or both ends

    19. Stamens 3, opposite the sepals

        20. Mature capsule slender, tapering gradually to a prominent beak 0.5—1 mm long, exceeding tips of sepals; heads densely globose or almost so . . . . . .
. . . . 18. **J. scirpoides**

        20. Mature capsule plump, shorter than the sepals (rarely much longer), ± abruptly terminating in a very short tip; heads globose or (in more common species) narrower

            21. Capsules much shorter than sepals; heads densely globose . . 19. **J. brachycarpus**

            21. Capsules about equalling sepals; heads usually hemispherical or narrower
. . . . 20. **J. acuminatus**

    19. Stamens 6, opposite the petals and sepals

        22. Heads ± spherical, the lower flowers reflexed; involucral bract exceeding the inflorescence; capsule slender, (3) 3.5 or more times as long as wide, thickest at base, tapering rather regularly to apex

            23. Sepals 4—5.5 (6) mm long, usually slightly exceeding the petals; largest heads often over 1 cm in diameter; auricles at summit of sheaths of cauline leaves over 1.5 mm long . . . . 21. **J. torreyi**

            23. Sepals 2.5—3.5 mm long, usually slightly shorter than the petals; heads under 1 cm in diameter; auricles shorter than 1.5 mm (very rarely longer)
. . . . 22. **J. nodosus**

        22. Heads hemispherical or narrower, the lower flowers not reflexed; involucral bract shorter than inflorescence (though cauline leaf may exceed it); capsules plump, less than 3 times as long as wide

            24. Plants stout, over 50 cm tall; cauline leaf 1, overtopping the inflorescence, a bladeless (or at most a very short-bladed) sheath on stem between inflorescence and sheath of the cauline leaf . . . . 23. **J. militaris**

---

*Specimens with well developed fruit and seeds are necessary for accurate identification in this group. Stamens are persistent even on older flowers and may be seen under the perianth parts. Flowers occasionally are reduced to sterile bulblets or are made unrecognizable by conversion into enlarged insect galls.

24. Plants more slender, very rarely as tall as 50 cm; cauline sheaths all with
blades, these not overtopping the inflorescence
25. Petals 1.5–2.5 mm long, ± blunt or rounded at apex, slightly but
noticeably shorter than sepals; branches of inflorescence ascending . . .
. . . . 24. **J. alpinus**
25. Petals 2–3 mm long, sharp-pointed at apex, about the length of the
sepals or often slightly longer; at least some branches of inflorescence
usually spreading . . . . 25. **J. articulatus**

1. **J. effusus** L.   Fig. 292

Map 572. Widespread in wet ground: marshes, shores, banks of ditches and
streams, bog borders and clearings, pastures, etc.; damp woods and thickets.

Very variable, several varieties having been described on the basis of sepal
texture, whether the sepals are longer than the petals and capsules, and the
degree of compactness or laxness of the inflorescence. While the extremes of
variation are well marked, too many intermediate plants occur to make it useful
to attempt to distinguish varieties here.

2. **J. filiformis** L.

Map 573. Rather local on sandy or peaty shores, gravelly river borders, and
moist edges of thickets in the Upper Peninsula.

Distinct from the other species with lateral inflorescence in its slender stature
and very long involucral bract.

3. **J. inflexus** L.

Map 574. A native of the Old World, very rarely established in North
America. First collected in Michigan by F. J. Hermann in 1936 and 1937 (*8289*,
MICH, AQC, NY; *9203*, MSC, MO) on a wet sandy and peaty hillside (Quincy
Hill) above the highway just north of Hancock; Farwell later visited the station
(*12734 & 12735* in 1940, MICH, BLH, MSC). The species was still thriving here
in the ditch and on the hillside in 1958 (*Voss 7889*, MICH, UMBS) and is
presumably there yet.

4. **J. balticus** Willd.

Map 575. Especially characteristic of wet sandy (to gravelly or marly) shores
and damp flats between dunes along the Great Lakes and inland lakes, but also
in marshy situations of all kinds, ditches, etc.

The straight rhizomes develop extensively in sand, with evenly spaced stems
arising in conspicuous rows. American plants belong to var. *littoralis* Engelm.
Most Michigan specimens appear to represent the f. *dissitiflorus* Fern. & Wieg.,
with the inflorescences 4–10 (17) cm long, elongate and rather remotely
flowered. A few collections are typical f. *littoralis*, with the inflorescence less
than 4 cm long, compact, and the flowers more or less crowded. Some plants,
especially less robust ones with small (though sometimes remotely flowered)
inflorescences, are intermediate in regard to these characters. Most of these

should probably be referred to f. *dissitiflorus* — if such a form merits recognition at all.

5. J. stygius L.

Map 576. Known in Michigan only from Isle Royale (*Stuntz & Allen 1096* in 1901, WIS) and a peat bog at Marquette (*E. J. Hill 169* in 1889, ILL, MICH). A circumboreal species ranging south in North America to Lake Superior.

American plants are referred to var. *americanus* Buch.

6. J. vaseyi Engelm.

Map 577. Very local in low sandy ground with pines, sandy excavations, etc.

7. J. greenei Oakes & Tuckerman

Map 578. Moist or dry sandy open ground: shores, swales, fields, clearings, dunes, and interdunal depressions.

See comments under *J. dudleyi*.

8. J. marginatus Rostk.

Map 579. Very local: wet banks, marshes, sandy lake margins, and probably other moist sites.

571. Luzula parviflora

572. Juncus effusus

573. Juncus filiformis

574. Juncus inflexus

575. Juncus balticus

576. Juncus stygius

9. **J. biflorus** Ell.

Map 580. Wet open often sandy ground, ditches, swales. Early lists of the state flora included *J. marginatus* for Macomb County on the authority of Cooley. Cooley's material (in 1839, MSC) is *J. biflorus*, but has no locality data other than "Mich." on the label; hence it has not been mapped although probably from Macomb County.

10. **J. bufonius** L.   Figs. 290a, 290b   Toad Rush

Map 581. Moist often sandy soil (e. g., shores) but also in clay and in disturbed ground of roadsides, old trails, etc.

The leaf sheaths lack auricles at the summit, thus distinguishing this species from rare individuals of *J. tenuis* or *J. dudleyi* which might resemble it, and from *J. pelocarpus*, dry specimens of which may have a similar aspect. The rare f. *congestus* (Schousboe) Wahlb. has the flowers aggregated into small heads of about 3; it was collected in 1895 at Muskegon (*C. D. McLouth*, MSC) and could be expected anywhere.

11. **J. gerardii** Loisel   Black-grass

Map 582. Collected in Michigan only by Dodge in 1911 and 1915 at Port Huron (MICH, BLH) and by Farwell at Wayne in 1931 and 1932 (MICH, BLH, GH). Found in ditches, railroad yards, and damp open ground. Quite probably not indigenous in Michigan

577. Juncus vaseyi

578. Juncus greenei

579. Juncus marginatus

580. Juncus biflorus

581. Juncus bufonius

582. Juncus gerardii

12. J. tenuis Willd.   Path Rush

Map 583. Widespread in dry to moist ground, especially along roadsides, trails, ditches, gravel pits, etc.; on shores and less commonly in boggy areas; in open woods, fields, and clearings.

Care must be taken to find well developed, unbroken auricles in determining *J. tenuis* and *J. dudleyi*. The sheath margins in *J. tenuis* are, like the auricles, white-hyaline and fragile.

J. interior Wieg., of which no Michigan material has been located despite reports from the state, differs from *J. tenuis* in having short, hardly prolonged auricles, and from *J. dudleyi* in having them thin; from both of these it differs in having the perianth appressed-ascending and the bracteoles ± acuminate rather than blunt or merely acute. Another similar taxon with auricles intermediate between those of our two common species is *J. dichotomus* var. *platyphyllus* Wieg.; it differs from *J. interior* in its more spreading tepals and acute bracteoles — as well as more eastern range (Atlantic states). A collection by Hermann and Hanes from Kalamazoo County (east side of Austin Lake, July 9, 1937, WMU) may be this poorly understood taxon, according to Hermann.

13. J. dudleyi Wieg.

Map 584. Especially frequent on sandy lake shores, but in almost all kinds of moist places, including edges of marshes, moist waste ground and roadsides, ditches, clearings in bogs, rock crevices along Lake Superior.

Sometimes treated as a variety of the preceding, in which case the correct name is, unfortunately, var. *uniflorus* (Farw.) Farw. (TL: Dearborn [Wayne County]). Specimens of *J. greenei* might be keyed here, but can be distinguished by the mature capsules clearly exceeding the perianth, more compact inflorescence, and leaves (though very narrowly channeled) so tightly rolled as to be essentially terete.

14. J. pelocarpus Meyer

Map 585. Spreading by slender rhizomes on sandy, peaty, or mucky shores (or drying beds) of soft-water lakes and ponds, old bogs, etc.

Completely submersed sterile plants may be locally abundant [f. *submersus*

583. Juncus tenuis   584. Juncus dudleyi   585. Juncus pelocarpus

Fassett]; their rather flattened rosettes of reddish leaves are typically associated with such plants as *Eriocaulon septangulare*, *Littorella uniflora*, and *Myriophyllum tenellum*. Emersed plants may have some or all of the flowers converted to sterile viviparous bulblets.

15. **J. canadensis** La Harpe   Fig. 293
Map 586. Sandy, muddy, marshy, or boggy shores, ditches, interdunal ponds, swales; drying mucky-sandy hollows.

The next two species were once considered varieties of *J. canadensis*, and are certainly closely related although distinct in their extremes. Identification must be largely comparative and is difficult until one is familiar with typical specimens of all three. A few collections with immature seeds but with flowers and inflorescences resembling *J. brevicaudatus* have been referred to *J. canadensis* by Hermann, presumably because of their stout and tall stature, which is that of the latter species. Specimens intermediate between *J. brachycephalus* and *J. brevicaudatus* also occur rarely. The stamens in all three species may be 6, at least on many of the flowers, although usually they are 3.

16. **J. brachycephalus** (Engelm.) Buch.
Map 587. Wet ground: sandy or marly shores and beachpools; less commonly bogs, stream banks, moist black soil, and other damp sites.

17. **J. brevicaudatus** (Engelm.) Fern.
Map 588. Moist ground of almost all types except bogs: sandy lake shores, excavated flats and clearings, beachpools and interdunal swales, borders of streams and ponds, rock crevices along Lake Superior.

18. **J. scirpoides** Lam.
Map 589. An Atlantic Coastal Plain species, disjunct with other Coastal Plain plants at a few inland stations including the Indiana Dunes and southwestern Michigan. Here it has been found by F. J. Hermann and C. R. Hanes on sand dunes and in sandy oak groves on the shore of Austin Lake, Kalamazoo County;

586. Juncus canadensis   587. Juncus brachycephalus   588. Juncus brevicaudatus

and by Gary Pierce on sandy shores of Daggett Lake in Barry County and Keeler Lake in Van Buren County.

19. J. brachycarpus Engelm.

Map 590. Very local in low ground.

20. J. acuminatus Michaux

Map 591. Moist, often sandy ground, but not in bogs.

In this species and in *J. scirpoides*, the sheaths of the lower leaves may be open and flat.

21. J. torreyi Cov.   Fig. 294

Map 592. In moist, usually rather open ground, at edges of creeks, rivers, cat-tail marshes, etc.; ditches, sandy excavations, and along railroads. Spreading by tuber-bearing rhizomes.

A few specimens, including the types of Farwell's f. *longipes* (TL: River Rouge [Wayne County]) and var. *paniculatus* (TL: Detroit [Wayne County]) have the inflorescence smaller, hemispherical or narrower, and the flowers for the most part reduced to sterile bulblets; they may be referred to *J. torreyi* by the prominent subulate sepal-tips, 6 stamens (when developed), large auricles, and general stature. Although difficult to describe in other than a comparative way, the prominent rigid subulate sepal-tips of *J. torreyi* contrast with the less stiff tips of *J. nodosus*, giving a decidedly bristly appearance to the former.

22. J. nodosus L.

Map 593. Moist ground generally, spreading by slender rhizomes with tuberous thickenings: shores (sand, gravel, clay, marl), edges of creeks and rivers, ditches, fields, edges of bogs and roads through boggy ground, wet sandy excavations.

23. J. militaris Bigelow   Plate 5-C

Map 594. Very local, known in Presque Isle County only from Lake Sixteen

589. Juncus scirpoides

590. Juncus brachycarpus

591. Juncus acuminatus

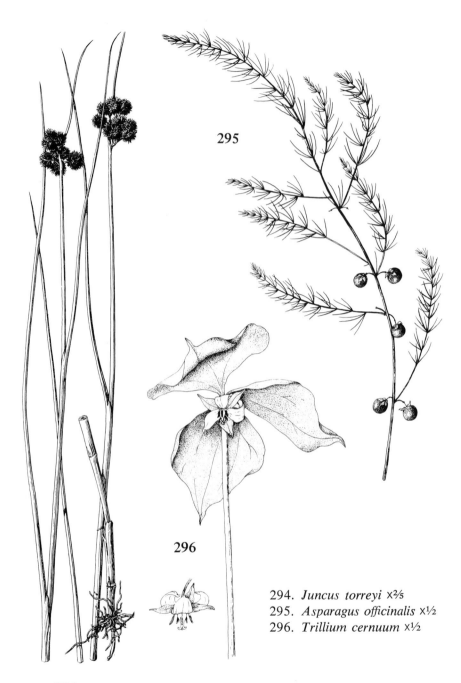

295

296

294. *Juncus torreyi* ×⅖
295. *Asparagus officinalis* ×½
296. *Trillium cernuum* ×½

294

and in Cheboygan County from another lake of the same name not far away (or "Little Lake 16") and from nearby smaller bogs and marshy mucky hollows which are the remnants of former bogs. Primarily a plant of the Atlantic coastal region, from Newfoundland to Delaware, with a few inland sites in New York, the Temagami Forest Reserve of Ontario, and Michigan apparently associated with features of earlier postglacial stages of the Great Lakes. Formerly known (as recently as 1955) from Goose Lake in the Indiana Dunes (Porter County) but the station has since been destroyed.

Flowers chiefly when water levels are down; usually remains sterile, with copious filiform submersed leaves from the rhizomes, when submersed (in up to about 2 feet of water). These tufts of leaves somewhat resemble *Scirpus subterminalis* and *Eleocharis robbinsii* – which may both be growing with *J. militaris* – but the hair-like leaves of the latter are easily distinguished by the typical nodulose character of this section of the genus.

24. **J. alpinus** Vill.

Map 595. Moist ground, especially wet sandy and gravelly (often calcareous) shores; not in bogs.

Many plants have one or more central flowers of some heads elevated on distinct pedicels; these have been called var. *rariflorus* (Hartman) Hartman.

See comments under the next species.

592. Juncus torreyi

593. Juncus nodosus

594. Juncus militaris

595. Juncus alpinus

596. Juncus articulatus

597. Asparagus officinalis

25. **J. articulatus** L.

Map 596. Moist ground, both sandy and boggy shores, streams and springy places, much less often encountered than *J. alpinus*.

This species and *J. alpinus* are often difficult to distinguish on the basis of any single character, some exceptions occurring to each item in the key. For example, certain collections from near Laurium (Houghton County) have shorter, rounder petals than usual for *J. articulatus*, but are referred to this species by the spreading inflorescence and the shape of the capsule. (The capsule of *J. articulatus* tends to taper throughout the terminal half, while that of *J. alpinus* tapers more rapidly nearer the end.) Some other plants have more acute sepals than usual for *J. alpinus*, but are referred to that species by the strongly ascending inflorescence and the petals, which are distinctly shorter than the sepals.

LILIACEAE   Lily Family

This family is traditionally separated from the Amaryllidaceae by the position of the ovary: superior in the Liliaceae, inferior in the Amaryllidaceae. However, the importance of this character is seriously questioned by modern taxonomists who are inclined to unite the two families, generally segregating the coarse often shrubby *Yucca* of the former and *Agave* of the latter into a separate family, the Agavaceae. *Smilax* may also warrant segregation as the Smilacaceae. *Hypoxis*, our only native genus of the traditional Amaryllidaceae, is sometimes made the type genus of another family, the Hypoxidaceae. Such reorientations, based on assemblages of characters rather than simply ovary position, seem more natural. In the local flora, *Aletris* and *Zigadenus* have the ovary apparently half-inferior.

In addition to the species here treated as native or as reasonably well established escapes, a number of ornamental garden plants in other genera of the Liliaceae are reported occasionally to be long persistent after cultivation or even to spread somewhat near the site of original cultivation. Works on garden plants, while not always satisfactory, may be consulted for their identification. Species of *Scilla* (Squill) and *Chionodoxa* (Glory-of-the-snow) with 1 or a few blue flowers appearing in early spring on scapes with a pair of leaves at or near the ground have been collected and reported as escaped in lawns, etc., in Houghton and Baraga counties and surely behave similarly elsewhere in the state. In *Scilla*, the tepals are separate nearly or quite to the base, while in *Chionodoxa* they are united for some distance. *Tulipa* (Tulip) would run in the key to genera near *Lilium*, but differs in having no more than 5 cauline leaves and filaments attached to the ends of the anthers. Specimens have been seen collected in Houghton, Keweenaw, Wayne, and Washtenaw counties, the labels for at least the latter two definitely stating that the plants were escaped; Farwell (Asa Gray Bull. 4: 53. 1896) considered the tulip hardy wherever it obtained a footing in Michigan. *Hosta ventricosa* (Salisb.) Stearn (Plantain-lily; Funkia) was collected by Farwell at Island Lake [Livingston County] in 1905, apparently as an escape. It would run in the key next to *Hemerocallis* and *Yucca*, but differs obviously in having blue flowers in a raceme and broad leaves (to 12 cm).

Veratrum viride Aiton (False Hellebore) has been reported from southern Michigan but no specimens have been located. It would run in the key near *Zigadenus* and *Stenanthium*, but differs in being pubescent on the axis of the inflorescence and summit of stem and in its broadly elliptic or ovate leaves. It is an acutely poisonous plant if eaten by man or beast. *V. virginicum* (L.) Aiton f. [*Melanthium virginicum* L.] has been found along a railroad in St. Joseph County, Indiana (just south of Berrien and Cass counties, Michigan); it has linear leaves.

KEY TO THE GENERA

1. Flowers or inflorescences lateral, arising from the axils of alternate cauline leaves or scales
  2. Leaves scale-like, mostly brownish or yellowish, those on the much-branched upper portion of the plant subtending short green filiform branches (often mistaken for leaves) (fig. 295) . . . . 1. **Asparagus**
  2. Leaves broad, flat, green (scale-like leaves or bracts may be present in addition to normal leaves)
    3. Leaves net-veined with long or short (but distinct) petioles; flowers unisexual, in umbels of several to many . . . . 2. **Smilax**
    3. Leaves parallel-veined, sessile, clasping, or perfoliate at base; flowers perfect, 1–5 at a node
      4. Flowers yellow, of separate tepals; fruit a 3-lobed capsule, green, becoming brownish when fully mature . . . . 13. **Uvularia**
      4. Flowers greenish to purple, of separate or united tepals; fruit a berry, usually spherical, red or blue-black when ripe
        5. Tepals united most of their length; ripe fruit blue to black, young fruit green; peduncles appearing axillary . . . . 3. **Polygonatum**
        5. Tepals separate; ripe fruit red, young fruit whitish green; peduncles appearing to arise *beneath* the axils . . . . 4. **Streptopus**
1. Flowers or inflorescences terminal on scapes or leafy (simple or branched) stems
  6. Leaves all withering before plant flowers . . . . 12. **Allium (tricoccum)**
  6. Leaves present at flowering time
    7. Leaves *all* in one or two whorls on the stem
      8. Flower solitary, with white to maroon petals over 1 cm long, the sepals green; stem with 1 whorl of leaves, glabrous . . . . 5. **Trillium**
      8. Flowers in a sessile umbel, the petals yellow or yellowish green, 1 cm or less, similar to sepals; stem with 2 whorls of leaves, ± covered when young with loose woolly pubescence (at least the upper internodes usually becoming glabrous in age) . . . . 6. **Medeola**
    7. Leaves alternate or basal, or if in whorls these more than two or alternate leaves also present
      9. Flowers more than 3.5 cm long
        10. Leaves perfoliate . . . . 13. **Uvularia (grandiflora)**
        10. Leaves not perfoliate
          11. Plant with only 2 basal leaves and 1 nodding flower . . . . 7. **Erythronium**
          11. Plant with several–many leaves and often more than 1 flower
            12. Principal leaves cauline, not crowded toward base of plant . . . . 8. **Lilium**

12. Principal leaves basal or nearly so; stem leafless above or with very small bracts

13. Flowers in an umbel or rather irregular corymb, orange or yellow; leaves without fibrous margins . . . . 9. **Hemerocallis**

13. Flowers in a panicle, white or creamy; leaves with curling fibers separating from their margins (fig. 299) . . . . 10. **Yucca**

9. Flowers less than 3.5 cm long

14. Flowers in an umbel on an unbranched stem or scape

15. Leaves over 3 cm broad, flat, present at flowering time; flowers yellowish, 10–20 mm long, 3–7 in an umbel; umbels not subtended by an involucral spathe, never bearing bulblets; plants not bulbous, nor with odor of onion or garlic; fruit a blue berry . . . . 11. **Clintonia**

15. Leaves narrower, sometimes terete (if broader, then withering before flowering time); flowers less than 10 mm long (or if slightly longer, then pink), more than 7 or the umbel bearing bulblets; umbel subtended by a papery involucral spathe; plants bulbous, with odor of onion or garlic when bruised; fruit a capsule . . . . 12. **Allium**

14. Flowers solitary or in a raceme, panicle, or corymb, on a simple or branched stem (if in a few-flowered umbel, then the stem branched or forked)

16. Plants with principal leaves clearly cauline; basal leaves (at least of current season) absent or at most apparently one

17. Stem forked or branched; perianth over 12 mm long

18. Perianth pale to deep yellow, unspotted; stem and pedicels glabrous; fruit a glabrous capsule . . . . 13. **Uvularia**

18. Perianth white or creamy, in one species purple-dotted; stem (at least when young) and pedicels pubescent; fruit a finely pubescent or glabrate berry . . . . 14. **Disporum**

17. Stem unbranched (above the ground and below the inflorescence); perianth usually less than 12 mm long

19. Ovary with a single style; fruit a berry; inflorescence a raceme (tepals up to 7 mm long) or if a panicle, the tepals less than 3 mm long; leaves broadly lanceolate to ovate

20. Perianth of 4 parts; leaves 3 or fewer (very rarely 4, usually 2), sometimes pubescent beneath . . . . 15. **Maianthemum**

20. Perianth of 6 parts; leaves often more than 3 (1–4 in one species, where completely glabrous) . . . . 16. **Smilacina**

19. Ovary with 3 styles (one on each lobe); fruit a capsule; inflorescence a panicle; tepals ca. 5–13 mm long; leaves very elongate

21. Tepals with a conspicuous dark gland below the middle on the inner side; flowers mostly on pedicels longer than the perianth (fig. 309) . .
. . . . 17. **Zigadenus**

21. Tepals glandless; flowers sessile or on pedicels shorter than the perianth (fig. 304) . . . . 18. **Stenanthium**

16. Plants with the principal leaves all basal (at ground level) or nearly so; cauline leaves absent, reduced to bracts, or at most much smaller or fewer than basal leaves

22. Flower solitary . . . . 7. **Erythronium**

22. Flowers 2 or more in an inflorescence

23. Tepals united for half or more of their length

24. Perianth blue, less than 6 mm long; leaves linear (longest leaves at least 20–40 times as long as wide, up to 8 mm wide); plants bulbous . 19. **Muscari**

24. Perianth white, 5–10 mm long when mature; leaves lanceolate to elliptic (longest leaves less than 20 times as long as wide or over 1 cm

wide, or both); plants not bulbous

25. Leaves elliptic, the widest 2–6 cm broad; stems up to 35 cm tall, about equalling or shorter than the leaves; flowers nodding, on pedicels longer than the subtending bracts; perianth smooth on outside; fruit (rare) a berry . . . . 20. **Convallaria**

25. Leaves narrowly lanceolate or oblanceolate, the widest less than 2 cm broad; stems over 40 cm tall, much surpassing the leaves; flowers ascending on pedicels shorter than the subtending bracts; perianth ± granular-roughened on the outside; fruit a capsule . . . . 21. **Aletris**

23. Tepals completely separate or united at base only

26. Ovary with 3 styles (1 on each lobe) in perfect or pistillate flowers (or flowers all staminate)

27. Plants dioecious (flowers unisexual); pedicels not bracted; perianth less than 4 mm long; inflorescence a long narrow spike-like raceme ca. 1–1.5 cm in diameter and 8–30 cm long (fig. 308) . . . 22. **Chamaelirium**

27. Plants with perfect flowers; pedicels with a bract at the base; inflorescence shorter than 6 cm or perianth over 6 mm long and the flowers in a broad panicle

28. Perianth 1.5–5 mm long, the tepals without glands within; inflorescence a compact raceme less than 6 cm long; leaves ± equitant; peduncle and pedicels glandular-sticky in common species . . . . 23. **Tofieldia**

28. Perianth 5–12 mm long, the tepals glandless or with a conspicuous dark gland below the middle on the inner side; inflorescence over 7 cm (rarely 5 cm) long, open and almost always paniculate at least at base; leaves not equitant; peduncle and pedicels smooth . . . . [go to couplet 21]

26. Ovary with a single style; flowers perfect

29. Leaves broad (over 3 cm); perianth yellowish; fruit a blue berry . . . . 11. **Clintonia**

29. Leaves long and narrow (less than 1.5 cm wide); perianth white to blue; fruit a capsule

30. Tepals white, with green median stripe on outer side; filaments broad and flat (winged), wider at the middle than the anthers; flowers fewer than 20, in a raceme or corymb . . . . . . 24. **Ornithogalum**

30. Tepals white to blue, without green stripe; filaments slender or somewhat flattened toward base only; flowers usually more than 20, in an elongate raceme . . . . 25. **Camassia**

1. Asparagus

1. **Asparagus officinalis** L.　Fig. 295　　　　　　　　Garden Asparagus

Map 597. A native of the Old World (originally the eastern Mediterranean area?), long cultivated for medicinal and food uses, now rather widely escaped and established in waste places, fencerows, roadsides near old garden sites, etc., spreading into woods, dunes, and other habitats.

The greenish yellow flowers are followed by bright red berries.

2. Smilax

Greenbrier; Catbrier

KEY TO THE SPECIES

1. Leaves glabrous beneath (sometimes roughened on main veins); stems prickly or smooth
    2. Stems herbaceous throughout, without prickles; umbels with more than 20 flowers; fruiting peduncles flattened or terete; fruit glaucous; ultimate branchlets terete or many-ridged (not sharply 4-angled) . . . . 1. **S. herbacea**
    2. Stems woody and prickly, at least at base; umbels with fewer than 25 (usually fewer than 20) flowers; fruiting peduncles strongly flattened; fruit not glaucous, or if so, the branchlets strongly 4-angled
        3. Youngest branchlets strongly 4-angled; fruit glaucous; prickles broad and flattened at base, green with (usually) dark tip; fruiting peduncles less than twice as long as the subtending petioles . . . . 2. **S. rotundifolia**
        3. Youngest branchlets terete or with more than 4 ridges; fruit dark when ripe, not glaucous; prickles slender, terete, dark (or when young, green) throughout; fruiting peduncles twice as long as the subtending petioles, or longer . . 3. **S. tamnoides**
1. Leaves finely puberulent, at least on the veins, beneath; stems never prickly
    4. Stem of full-grown plants over 1 m long, the main stem or elongate branches climbing (or resting on other objects for support); plant almost always branched, with total of more than 25 leaves; tendrils conspicuously curled, present at most nodes, including those from which peduncles arise; peduncles longer than petioles (sometimes several times as long), all or most arising from axils of foliage leaves; flowers (at least on main stem) more than 25 in an umbel (but not all develop into fruit) . . . . 4. **S. lasioneura**
    4. Stem under 1 m tall, stiffly erect much of its length; plant unbranched, with fewer than 25 leaves (in *S. illinoensis* rarely more); tendrils absent or at most poorly developed and limited to uppermost nodes (never at the lower nodes from which peduncles arise); peduncles longer or, more often, shorter than petioles, at least the lowest ones usually arising from scale-like bracts on the stem below the foliage leaves; flowers more or fewer than 25 in an umbel
        5. Pistillate (and usually also staminate) flowers fewer than 25 in an umbel; leaves fewer than 20 (usually 7–9) on a plant; stems under 50 cm tall; peduncles usually shorter than the petioles or slightly longer; tendrils completely absent (rarely on upper 2–3 nodes) . . . . 5. **S. ecirrata**
        5. Pistillate and staminate flowers usually more than 25 in an umbel and plants with one or more other exceptions to the above (i.e., leaves more than 20, stems over 50 cm tall, peduncles more than 2 cm longer than petioles, tendrils present on several upper nodes) . . . . 6. **S. illinoensis**

598. Smilax herbacea   599. Smilax rotundifolia   600. Smilax tamnoides

1. S. herbacea L. Carrion-flower

Map 598. Thickets, low ground, and wooded banks.

Most reports of this species from Michigan should be referred to *S. lasioneura.*

2. S. rotundifolia L. Common Greenbrier

Map 599. Dry sandy woodland and dunes, often under oaks, sometimes covering large areas with a tangled thicket; less often in rich deciduous woods.

3. S. tamnoides L. Bristly Greenbrier

Map 600. Thickets, riverbanks, fencerows, and woods (oak–hickory, beech–maple, swamp forests on floodplains).

Plants in this region belong to var. *hispida* (Torrey) Fern., and in view of some doubt surrounding application of the name *S. tamnoides* might best be called *S. hispida* Torrey.

4. S. lasioneura Hooker Carrion-flower

Map 601. Edges of woods and clearings, fencerows, stream banks and floodplains, sandy oak woods and ridges.

The peduncles may be as long as 27 cm, with 150–250 pedicels, although usually the umbels are less robust. Often treated as *S. herbacea* var. *lasioneura* (Hooker) DC.

5. S. ecirrata (Kunth) S. Watson Carrion-flower

Map 602. Rich deciduous woods, low woods and thickets along riverbanks and floodplains, oak and oak–hickory woods.

A distinctive erect *Smilax* with a few large leaves and no tendrils. The specific epithet is often incorrectly spelled *ecirrhata.*

6. S. illinoensis Mangaly Plate 6-A Carrion-flower

Map 603. Woods and thickets, on floodplains and riverbanks, under oaks, and in rich deciduous woods.

For the plants which will run here in the key, the name published by Mangaly

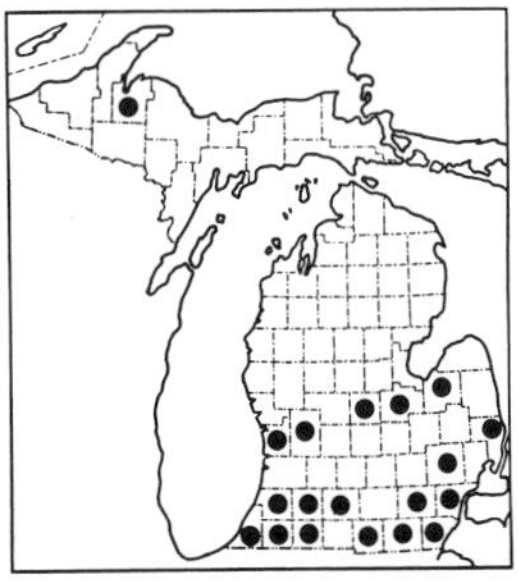

601. Smilax lasioneura   602. Smilax ecirrata

603. Smilax illinoensis

in 1968 is tentatively applied. Previously these plants had been treated by me merely as "intermediate forms, close to *S. ecirrata*," and almost every one of these which has been examined by Mangaly he has annotated as *S. illinoensis*. He points out (Rhodora 70: 264. 1968) that such plants may have arisen from hybridization. The leaf characters he cites do not appear to be consistent in our material, but the taxon is nevertheless quite well distinguished. These plants (and a few which will not fit anywhere in the key) have the general aspect of *S. ecirrata* but differ, usually, in the more densely flowered umbels, taller and more leafy stems with tendrils at uppermost nodes, etc., thus approaching *S. lasioneura*, although the distinction from the latter species is more clearcut than from *S. ecirrata*. For comments on some of the problems in the herbaceous species of *Smilax*, see also Deam (1940, p. 326) and Hanes & Hanes (1947, pp. 72–73).

3. **Polygonatum**   Solomon-seal

REFERENCES

Kawano, Shoichi, & Hugh H. Iltis. 1963. Cytotaxonomy of the Genus Polygonatum (Liliaceae) I. Karyotype Analysis of Some Eastern North American Species. Cytologia 28: 321–330.
Ownbey, Ruth Peck. 1944. The Liliaceous Genus Polygonatum in North America. Ann. Missouri Bot. Gard. 31: 373–413.

KEY TO THE SPECIES

1. Leaves completely glabrous . . . . 1. **P. biflorum**
1. Leaves finely pubescent on the veins beneath . . . . 2. **P. pubescens**

1. **P. biflorum** (Walter) Ell.

Map 604. Woods and borders (such as roadsides): rather open and dry (oak–hickory, sassafras) or floodplain swamp forest; sometimes relic in fields; and wet prairies. Farwell described var. *melleum* from "sandy grounds in open thickets at Algonac".

Generally a much stouter and taller plant than the next species. The peduncles arch throughout their length, at least some on a plant usually bearing 3–7 flowers. This name was originally restricted by Ownbey to diploid plants ($n = 10$) with a range largely beyond the Wisconsin glacial border but represented in this area by var. *melleum* (Farwell) Ownbey, known only from a few collections from the vicinity of Algonac (St. Clair County) and neighboring Ontario (TL: Algonac). She referred all other glabrous Michigan plants to *P. commutatum* (Schultes f.) A. Dietr. [often erroneously called *P. canaliculatum*, a synonym of the diploid *P. biflorum*]. *P. commutatum* was considered a tetraploid ($n = 20$), smaller plants of which could positively be distinguished only by the chromosome number. Subsequent discovery of additional chromosome numbers in both *P. biflorum* and *P. commutatum*, and the repeated difficulty botanists in various parts of the country have had in separating the two species morphologically, support the recognition of the latter taxon as, at best, *P. biflorum* var. *commutatum* (Schultes f.) Morong. Much of the older wildflower literature is confusing because the name *P. biflorum* was once erroneously applied to what now is called *P. pubescens*.

The longest leaves subtending peduncles in var. *melleum* are less than 3 times as long as wide, whereas in var. *commutatum* they are often (although not always) longer in proportion to width. The filaments in var. *melleum* are ± hairy, while in var. *commutatum* they are glabrous or minutely roughened or papillose. The flowers of var. *melleum* are said to be "honey-yellow" and those of var. *commutatum* "yellowish-green to greenish white." In var. *commutatum* the leafy portion of the stem is often longer than the lower naked portion, while these regions are of about equal length in var. *melleum*. Collectors in the region of Lake St. Clair should watch for var. *melleum*.

2. **P. pubescens** (Willd.) Pursh   Fig. 297

Map 605. In the same diversity of wooded habitats as the preceding, but more characteristic of beech—maple woods (both moist and dry), especially northward, where neither oak—hickory forests nor *P. biflorum* occurs.

A diploid species, usually smaller than most plants of *P. biflorum*. The peduncles are reflexed from their bases and bear 1—2 (rarely some on a plant 3—4) flowers. The rhizomes of both species bear conspicuous round scars (source of the common name) where the stems of previous years were borne.

4. Streptopus   Twisted-stalk

REFERENCE

Fassett, Norman C. 1935. Notes from the Herbarium of the University of Wisconsin – XII. A Study of Streptopus. Rhodora 37: 88–113.

KEY TO THE SPECIES

1. Leaves entire or minutely denticulate, strongly clasping at the base, glaucous beneath; nodes and upper internodes glabrous (lower internodes sometimes hispid); tepals spreading or curving from near the middle; flowers whitish green . .
. . . . 1. **S. amplexifolius**
1. Leaves prominently ciliate on the margins, the cilia usually visible to the naked eye (plate 7-A), sessile or slightly clasping (the larger ones subtending branches more strongly clasping), sometimes paler but not glaucous beneath; nodes and upper internodes ± pubescent or sparsely hispidulous; tepals spreading or recurved only at the tips; flowers usually pinkish (or even maroon) . . . . 2. **S. roseus**

1. **S. amplexifolius** (L.) DC.

Map 606. Rich deciduous or mixed woods, often in moist ravines, on seepy slopes, or along streams; also in cedar swamps.

The typical variety of this species is European. Our specimens have been placed in two weakly distinguished varieties based largely on the leaf margins. Most Michigan material appears to be var. *denticulatus* Fassett, with more than 10 tiny teeth per centimeter on at least some portions of the leaf margins. In var. *americanus* Schultes, the margins are entire or have fewer than 10 denticulations per cm. Both are distributed throughout the state, and intermediate forms difficult to place are not uncommon.

2. **S. roseus** Michaux   Plate 7-A
Map 607. In all northern woods: mixed (e. g., birch–fir), hemlock–hardwoods, or coniferous (especially cedar swamps); often becoming abundant along roadsides or powerline clearings through woods.

The typical variety occurs in the southern Appalachians. Michigan plants represent two varieties usually difficult to distinguish unless rhizomes are available. The rhizome of var. *longipes* (Fern.) Fassett has conspicuous elongate internodes several cm long, while var. *perspectus* Fassett has the matted internodes of the rhizome very short and mostly hidden by the roots. The latter also tends to have more than 30 cilia per cm of leaf margin and 7–11-nerved sepals; while var. *longipes* usually has fewer than 30 cilia per cm and 3–5-nerved sepals. *S. roseus* tends to be a smaller plant, but no clear distinction can be made from *S. amplexifolius* on size alone.

5. **Trillium**   Trillium; Wake-robin

Fortunately, the trilliums as a group are familiar to most people, so that the frequent "sports" which would cause trouble in the keys to family and genus are likely to be recognized as aberrant trilliums. *T. grandiflorum* is our most variable species – perhaps because it is our most common – but other species, such as *T. cernuum* and *T. flexipes*, are also known (though rarely) to have variously deformed or green-striped petals. The conservation laws of Michigan include all trilliums on the protected list. Several of our species are very rare; all of them suffer from picking because one generally picks the leaves and no part remains to nourish the underground portions.

604. Polygonatum biflorum   605. Polygonatum   606. Streptopus
                              pubescens   amplexifolius

The stigmas of our species are usually described as sessile. They lack a common style, and the stigmatic surface extends along the inner side of each of the 3 separate styles.

REFERENCE

Case, Frederick W., Jr., & George L. Burrows, IV. 1962. The Genus Trillium in Michigan. Some Problems of Distribution and Taxonomy. Pap. Mich. Acad. 47: 189–200.

KEY TO THE SPECIES

1. Flower sessile; petals maroon or yellow; leaves usually mottled when fresh
   2. Leaves with distinct (though sometimes short) petioles; sepals becoming strongly reflexed; petals ± abruptly narrowed basally to a slender claw . . . . . 1. **T. recurvatum**
   2. Leaves sessile; sepals ascending or somewhat spreading; petals convex or gradually narrowed basally, without a slender claw
      3. Petals more than 4 cm long, at least 3 times as long as the stamens, yellow or maroon . . . . 2. **T. viride**
      3. Petals not over 4 cm long, about twice as long as stamens or shorter, maroon . .
      . . . . 3. **T. sessile**
1. Flower peduncled; petals maroon to white (or green in aberrant forms); leaves not mottled
   4. Ovary 3-angled to obscurely 3-lobed; leaves definitely petiolate
      5. Plant usually less than 10 cm tall (rarely up to 17 cm), flowering usually in April; leaf blades less than 4 cm long, obtuse or rounded at apex; petals white or sometimes pinkish at base . . . . 4. **T. nivale**
      5. Plant over 10 (usually over 20) cm tall, flowering in May or June; leaf blades more than 4 cm long, acuminate at apex; petals white, streaked or blotched with purple at base . . . . 5. **T. undulatum**
   4. Ovary strongly 6-angled or -winged and the leaves sessile or subsessile (except in aberrant forms with petioled leaves and/or 3-lobed ovary; these usually also have the petals ± marked with green and are obvious sports)
      6. Petals white to pink (never maroon), 3.5–8 (9.5) cm long, distinctly longer than the sepals, ± obtuse (occasional small plants with shorter petals – though still longer than sepals – may be recognized by the straight styles and broad obovate petals); stigmatic styles straight (though sometimes spreading) or slightly curved at very tip, uniform in diameter; peduncles held above the leaves . . 6. **T. grandiflorum**
      6. Petals white to maroon, usually less than 3.5 cm long (if longer, maroon in color and/or narrowly acute at apex), seldom much longer than sepals; stigmatic styles spreading, thick at base, tapering, and recurved; peduncles in white-flowered plants (and often also in maroon ones) usually reflexed or declined below the leaves
         7. Peduncles erect or slightly bent (rarely nearly horizontal), the flowers held above the leaves; petals and ovary (except sometimes in white forms) deep maroon; petals 2.5–6 (usually 3–4) cm long; filaments usually a fourth to half as long as the anthers (though occasionally longer or shorter); anthers (4.5) 6–12 mm long . . . . 7. **T. erectum**
         7. Peduncles bent from the base, usually nearly horizontal or reflexed below the leaves; petals white to maroon, less than 3 cm long (rarely 3.5–4); ovary white or rose; filaments usually less than a fourth or more than a half as long as the anthers

8. Filaments very short, almost always less than 2 mm long and less than a fourth the length of the anthers; anthers 6–15 mm long, yellowish (to pink in maroon-flowered forms); petals white to maroon, (1.5) 2–3 (4) cm long . . .
. . . . 8. **T. flexipes**
8. Filaments about as long as the anthers or occasionally as short as half as long; anthers 3–7 mm long, usually pink when fresh; petals white (rarely rosy), less than 2.5 (rarely 3 or very rarely 3.5) cm long . . . . 9. **T. cernuum**

1. **T. recurvatum** Beck

Map 608. Rich floodplains and upland beech–maple woods; very rarely seen in Michigan except in Warren Woods, Berrien County.

2. **T. viride** Beck

Map 609. Referred here are two collections which some authorities would treat as other species. In view of the perplexing taxonomic status of this group of *Trillium* and the uncertain status of the rare Michigan collections, they are included here under the broad concept of *T. viride*. Plants with yellow petals would by some botanists be referred to *T. luteum* (Muhl.) Harbison. This is a taxon of the southeastern United States, which suddenly appeared in 1950 in a mature beech–maple woods in Saginaw County, where it had not been noted in previous botanizing in the same woods (*Case & Burrows* in 1957, MICH; see Case & Burrows, 1962). A similar plant with maroon petals and ovary is labeled as having been collected in northwestern Gratiot County by a student at Alma College (*G. Gopoian* in 1968, ALM). This would presumably be referred by some to *T. hugeri* Small or *T. cuneatum* Raf., another taxon of the southeastern and southern states.

Such occurrences as these – and reports from Berrien County – may represent long-dormant escapes from wildflower gardens, as suggested near Kingston, Ontario (Blue Bill 17: 23–27. 1970), although we have no evidence regarding the possible origin of the Michigan plants.

3. **T. sessile** L.   Plate 6-B   Toadshade

Map 610. Woods and thickets, very local; Michigan specimens almost all lack habitat data.

607. Streptopus roseus   608. Trillium recurvatum   609. Trillium viride

Varies greatly south of our region, but apparently so rare in Michigan that the expected range of forms has not yet been found.

4. **T. nivale** Riddell   Snow Trillium
    Map 611. Very rare and local; alluvial woods and riverbanks, apparently in ± calcareous sites. Blooming so early in the spring it is apt to be overlooked.

5. **T. undulatum** Willd.   Painted Trillium
    Map 612. Very local, under hemlocks and in damp coniferous or mixed woods.

6. **T. grandiflorum** (Michaux) Salisb.   Plate 6-C   Common Trillium
    Map 613. Moist to rather dry deciduous woods, most luxuriant and often forming large colonies in rich beech—maple forest, but also, less frequently, found in oak—hickory or aspen woods, swampy woods and thickets, mixed conifer—hardwoods, and even coniferous swamps. Apparently very local in Houghton and Keweenaw counties and Isle Royale — a single collection seen from each of these areas.
    The petals tend to turn pink with age; in other species, the color, whether white, maroon, or intermediate, appears to be more constant from the bud to maturity.
    Forms with the petals variegated with green are not uncommon and are sometimes locally abundant (as in Alpena, Lapeer, Leelanau, and Oakland counties). Long- or short-petioled leaves often occur on plants with abnormal flowers. Parts of the flower are sometimes changed toward more leaf-like states, and are sometimes multiplied in number. Plants with parts in 2's or 4's are known from Michigan, and I have collected a specimen in Iron County (*4126*, MICH) in which a stem bearing 3 normal leaves and a 3-merous flower arose from the same root as one bearing 4 leaves and a 4-merous flower.
    While the variations are striking, most of them result from a mycoplasma-like infection (Hooper, Case, & Myers, Pl. Disease Rep. 55: 1108–1110. 1971). Such diseased states are scarcely worthy of recognition by formal scientific names. Names for some of them may be found in *Gray's Manual*; one of them, f.

610. Trillium sessile   611. Trillium nivale   612. Trillium undulatum

chandleri (Farw.) Vict., with leaves absent, was originally described from Michigan (TL: Farmington Tp., Oakland County), but that name is illegitimate in the rank of form. Discussions of some of the more striking populations in Michigan are in the following articles:

Farwell, Oliver Atkins. 1919. The Trillium grandiflorum Group. Farmington Township, Oakland County, Michigan. Rep. Mich. Acad. 20: 155–159.
Hall, Marion Trufant. 1961. Teratology in Trillium grandiflorum. Am. Jour. Bot. 48: 803–811.
Smith, Erwin T. [sic, for 'F.'] 1879. A Michigan Trillium. Bot. Gaz. 4: 180–181.

7. **T. erectum** L.   Stinking Benjamin
Map 614. Usually in rich, even swampy, deciduous woods.

Frequently said to be distinguished in the field from *T. flexipes* (and other species) by the strong fetid "wet-dog" odor. However, I have noted this odor in plants which otherwise have all the characteristics of *T. flexipes*. The white-flowered form (with dark or pale ovary) [f. *albiflorum* R. Hoffm.] and the yellow-flowered form [f. *luteum* Louis-Marie] are known from Michigan. The form with petals colored at the base and pale terminally [f. *cahnae* (Farw.) Louis-Marie] was originally described from Michigan (TL: Clawson [Oakland County]). See also discussion under the next species.

8. **T. flexipes** Raf.
Map 615. Woods: rich beech—maple, oak—hickory, mixed swampy and floodplain forests.

The very short filaments and long anthers usually make this species easily recognizable; the filaments are rarely as long as 4 mm. Nearly erect peduncles are occasionally reported in this species, and such plants are close to *T. erectum*. The two species appear to intergrade in Michigan, but most plants of this group with erect peduncles also have deep maroon petals and ovary, longer filaments, and larger petals, and thus are referred to *T. erectum* as here treated. Case & Burrows (1962) suggest that f. *walpolei* and intermediate forms may represent a hybrid swarm derived from *T. erectum* and white-flowered *T. flexipes*.

613. Trillium grandiflorum   614. Trillium erectum   615. Trillium flexipes

The petals may be white [f. *flexipes*], maroon [f. *walpolei* (Farw.) Fern. (TL: Ypsilanti [Washtenaw County])], intermediate in shade, or reddish basally and white terminally. Different color forms may be found growing together. The odor may apparently be either sweet or foul; this varies somewhat with the age of the flowers. In some manuals, this species is called either *T. declinatum* (Gray) Gl. or *T. gleasonii* Fern.

9. **T. cernuum** L.   Fig. 296   Nodding Trillium
    Map 616. Coniferous swamps and borders of bogs, moist or swampy mixed woods especially where there is birch, thickets along streams; less often in rich hardwoods, such as beech–maple woods, and floodplain forests. Seldom common where found.

This species is often said to have shorter peduncles than the preceding, but Michigan specimens average just the opposite, if anything. Peduncles of our flowering specimens of both species range from (1) 1.5 to 5.5 (6) cm in length (one *T. flexipes* seen at 10 cm), the majority falling between 3 and 4.5 cm in *T. cernuum* and between 2.5 and 3.5 cm in *T. flexipes*. There is much variation, however, and since similar extremes occur in both species, measurements are of little help in identification.

Most Michigan specimens are var. *macranthum* Eames & Wieg. (larger flowers, with anthers 4–7 mm long); a few collections, from both the Upper and Lower Peninsulas, may be referred to typical var. *cernuum* (smaller flowers, with anthers 3–4 mm long). Leaves of this species tend to be subsessile or almost petiolate.

6. Medeola

1. **M. virginiana** L.   Indian Cucumber-root
    Map 617. Usually in moist or swampy woods: beech–maple forests (especially at margins of pools), hemlock knolls, cedar bogs, less often in oak or pine woods.

The crisp white tuberous rhizomes have a decidedly cucumber-like flavor.

616. Trillium cernuum   617. Medeola virginiana   618. Erythronium
                                                                 americanum

7. Erythronium   Trout-lily; Dogtooth-violet; Adder's-tongue

The characteristic mottled leaves (shades of green and reddish) make the genus easily recognizable vegetatively. Both species flower in April and May, the leaves withering soon afterward, so that by mid-summer there is no trace above ground of plants which may have carpeted an area in the spring.

Although these plants are not true violets of any sort, the common name "dogtooth-violet" is in wide use, derived from the European species, *E. dens-canis* L.

KEY TO THE SPECIES

1. Flowers yellow . . . . 1. **E. americanum**
1. Flowers white . . . . 2. **E. albidum**

1. E. americanum Ker   Plate 6-D

Map 618. Rich deciduous woods and moist thickets, at least the leaves, if not always flowering plants, often abundant.

The stigmas are generally straight, very short, not recurved. However, *Gray's Manual*, ed. 5 (1867) noted a yellow-flowered *Erythronium* like *E. albidum* from Lake Superior. This report apparently was based on a collection (*J. W. Robbins* in 1863, GH) from Houghton County at Portage Lake, recorded as having unspotted leaves, a 3-parted stigma, and yellow flowers. In 1938 Farwell published a description of *E. americanum* var. *rubrum* (TL: Lake Linden vicinity, Houghton County) — a larger plant with red anthers and 3-lobed stigmas (but these hardly spreading).

The anthers in this species may be either yellow or rich reddish brown. Rarely the perianth is reddish brown basally. An unusual form in which the scapes bear a leaf-like bract (ca. 4.5 cm long) has been collected on Sugar Island (Chippewa County) (*Hiltunen 2787*, WUD).

2. E. albidum Nutt.

Map 619. Rich deciduous woods, especially in low areas and along flood-plains, sometimes with the preceding species, but much more local, even in the

619. Erythronium albidum   620. Lilium   621. Lilium michiganense
                             philadelphicum

southern part of the state. Northward, known only along the Sturgeon River in Houghton County (*C. D. Richards 2889* in 1950, MICH, ILL; *Bourdo 4224* in 1962, MCT-F, MICH).

The stigmas are spreading or recurved and short, but tend to be slightly longer and more distinctly separated than in the preceding species.

8. **Lilium**   Lily

REFERENCES

Ingram, John. 1968. Notes on the Cultivated Liliaceae 7. Lilium lancifolium Thunb. vs. L. tigrinum Ker. Baileya 16: 14–19.
Voss, Edward G. 1964. The Michigan Lily. Mich. Bot. 3: 44–47.
Wherry, Edgar T. 1947 ["1946"]. A Key to the Eastern North American Lilies. Bartonia 24: 5–8.

KEY TO THE SPECIES

1. Flowers erect; tepals distinctly narrowed at the base to a slender claw; leaves up to 8 (14) mm wide, smooth on veins beneath but sometimes slightly roughened on margins . . . . 1. **L. philadelphicum**
1. Flowers nodding (fruit becoming erect); tepals narrowed gradually toward base, not clawed; widest leaves 8–25 (28) mm wide, the margins and usually veins of most leaves finely scabrous or papillose beneath (uppermost leaves sometimes smooth)
   2. Tepals glabrous within; stem glabrous above; leaves mostly whorled . . . .
   . . . . 2. **L. michiganense**
   2. Tepals with pubescent strip basally within; stems with cobwebby pubescence, especially above; leaves alternate but crowded . . . . 3. **L. lancifolium**

1. **L. philadelphicum** L.   Wood Lily

Map 620. Open sandy or rocky ground, dunes, sandy open woods (jack pine, aspen), edges of coniferous woods, in bogs; often thriving along grassy roadsides and in clearings.

The typical variety, largely found east of Michigan, has the leaves mostly whorled. Our plants often have the leaves mostly scattered except for one or two nodes [var. *andinum* (Nutt.) Ker, sometimes called *L. umbellatum* Pursh]. But in our area, so many plants intermediate with this western variety occur that they cannot be clearly distinguished.

Plants may have as many as 5 flowers, although fewer (often only 1) are usual. The flowers are normally orange to nearly red, but the yellow-flowered form [f. *flaviflorum* Williams] has been found very rarely in Michigan, including on Drummond Island (*R. McVaugh 9103* in 1948, MICH) and at Hammond Bay, Presque Isle County (*M. A. Piehl 6060* in 1960, MICH). A form with the tepals unspotted as well as yellow [f. *immaculatum* Raup] has been collected at West Bluff in Keweenaw County (*N. C. Fassett 22009* in 1938, WIS).

2. **L. michiganense** Farw.   Plate 6-E   Michigan Lily

Map 621. Wet meadows, thickets, borders of streams and woods, sometimes in partial or deep shade in swamp forests (floodplains, etc.) and boggy woods with tamarack and poison sumac. Old reports from the Mackinac Straits region (Burt Lake, Cheboygan County; Bois Blanc Island, Mackinac County) are not supported by specimens but presumably are reliable.

Originally described from east of Ypsilanti, Washtenaw County (along with some variants of little significance). There is some doubt as to whether this midwestern plant is sufficiently distinct from the eastern *L. canadense* L. to warrant recognition as a separate species. The chief — perhaps only — consistent difference is in the tepals, which are slightly curved or arching in *L. canadense*, whereas in *L. michiganense* (when fully open) the tepals are strongly recurved, the tips being turned back beyond the base of the perianth. A strongly recurved perianth is also characteristic of *L. superbum* L., which has sometimes been reported from Michigan. This, however, is an eastern and southern species, with longer anthers and smooth leaves at most slightly papillose on the margins. (The anthers of our *L. michiganense* are 4.5–12 (17) mm long.) A very few specimens from Michigan do not appear to have the strongly recurved perianth of *L. michiganense*, but are doubtless young or variable individuals of this species.

3. **L. lancifolium** Thunb.   Tiger Lily

Map 622. Rarely escaped from cultivation but probably somewhat more frequent than the single Michigan collection would indicate; roadside in Cleveland Tp., Leelanau County (*Thompson L-1650* in 1951, BLH).

Long known under the later name of *L. tigrinum* Ker.

9. Hemerocallis   Day-lily

KEY TO THE SPECIES

1. Flowers orange . . . . 1. **H. fulva**
1. Flowers yellow . . . . 2. **H. lilio-asphodelus**

622. Lilium lancifolium   623. Hemerocallis fulva   624. Hemerocallis
                                                              lilio-asphodelus

1. H. fulva (L.) L.   Fig. 298   Orange Day-lily

Map 623. A cultivated plant rather freely spreading by rhizomes and tuberous roots to roadsides and dumps; much more widely established in southern Michigan than extant collections would suggest.

2. H. lilio-asphodelus L.   Yellow Day-lily; Lemon-lily

Map 624. Another cultivated plant, less commonly escaping. The yellow day-lily is a fragrant spring- to summer-flowering species, while the orange one is summer-flowering and without fragrance. Usually known as *H. flava* (L.) L., an illegitimate name.

10. Yucca

1. Y. filamentosa L.   Fig. 299   Yucca; Adam's-needle

Map 625 A locally established escape from cultivation in the southwestern part of the state. Particularly well established in sandy barren open ground at places in the Allegan State Forest, where it presents a handsome sight in association with the prickly-pear cactus, *Opuntia humifusa.*

Our plants are apparently the narrowed-leaved segregate designated by Fernald as *Y. smalliana*, although some have smooth and glabrous panicles as in true *Y. filamentosa.* Yucca moths, both "true" (*Tegeticula yuccasella* Riley) and "false" (*Prodoxus quinquepunctellus* Chambers), are found in the flowers, the former effecting pollination after laying their eggs in the ovary where the caterpillars will be assured of a developing food supply.

11. Clintonia

1. C. borealis (Aiton) Raf.   Plate 5-D   Corn-lily; Bluebead-lily

Map 626. At the edge of its range in southern Lower Michigan, usually restricted to bogs and coniferous woods; northward, widespread in coniferous, deciduous, and mixed woods on sandy and richer soils; particularly luxuriant under cedar and in birch–fir woods near the Great Lakes.

The flowers are usually in an umbel, although occasionally one or more of

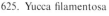

625. Yucca filamentosa   626. Clintonia borealis   627. Allium tricoccum

them may be in a separate little umbel or solitary, giving the inflorescence a racemose tendency. The perianth is pale yellow and the fruit a bright blue berry. The margins of the leaves are ± ciliate with loose fine hairs, glabrescent in age. The leaves, often consipicuous in the woods, may thus be readily distinguished from those of similar plants.

12. Allium

REFERENCE

Moore, Harold E., Jr. 1954–1955. The Cultivated Alliums. Baileya 2: 103–113; 117–123. 3: 137–149; 156–167. [This series includes keys to over 125 species reported to be in cultivation, including all native Michigan ones.]

KEY TO THE SPECIES

1. Leaves usually over 2 cm broad, flat, petiolate, withering before the plant flowers .
. . . . 1. **A. tricoccum**
1. Leaves linear, flat or terete, less than 2 cm broad (usually less than 1 cm), not petiolate, present at flowering time
  2. Umbel nodding, on bent or reflexed tip of scape; leaves flat . . . . 2. **A. cernuum**
  2. Umbel erect on straight tip of scape; leaves flat or terete
    3. Leaf blades terete, at least most of their length [may be flattened where pressed in drying, but base of blade, just above summit of sheath, will not show 2 distinct surfaces]
      4. Stem stout, over 5 mm in diameter for most or all of its length, distinctly inflated below the middle . . . . 3. **A. cepa**
      4. Stem 5 mm in diameter or less (very rarely as stout as 7 mm), without inflated section
        5. Pedicels equalling or shorter than the flowers; inflorescence without bulblets; filaments unappendaged . . . . 4. **A. schoenoprasum**
        5. Pedicels longer than mature flowers; umbels bearing bulblets in addition to 0–many flowers; inner filaments (if flowers present) with a long slender appendage on each side . . . . 5. **A. vineale**
    3. Leaf blades flat (sometimes keeled)
      6. Umbels bearing bulblets (flowers few or none)
        7. Involucral spathe composed of one bract with a beak usually 2–10 cm long; outer coverings of bulb membranous . . . . 6. **A. sativum**
        7. Involucral spathe composed of 2–3 bracts, with beaks less than 1.5 cm long; outer coats of bulb strongly fibrous . . . . 7. **A. canadense**
      6. Umbels not bearing bulblets
        8. Leaves all at base of plant; filaments slender, unappendaged; ovary with thin erect projections or crests on top. . . . . 8. **A. stellatum**
        8. Leaves scattered on lower half of stem; inner filaments broad and flat, trifid, with a long appendage on each side extending beyond anther-bearing central filament; ovary without crests . . . . 9. **A. rotundum**

1. **A. tricoccum** Aiton   Fig. 300   Wild Leek; Ramps
    Map 627. Rich deciduous woods, both on upland and floodplain sites, especially characteristic in moist areas of beech–maple–hemlock stands.

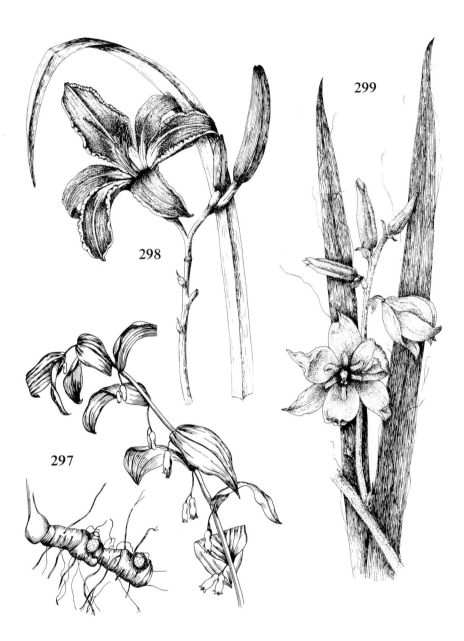

297. *Polygonatum pubescens* x½ ; rhizome x½
298. *Hemerocallis fulva* x½
299. *Yucca filamentosa* x½

The broad green leaves appear in May and often carpet extensive areas; the umbels of white flowers appear in late June or July, when the leaves are nearly or quite gone. All parts of the plant, but especially the bulbs, have a strong odor of onion. The plants are called "ramps" in some regions and festivals on their behalf (e. g., in West Virginia) consume more bulbs than Michigan stomachs are accustomed to tolerate. A bulb or two will add plenty of flavor to a stew, and the broad leaf may be placed in a sandwich.

On the basis of his extensive observations in Kalamazoo County, Hanes described var. *burdickii*, which differs from the typical plant in having the sheaths and petioles white or greenish instead of red or reddish. This variety is also said to appear later in the spring but to flower earlier, and to be smaller than typical var. *tricoccum* (see Rhodora 48: 61–63. 1946; 55: 243. 1953). Such plants are known from Kalamazoo and Cass counties. A clump lacking red color in Gogebic County (*Voss 6258*, May 31, 1958, MICH) appeared in all other respects similar to reddish plants, among which it was growing.

2. **A. cernuum** Roth   Nodding Wild Onion
Map 628. Marshy ground, swales and meadows, grassy wooded banks, spreading along railroad embankments and roadsides.

The flowers range from white to rose. The umbels may tend to appear erect in old fruiting material, but the tip of the scape is still usually bent. Plants with the flowering umbel apparently erect were collected by Farwell in Washtenaw County (Am. Midl. Nat. 12: 114–115; 119–120. 1930). Such plants would run to *A. stellatum* in the key, but differ in the flowering pedicels being more flexuous (they become stiffer in fruit; are stiff but arching in flower in *A. stellatum*), and the leaves are soft and flat (in *A. stellatum*, the leaves are stiffish and rounded on the back).

3. **A. cepa** L.   Fig. 301   Onion
Map 629. Occasionally established as "escaped" — at least persisting and spreading somewhat where old bulbs have been dumped, and doubtless more often than the map indicates.

628. Allium cernuum   629. Allium cepa   630. Allium
                                                      schoenoprasum

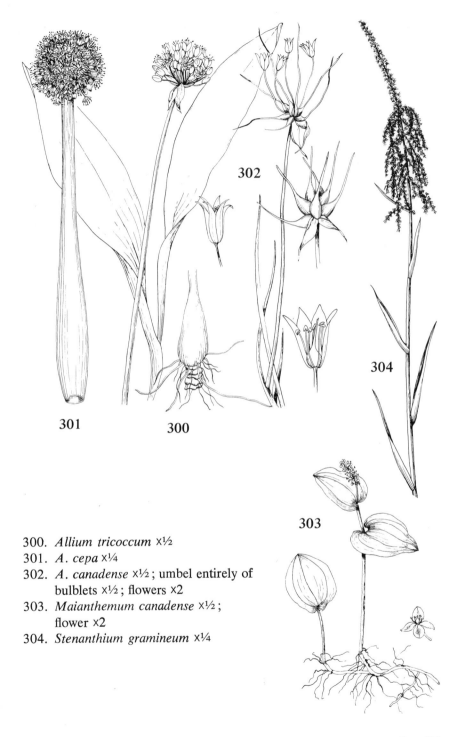

300. *Allium tricoccum* ×½
301. *A. cepa* ×¼
302. *A. canadense* ×½ ; umbel entirely of bulblets ×½ ; flowers ×2
303. *Maianthemum canadense* ×½ ; flower ×2
304. *Stenanthium gramineum* ×¼

4. **A. schoenoprasum** L.   Chives

Map 630. The native wild chives of the Lake Superior region (stream banks and rock crevices) has usually been referred to var. *sibiricum* (L.) Hartman, supposedly larger and somewhat coarser than the chives of the garden, which may occasionally escape in waste ground. Southern Michigan and Baraga County records are not from indigenous stands.

5. **A. vineale** L.   Field Garlic

Map 631. Dry sandy open ground, oak—sassafras woods, and waste ground. A native of Europe, locally well established.

Sometimes confused with *A. canadense*, especially when specimens are dry or incomplete. In *A. vineale* the uppermost leaf arises near the middle of the stem, while in *A. canadense* the leaves are all on the lower 20% of the stem.

6. **A. sativum** L.   Garlic

Map 632. Roadsides in Kalamazoo County, and probably elsewhere; an Old World species, escaped from cultivation.

7. **A. canadense** L.   Fig. 302   Wild Garlic

Map 633. Usually in low rich woods, floodplains, stream banks, or wet meadows.

631. Allium vineale

632. Allium sativum

633. Allium canadense

634. Allium stellatum

635. Allium rotundum

636. Uvularia sessilifolia

8. **A. stellatum** Ker   Wild Onion

Map 634. Collected only once in Michigan, on low dunes near Manistique (*Ehlers 4072* in 1929, MICH). Possibly not native quite this far east.

9. **A. rotundum** L.

Map 635. Fields, roadsides, etc.; a locally established escape from cultivation, native to the Old World.

13. Uvularia

REFERENCES

Voss, Edward G. 1966. The Little Bellwort. Mich. Audubon Newsl. 13(5): 3–4.
Wilbur, Robert L. 1963. A Revision of the North American Genus Uvularia (Liliaceae). Rhodora 65: 158–188.

KEY TO THE SPECIES

1. Leaves sessile, glaucous but glabrous beneath; rhizome elongate, bearing scattered small roots; mature capsule over 15 mm long . . . . 1. **U. sessilifolia**
1. Leaves perfoliate, finely puberulent (rarely almost glabrous) and usually light or dark green but not glaucous beneath; rhizome short with many crowded roots; mature capsule less than 14 (usually 8–10) mm long . . . . 2. **U. grandiflora**

1. **U. sessilifolia** L.   Merrybells

Map 636. Much less common than the next species, but in similar habitats; both species seem to thrive in openings and borders of woods. Apparently absent from the western and northern Lower Peninsula and the central and eastern Upper Peninsula. From the Porcupine Mountains (and Minnesota) it ranges south through Wisconsin to the southern states, and from Saginaw Bay southward to northeastern and southern Ohio and beyond, eastward to New England. This essentially circular pattern nearly surrounds a large area from northern and eastern Lake Michigan southward.

The flowers of *U. sessilifolia* are distinctly paler (cream-colored) than those of *U. grandiflora*, and the whole plant is more delicate in general appearance.

2. **U. grandiflora** Sm.   Plate 5-E   Bellwort

Map 637. Usually in rich deciduous woods, often on hillsides, ranging from upland beech—maple and hemlock—hardwoods to low floodplain woods and streamside thickets, but also in oak woods and rarely under pine or in moist ground under other conifers (such as cedar—hemlock).

Plants of this species are not fully developed vegetatively at the time the bright yellow flowers are open; the arched or drooping aspect of the upper part of the stem helps to give the whole plant a somewhat wilted look. The leaves expand and the stem straightens considerably as the distinctive triangular fruit ripens.

U. perfoliata L. is an eastern species which does not range as far west as Michigan although often erroneously reported from the state. It has the perfoliate leaves glabrous beneath; the tepals are papillate within, not smooth and glabrous as in *U. grandiflora*.

14. Disporum

REFERENCE

Jones, Quentin. 1951. A Cytotaxonomic Study of the Genus Disporum in North America. Contr. Gray Herb. 173. 39 pp.

KEY TO THE SPECIES

1. Tepals copiously purple-dotted; ovary ± obovoid, densely glandular-pubescent; cilia of leaf margin ± spreading . . . . 1. **D. maculatum**
1. Tepals unspotted; ovary ± ellipsoid, somewhat hispidulous especially toward the summit; cilia of leaf margin mostly pointing forward . . . . 2. **D. hookeri**

1. **D. maculatum** (Buckley) Britton   Fig. 305   Nodding Mandarin

Map 638. A southern species, the nearest location to Michigan being in southern Ohio. Known only from Farwell's 1921 and 1922 collections (BLH) from near Farmington. He described the plants as a new species (*D. cahnae*), named for Mrs. Cahn, who led him to the place [undescribed] where they were growing. A remarkable disjunction in range, and the status of the Michigan plants somewhat mysterious. Reported from Washtenaw County by Jones on the basis of an 1884 collection which has not been located despite intensive search.

2. **D. hookeri** (Torrey) Nicholson   Fairy Bells

Map 639. This western species was unknown east of western Montana until a flowering specimen of var. *oreganum* (S. Watson) Q. Jones was collected in woods near Carp River in the Porcupine Mountains State Park (*Thompson M-1593*, May 30, 1968, BLH).

637. Uvularia grandiflora   638. Disporum maculatum   639. Disporum hookeri

15. Maianthemum

REFERENCE

Weller, Stephen G. 1970. A Preliminary Report on the Varieties of Maianthemum canadense in Northern Michigan. Mich. Bot. 9: 48–52.

1. **M. canadense** Desf.   Fig. 303   Wild or False Lily-of-the-valley; Canada Mayflower

Map 640. Almost everywhere, in dry to moist sites: deciduous and coniferous woods, dunes, bogs, and swamps.

Plants with the stems and under surfaces of the leaves more or less pubescent are var. *interius* Fern., while var. *canadense* is completely glabrous. Both varieties occur throughout the state, and intermediates are not uncommon. There seem to be no habitat distinctions between the varieties, nor is there consistency in which blooms first in a given area and season.

16. Smilacina   False Solomon-seal

KEY TO THE SPECIES

1. Inflorescence a panicle; perianth 1–2.5 mm long, the stamens up to 3 mm long . .
. . . . **1. S. racemosa**
1. Inflorescence a raceme; perianth 2.5–9 mm long, exceeding the stamens
  2. Cauline leaves 1–4, completely glabrous; inflorescence almost always overtopping
    leaves . . . . **2. S. trifolia**
  2. Cauline leaves more than 6, finely pubescent beneath (rarely almost glabrous);
    uppermost leaves surpassing the top of the inflorescence . . . . **3. S. stellata**

1. **S. racemosa** (L.) Desf.   Fig. 306   False Spikenard

Map 641. In moist to dry woods, most often in rich beech–maple stands but also on floodplains and wooded dunes and in oak–hickory and other drier woods; often thrives in open second-growth stands; rarely in moist coniferous woods.

The stem is not erect, but arches gracefully. The greenish white flowers

640. Maianthemum   641. Smilacina racemosa   642. Smilacina trifolia
canadense

develop into red berries which may be conspicuous in the fall. This is generally a large plant, taller and with many more flowers than the other two species. It is perhaps unfortunate that the only species of this genus which does not have a simple raceme bears the epithet *racemosa*.

2. **S. trifolia** (L.) Desf.

Map 642. Bogs and coniferous swamps, especially along brooks, springs, and wet areas.

The flower is white, the berry red (although mature fruit is less often seen in this species than the others). Sometimes confused with *Maianthemum canadense* but readily distinguished from that species, even in the absence of flowers, by the more tapering leaf bases. In *Maianthemum* the leaves, especially the lower ones, are definitely cordate; var. *interius*, furthermore, differs from *S. trifolia* in being pubescent.

3. **S. stellata** (L.) Desf.

Map 643. Particularly characteristic of sand ridges, dunes, and shores (and neighboring thickets) near the Great Lakes, where the plants are stiffly erect with the leaves strongly ascending and somewhat folded or trough-like. In rich hardwoods, thickets along riverbanks, floodplains and swampy woods, coniferous swamps, and other shaded sites, plants are sometimes much like *S. racemosa* in aspect, with an arching stem and flat blades.

The flowers are whiter than in *S. racemosa*, and the fruit is a larger berry, red or greenish with conspicuous dark bands often strikingly developed.

17. Zigadenus

1. **Z. glaucus** (Nutt.) Nutt.   Fig. 309   White Camas

Map 644. Dunes and sandy or rocky shores of the Great Lakes; also inland on calcareous soils and banks, in bogs and low ground.

The plant, especially the bulb, contains a very poisonous alkaloid and should never be eaten by man or domestic animals. The flowers have a distinctive and strongly unpleasant odor. The leaves are long and narrow, less than 12 mm broad (rarely to 20 mm), mostly crowded toward the base of the plant. Probably not really distinct from *Z. elegans* Pursh, a species considered to occur west of the Great Lakes.

18. Stenanthium

1. **S. gramineum** (Ker) Morong   Fig. 304   Featherbells

Map 645. A local waif well established in marshy ground between the highway (M–28) and railroad on the Baraga-Houghton county line (*Bourdo* in 1958, 1960, 1965, 1967, MCT-F, MICH, BLH). Native from Pennsylvania to Indiana, southward to Florida and Texas.

19. Muscari   Grape-hyacinth

KEY TO THE SPECIES

1. Leaves 1–2 (3) mm wide, nearly terete; perianth cylindrical, about twice as long as
wide . . . . 1. **M. atlanticum**
1. Leaves flat, 2.5–8 mm wide (or wider); perianth subglobose, only slightly longer
than wide . . . . 2. **M. botryoides**

1. **M. atlanticum** Boiss. & Reuter   Fig. 307

Map 646. Fields, banks, and roadsides – escaped from cultivation; throughout much of Berrien County but only rarely collected elsewhere.

Not *M. racemosum* Miller, a name long applied erroneously.

2. **M. botryoides** (L.) Miller

Map 647. Like the preceding, an escape from cultivation along roadsides and
neighboring woods.

20. Convallaria

1. **C. majalis** L.   Lily-of-the-valley

Map 648. Another ornamental locally established as an escape from cultivation spreading in woods near cemeteries, gardens, etc.

643. Smilacina stellata

644. Zigadenus glaucus

645. Stenanthium
gramineum

646. Muscari atlanticum

647. Muscari botryoides

648. Convallaria majalis

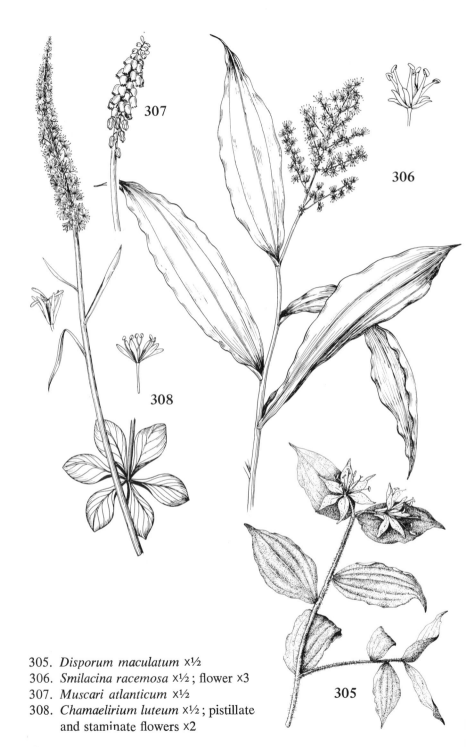

305. *Disporum maculatum* ×½
306. *Smilacina racemosa* ×½; flower ×3
307. *Muscari atlanticum* ×½
308. *Chamaelirium luteum* ×½; pistillate
and staminate flowers ×2

309. *Zigadenus glaucus* ×½; flower ×1
310. *Aletris farinosa* ×½
311. *Tofieldia glutinosa* ×½; flower ×2; fruit ×2
312. *Camassia scilloides* ×½; flower ×1

21. Aletris

1. **A. farinosa** L.   Fig. 310   Colic-root; Stargrass
Map 649. In moist or sometimes dry, usually sandy or sandy-mucky soil, on lake shores and in swales, meadows, clearings, and abandoned fields.

22. Chamaelirium

1. **C. luteum** (L.) Gray   Fig. 308   Blazing-star
Map 650. Despite other reports from the state, the only Michigan collection seen was made by Farwell at Detroit in 1917 (BLH). Since he collected many cultivated plants the same day, it is possible that this one was not native.

23. Tofieldia   False Asphodel

KEY TO THE SPECIES

1. Perianth ca. 2–3 mm long, not subtended by bractlets; peduncles and pedicels
   smooth . . . . 1. **T. pusilla**
1. Perianth 3–5 (6) mm long, subtended by tiny bractlets (in addition to the bracts
   at the bases of the pedicels); peduncle and pedicels glandular-sticky . . . . 2. **T. glutinosa**

1. **T. pusilla** (Michaux) Pers.
Map 651. In Michigan known only from Isle Royale, where collected at Scoville Point and on Passage Island and presumably occurring elsewhere; local in rock crevices and pools on the north shore of Lake Superior in Ontario.

Closely resembling the next species in habit, but smaller in every way and with a smooth scape.

2. **T. glutinosa** (Michaux) Pers.   Fig. 311
Map 652. Common on wet sandy and marly shores, beach pools, interdunal sand flats, etc., around northern Lakes Michigan and Huron; rock crevices along Lake Superior; on boggy shores and in ditches and bogs, especially marly ones, throughout the state.

649. Aletris farinosa

650. Chamaelirium luteum

651. Tofieldia pusilla

The red capsules are as conspicuous in late summer as the white flowers are in early summer.

24. Ornithogalum   Star-of-Bethlehem

1. **O. nutans** L.   Fig. 313
   Map 653. Like the next, a native of Europe grown as an ornamental, but less often escaped. The only collection seen was made at Ypsilanti (*Farwell 1230* in 1892, BLH).

2. **O. umbellatum** L.   Fig. 314
   Map 654. Rarely escaped from cultivation along roadsides, woods, etc. A garden plant native to Europe.

5. Camassia

1. **C. scilloides** (Raf.) Cory   Fig. 312   Wild-hyacinth
   Map 655. River-bottom flats and banks; in Michigan collected mainly along the Huron River near Lake Erie (both Monroe and Wayne county sides) in 1914–1918, and along the Raisin River in Lenawee County, as recently as 1946.

DIOSCOREACEAE   Yam Family

1. Dioscorea

1. **D. villosa** L.   Plate 7-B   Wild Yam
   Map 656. A vine in rich dense or dry open woods, fencerows, and thickets; often along pond and marsh borders, creek bottoms, and railroads.

AMARYLLIDACEAE   Amaryllis Family

There is a growing tendency among systematists not to recognize this family as distinct from the Liliaceae. We have only a single small-flowered species native in Michigan. Some larger flowered ornamentals in this family have rarely been collected as doubtfully established escapes from cultivation. *Galanthus nivalis* L. (Snowdrop) was collected by Farwell May 2, 1933 (*9372*, BLH), on the banks of the Rouge River at Dearborn [Wayne County], long after flowering; it was said by him to be spreading. Several species of *Narcissus* are cultivated, inclu ling the daffodil. They rarely escape to roadsides, and are found on dump h ?aps, in abandoned gardens, cemeteries, etc., though hardly established.

1. Hypoxis

1. **H. hirsuta** (L.) Cov.   Plate 7-C   S ar-grass
Map 657. Sandy open ground and oak woods, more often in damp ℩ɔ wet meadows, swamp borders, and shores, often in calcareous boggy places.

The attractive yellow to orange flowers are less than 12 mm long and ± pilose on the outside.

IRIDACEAE   Iris Family

As in the Liliaceae and Amaryllidaceae, species of at least one genus of garden ornamentals in this family have sometimes been collected and reported as barely escaped from cultivation. Plants of *Crocus*, with their familiar bright early spring flowers at ground level, may appear to be established in lawns; specimens have been seen from Houghton County and various species are doubtless of similar status elsewhere in the state.

652. Tofieldia glutinosa

653. Ornithogalum nutans

654. Ornithogalum
     umbellatum

655. Camassia scilloides

656. Dioscorea villosa

657. Hypoxis hirsuta

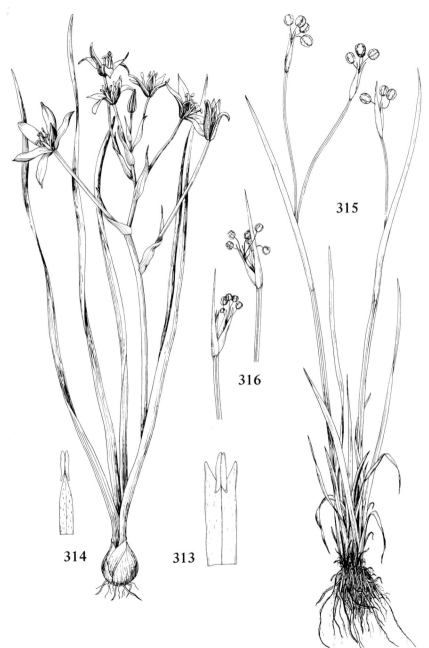

313. *Ornithogalum nutans,* stamen ×2
314. *O. umbellatum* ×½ ; stamen ×2
315. *Sisyrinchium angustifolium* ×½
316. *S. albidum* ×½

1. Perianth less than 1.5 cm long; styles filiform; sepals and petals alike . . . 1. **Sisyrinchium**
1. Perianth larger (over 1.8 cm long); styles expanded or club-shaped; sepals and petals alike or dissimilar
    2. Style branches broad and petal-like, concealing the stamens; sepals and petals dissimilar (the sepals larger, ± recurved, though petaloid in color and texture) . . . 2. **Iris**
    2. Styles club-shaped, not concealing the stamens; sepals and petals alike . 3. **Belamcanda**

1. Sisyrinchium   Blue-eyed-grass

A difficult genus of variable plants. Occasional specimens with peduncled spathes occur in species in which these are normally sessile; similarly, sessile spathes may occur in normally peduncled species. One should always collect a fair sample of a population, and be sure to include mature fruit. The flowers of our species range from violet to blue or white (or, in the case of *S. hastile*, are unknown); in some species beyond our range they may be yellow. The ovaries are usually sparsely to rather densely glandular-puberulent.

Measurements of stem width in the key always refer to the approximate middle of the stem — the widest portion. Measurements of the fused portion of the spathe bracts are from the end of the central portion of the stem (not the wing, which when broad may extend over the base of the spathe) to the end of the fused portion. Measurements of the length of the spathe are also from the end of the central or proper stem, not the wing.

REFERENCES

Bicknell, Eugene P. 1899. Studies in Sisyrinchium, II: — Four New Species from Michigan. Bull. Torrey Bot. Club 26: 297—300.

Bicknell, Eugene P. 1899. Studies in Sisyrinchium.—III: S. angustifolium and Some Related Species New and Old. Bull. Torrey Bot. Club 26: 335—349.

Fernald, M. L. 1946. The Identity of Sisyrinchium angustifolium. Rhodora 48: 152—160.

Ward, Daniel B. 1968. The Nomenclature of Sisyrinchium bermudiana and Related North American Species. Taxon 17: 270—276.

KEY TO THE SPECIES

1. Spathes on peduncles arising from a leaf-like bract, usually more than one, the upper portion of the stem thus appearing branched (fig. 315)
    2. Base of plant with very prominent dense tufts of light brown, rather straight fibers, sometimes as long as 6—7 cm but very fragile and breaking back to shorter tufts resembling a worn broom; corolla less than 7 mm long; spathes 15—21 mm long, the bracts subequal, the outer one with margins fused ca. 2.5—4 mm; capsule pale brown or tan, 3—5 mm long . . . . 1. **S. farwellii**
    2. Base of plant without dense fibers as described above — at most with a few rather lax fibers prominent only on lower 1—3 cm of plant; corolla often over 7 mm long; spathes various; capsule dark or pale, 2.5—5 mm long
        3. Stems 2 mm wide or less; spathe 16—22 mm long, the bracts subequal; pedicels (at least some of them) wing-margined basally more than half their length; capsule pale whitish tan (sometimes flushed with purplish apically) . . . 2. **S. strictum**

3. Stems 0.5–5.5 mm wide, if less than 2 mm then the inner spathe bract less than 15 mm long; pedicels merely 2-edged or slightly winged basally; capsules dark

    4. Stems (2) 2.5–3.5 (5.5) mm wide, with broad wings each wider than the central portion; inner bract of spathe 13–22 mm long, the outer bract ± longer or even leaf-like; margins of outer bract fused for (2.5) 3.5–6 mm; stems usually very minutely denticulate on margins [use strong lens!]; plants nearly always turning dark green in drying . . . . 3. **S. angustifolium**

    4. Stems 0.5–2 mm wide, the wings distinctly narrower than the central portion; inner bract of spathe 10–13 mm long, the outer scarcely longer; margins of outer bract fused for 2.5–3.5 mm; stems completely smooth-margined; plants drying pale green . . . . 4. **S. atlanticum**

1. Spathes sessile or nearly so at the end of a simple stem (see fig. 316)

  5. Spathes 2, surrounded at base by an outer leaf-like involucral bract with margins not fused beyond the wing (if any) of stem

    6. Stem slender, wiry, terete or compressed and somewhat 2-edged, but not winged (except at base); inner spathe bracts 2–3 cm long; anthers over 2.5 mm long . . . . 5. **S. hastile**

    6. Stem flattened and conspicuously winged; inner spathe bracts usually shorter; anthers up to 2.5 mm long . . . . 6. **S. albidum**

  5. Spathe 1, the outer (usually ± leaf-like) bract with the margins slightly to moderately fused at the base beyond wing of stem

    7. Plant very slender, the stem 0.5–1.75 mm wide (usually 1 mm or less) and barely margined or narrowly winged; largest leaves at most 1.5 mm wide; capsule ca. 2.5–4 mm long . . . . 7. **S. mucronatum**

    7. Plants stouter, the stem usually 1.75–3 (3.5) mm wide (most often 2–2.5 mm), winged; largest leaves usually 2–3 mm wide; mature capsule ca. 5–7 mm long . . . . 8. **S. montanum**

1. S. farwellii Bickn.

Map 658. Known only from the type collection from Birmingham (*Farwell 1625*, Sept. 27, 1898, NY, BLH, GH), one sheet of which gives the habitat as "fields." Said by Farwell to be rare.

The east coast *S. arenicola* Bickn. may not be different from this species, but in any event the epithet *farwellii* has priority. However, the Michigan plant also resembles some southeastern species which have similar fibrillose bases. The plant should certainly be looked for and studied further if it can be found. By Alexander (in Gleason, 1952; Gleason & Cronquist, 1963) this is considered a form of *S. atlanticum*, from which it differs in its pale, slightly larger (on the average), and non-apiculate capsules; its strongly fibrose bases; and its minutely denticulate stems.

2. S. strictum Bickn.

Map 659. Gravelly and sandy soil; edge of woods in rocky ground (Baraga County); very local.

The type collection is from near Vestaburg, Montcalm County (*Wheeler*, June 22, 1898, NY, MICH), and the species has been found subsequently in that county and a few other places. This is considered by Alexander (in Gleason, 1952; Gleason & Cronquist, 1963) to be "merely a form" of *S. montanum* with branched scapes. It differs from the latter species, however, also in its narrower stems, slightly smaller capsule, and subequal spathe bracts (with margins noticeably fused for 4–5 mm basally). Furthermore, the spathes are often distinctly geniculate (bent abruptly at summit of stem), unlike *S. montanum*. The margins of the leaves and stems are minutely denticulate or scabrous. *S. strictum* was accepted by Robinson and Fernald in *Gray's Manual*, ed. 7, but is completely ignored by Fernald in ed. 8. (See Mich. Bot. 6: 18–19. 1967.)

3. S. angustifolium Miller   Fig. 315

Map 660. Moist shores, meadows, fields, thickets, and swales; oak–hickory woods, forest borders, grassy clearings.

This is the plant long known as *S. gramineum* Curtis or *S. graminoides* Bickn. (and more recently as *S. bermudiana* L., sensu Shinners); most previous records of *S. angustifolium* in Michigan are referable to *S. montanum*. Some recent works continue to use the name *S. graminoides*. Plants with simple stems and sessile spathes belong here if the stems are very broadly winged.

4. S. atlanticum Bickn.

Map 661. Very local: field at cabin on Chippewa River, Midland County (*Dreisbach 6458a* in 1929, MICH) and several localities in Muskegon County, where apparently occurring on moist sandy shores.

Plants of this coastal species from Michigan and Indiana have been called *S. apiculatum* Bickn. (TL: Muskegon [Muskegon County]), distinguished from *S. atlanticum* by shorter spathes and more prominently apiculate capsule, but it is doubtful whether recognition of two species is warranted. The peduncles are generally strongly geniculate at the base, as are most of the spathes also.

5. S. hastile Bickn.

Map 662. Another species of doubtful relationships, known only from the type collection from sandy shores of Belle Isle in the Detroit River (*Farwell 867*, June 2, 1896, NY, BLH, GH). Quite possibly an escape from cultivation, although Farwell's notes for that day do not suggest any collecting of cultivated plants, and on the label for one of his sheets (NY) he states: "Perhaps only a stray immigrant." Shortly after discovery of the species, Farwell (Rep. Mich. Acad. 2: 32–33. 1902) observed that the site, on "the northeast corner of Belle Isle," had been covered by an artificial lake.

The specimens lack mature flowers or fruit, but closely resemble the Mexican *S. pringlei* Robinson & Greenman, as noted by Britton & Brown (Ill. Fl., ed. 2, i: 543. 1913). Said by Alexander (in Gleason, 1952) to be *S. junceum* E. Meyer, but the latter species has a stouter stem and long-exserted pedicels (contrasted with the scarcely exserted ones of *S. hastile*). The name is tentatively retained here in the hopes that more material can be found so that its status in our flora can be ascertained and its correct name (if truly an earlier described species) determined with the aid of flowers and fruit.

6. **S. albidum** Raf. Fig. 316

Map 663. Dry often sandy open fields, railroad embankments, oak–hickory woods; grassy, sometimes moist banks, shores, and pastures, even marshy ground. The Keweenaw County record (*Farwell 461* in 1886, BLH, GH) is far enough out of range to be suspicious.

Plants rarely occur with some spathes ± peduncled but these may be distinguished from *S. atlanticum* by the larger spathes, usually very unequal bracts, and little or no fusion of margins of spathe and involucral bracts.

7. **S. mucronatum** Michaux

Map 664. More or less open ground and fields, typically in moist calcareous flats and open boggy thickets near the north end of Lake Huron.

Ordinarily a distinctive, delicate, wiry plant with small capsules and purplish spathes; some forms are larger and approach *S. montanum* in stature, but may be placed here by smaller capsules and slightly narrower stems and foliage. The plant is rather dark green when dry (though usually slightly glaucous), the leaves and stems in Michigan specimens apparently being entirely smooth. The spathes are often ± geniculate at the base, while in the next species they are normally straight.

658. Sisyrinchium
farwellii

659. Sisyrinchium
strictum

660. Sisyrinchium
angustifolium

661. Sisyrinchium
atlanticum

662. Sisyrinchium hastile

663. Sisyrinchium albidum

8. **S. montanum** Greene

Map 665. Moist open, often grassy places; sandy, gravelly shores (or in rock crevices); mixed woods, especially in disturbed areas and clearings; old railroad beds, banks of ditches, and roadsides through wet ground.

Lighter green plants than the preceding, with a more glaucous aspect; the margins of stems and leaves are usually minutely denticulate. The margins of the outer spathe bracts are fused basally up to about 4 or 5 mm. Fernald recognizes two varieties. The common plant in Michigan is var. *crebrum* Fern. (formerly widely known as *S. angustifolium* and still called this by Alexander in Gleason, 1952; Gleason & Cronquist, 1963). It has deep green foliage which dries rather dark, and green or purple-tinged capsules, dark when ripe. The typical var. *montanum* seems to occur rarely in Michigan. It is a paler plant, with capsules pale straw-color even when ripe. The spathes of var. *crebrum* are often tinged with purple. A few plants from Sugar Island (Chippewa County) resembling this species but most culms bearing two peduncled spathes have been seen (*Hiltunen 934* in 1957, WUD); they would run in the key to *S. angustifolium*, from which they differ in generally narrower stems, failure to darken in drying, and purple tinge of spathes and capsules. They are presumably either hybrids or unusual branched forms of *S. montanum* var. *crebrum*. (They differ from *S. strictum* in often slightly wider stem, leaf-like outer bract of spathe, and absence of wing-margin on pedicels.)

2. Iris

Iris; Flag

REFERENCES

Anderson, Edgar. 1928. The Problem of Species in the Northern Blue Flags, Iris versicolor L. and Iris virginica L. Ann. Missouri Bot. Gard. 15: 241–332.
Anderson, Edgar. 1936. The Species Problem in Iris. Ann. Missouri Bot. Gard. 23: 457–509.

KEY TO THE SPECIES

1. Plants dwarf, the flowering stems less than 15 (usually less than 10) cm tall
  2. Sepals with a prominent beard above; rhizomes stout (much more than 5 mm thick) . . . . 1. **I. pumila**
  2. Sepals beardless; rhizomes slender (less than 5 mm thick most of their length) . .
. . . . 2. **I. lacustris**
1. Plants much taller
  3. Sepals with a prominent median beard above . . . . 3. **I. germanica**
  3. Sepals without a prominent beard, at most minutely pubescent
    4. Flowers yellow. . . . . 4. **I. pseudacorus**
    4. Flowers blue (white in albinos)
      5. Base of expanded portion of sepal with a bright yellow spot, finely pubescent with hairs as long as the thickness of the sepal; outer spathe bracts of uniform texture and color; seeds round to D-shaped, irregularly (but shallowly) pitted
. . . . 5. **I. virginica**

5. Base of expanded portion of sepal at most with a greenish yellow spot, with papillae shorter than thickness of the sepal; outer spathe bracts with the margins generally darker and more shiny than the rest of the dull surface; seeds D-shaped, with a ± regularly pebbled surface . . . . 6. **I. versicolor**

1. **I. pumila** L.

Map 666. Referred here are collections of dwarf irises from waste ground at Alma (*Davis* in 1897, MSC) and "prairie" in Newaygo County (*T. H. P. Marshall 1531* in 1951, MSC), apparently escapes from cultivation or perhaps only persisting where dumped.

2. **I. lacustris** Nutt.   Frontispiece   Dwarf Lake Iris

Map 667. Moist sands, gravel, and limestone crevices, usually slightly shaded, at edges of conifers (cedar, fir) along the northern shores of Lakes Michigan and Huron. Particularly well developed on the rubble of old beach ridges at such places as Wilderness State Park. The type locality is Mackinac Island, where this *Iris* was found by Thomas Nuttall in 1810. An endemic Great Lakes species (unless one treats it as a variety of the southern *I. cristata* Aiton), it is known outside of Michigan at present only from Manitoulin Island and the Bruce Peninsula of Ontario and the Door Peninsula of Wisconsin. Formerly known from the vicinity of Milwaukee, Wisconsin, and Windsor, Ontario. Reports from Lake Superior shores have defied confirmation, but the species should continue to be sought in the Whitefish Bay area, which would seem the most likely place for it if indeed it is anywhere around Lake Superior. (Perhaps old collections from Whitefish Bay of Lake Michigan, in Wisconsin, have been erroneously attributed to Whitefish Bay of Lake Superior.) Only two significantly inland localities are known: Menominee County, sandy gravel ridge among aspens at Koss (not far from Menominee River) (*Davis* in 1905, MICH, GH; *Voss 4044* in 1957, MICH, MSC); Delta County, calcareous bank of Escanaba River near Cornell (*J. H. Beaman 1797* in 1958, MSC; *Voss 5982* in 1958, MICH, BLH; presumably the same site, *Wheeler* in 1892, MSC). (See Mich. Bot. 2: 100–101. 1963.)

664. Sisyrinchium mucronatum

665. Sisyrinchium montanum

666. Iris pumila

Usually reaches the peak of blooming in late May; rarely some flowers as late as mid-June or even mid-July. Plants in cultivation have been known to bloom in September. Late-blooming plants have a quite different aspect, for the leaves have grown well beyond the flowers. White flowers are very rarely found.

3. **I. germanica** L.   Flag; Fleur-de-lys

Map 668. The common garden *Iris* rarely (or one might say "barely") escapes from cultivation, but could doubtless be found in more counties than those mapped; collectors are not often inspired to make specimens of plants such as these from roadsides, when they are so dubiously a part of the established flora.

Not all garden irises should be called *I. germanica* and the name is not used here in a strict sense. Works on cultivated plants should be consulted by those desiring more understanding of these plants.

4. **I. pseudacorus** L.   Yellow Flag

Map 669. A European species, locally spread from cultivation into wet lake shores, river edges, marshes, and ditches.

5. **I. virginica** L.   Southern Blue Flag

Map 670. Ponds and lake shores, marshes and swales, ditches, stream sides, riverbanks and thickets, swamp forests, and rarely bogs.

Plants from north of the Coastal Plain are referred to var. *shrevei* (Small) Anderson — recently associated with the next species as *I. versicolor* var. *shrevei* (Small) Boivin, and occasionally recognized as a distinct species, *I. shrevei* Small. Dry herbarium specimens, in particular, are often difficult to distinguish from *I. versicolor*, from which this taxon was once not separated in manuals.

The cauline leaves of *I. versicolor* are shorter than the tops of the inflorescence, while in *I. virginica* the cauline leaves frequently overtop the flowers. The ovaries of *I. versicolor* (at anthesis) are somewhat shorter (1–2 cm long), at least one of them frequently exserted on the pedicel beyond the tip of the spathe; while in *I. virginica* the ovaries (before forming fruit) are 1.5–3 cm long and usually are not exserted. The bases of plants of *I. versicolor* are more frequently flushed with purple than are those of *I. virginica*, which are generally

667. Iris lacustris

668. Iris germanica

669. Iris pseudacorus

brown. But all of these characters are variable, and several must often be considered before identification can be made. A hybrid between the two species has been described as *I.* × *robusta* Anderson, on the basis of two colonies studied near St. Ignace and at Engadine [Mackinac County]. Anderson also reports an albino form from Lawrence [Van Buren County] and a lavender or red-purple form from Otisville [Genesee County]. Partial albinos (or pale blue forms) are more common in this species than in the next.

6. **I. versicolor** L. Wild Blue Flag
   Map 671. Wet places generally: lake shores, marshes, river borders, stream bottoms, meadows, ditches, swamps, and sphagnum bogs.
   See discussion under the preceding species.

3. Belamcanda

1. **B. chinensis** (L.) DC. Fig. 317 Blackberry-lily
   Map 672. An orange-flowered Asiatic species, frequently cultivated and escaped south of our region. Collected by Farwell in waste places at Detroit (*1496* in 1894, BLH) and said by him to be adventive; and by H. R. Becker and C. R. Hanes (*1718* in 1938, WMU) along a roadside in Charleston Tp., Kalamazoo County.

ORCHIDACEAE Orchid Family

This familiar and popular family has been well treated for this region by Case (1964), whose work should be consulted by anyone desiring more discussion than the very brief consideration presented here. The family includes some of our showiest wildflowers as well as some of our most inconspicuous and rare ones. All native orchids are protected by the conservation laws of Michigan.

Some authors have preferred to split *Orchis* and *Habenaria* into smaller genera, and other genera may warrant similar treatment. However, lacking a comprehensive review of generic characters for our native orchids, the more

670. Iris virginica

671. Iris versicolor

672. Belamcanda chinensis

conservative genera traditionally recognized in recent works on the North American flora are here maintained.

REFERENCES

Case, Frederick W., Jr. 1964. Orchids of the Western Great Lakes Region. Cranbrook Inst. Sci. Bull. 48. 147 pp.

Correll, Donovan Stewart. 1950. Native Orchids of North America. Chronica Botanica, Waltham, Mass. 399 pp.

Fuller, Albert M. 1933. Studies on the Flora of Wisconsin Part I: The Orchids; Orchidaceae. Bull. Publ. Mus. Milwaukee 14: 1–284.

KEY TO THE GENERA

1. Lip a showy inflated pouch 1–5 cm long
   2. Plants with leafy stems; lip a closed pouch (i.e., open only at base above) . . . .
   . . . . 1. **Cypripedium**
   2. Plants with leaves basal; lip split down middle above or open at base about half its length
      3. Basal leaf single, petiolate, the blade less than 7 cm long, produced in late summer and withering after the plant blooms the following spring; lip ca. 1.5–2 cm long, open about half its length basally above; plants less than 20 cm tall . . . . 2. **Calypso**
      3. Basal leaves 2, longer, tapered to sheathing bases and not distinctly petiolate, present throughout the summer (but not winter); lip ca. 4–5 cm long, split down upper side; plants taller . . . . 1. **Cypripedium (acaule)**
1. Lip showy or inconspicuous, but not an inflated pouch, usually ± flat with or without a slender basal spur (or if somewhat saccate, hardly showy and less than 1 cm long)
   4. Flower solitary (rarely plants with 2 flowers in a population of normal ones)
      5. Leaves 4–6, whorled; lip whitish or greenish yellow streaked with green or purple . . . . 3. **Isotria**
      5. Leaf solitary or undeveloped at flowering time; lip light or deep pink
         6. Leaf linear, at most up to 7 (very rarely 10) mm wide, often poorly developed at flowering time, ± folded or plicate longitudinally, sheathing stem at base; plant from a small bulbous corm . . . . 4. **Arethusa**
         6. Leaf ± elliptic or lanceolate, usually over 7 mm wide and well developed at flowering time, flat, arising near middle of stem, sessile but not sheathing at base; plant from slender roots and rhizome . . . . 5. **Pogonia**
   4. Flowers 2 or more on a plant
      7. Lip produced into a distinct (usually slender and elongate) spur at base (in one species, this pouch-like and as short as 2 mm)
         8. Leaves cauline . . . . 8. **Habenaria**
         8. Leaves all basal or nearly so, or absent at flowering time (bracts subtending flowers may be leaf-like)
            9. Leaf solitary, absent or withered at flowering time, often dark-spotted above and purplish beneath; bracts absent or very minute at base of pedicels; perianth ca. 5–7 mm long (excluding the long spur), slightly asymmetrical
            . . . . 6. **Tipularia**
            9. Leaves 1 or 2 (3), well developed at flowering time (or perianth less than 5 mm long), green; bracts definite at base of pedicels; perianth of various sizes, bilaterally symmetrical

10. Flowers with white lip (spotted or not) broadly ovate to oblong, often crenate or lobed; lateral petals (and in one species also sepals) connivent or fused with dorsal sepal to form a pink to purple hood . . . . 7. **Orchis**

10. Flowers entirely white and/or green, the lip lanceolate to narrowly linear, entire; lateral petals free . . . . 8. **Habenaria**

7. Lip at most somewhat swollen or saccate (but not with a spur 2 mm or more long)

11. Plants lacking green color (except sometimes in fruit), leafless with red, yellow, brown, or purplish stems arising from a coralloid rhizome . . 9. **Corallorhiza**

11. Plants with normal green color, bearing leaves at some time in the year (if leaves absent at flowering time or plants apparently lacking green, arising from tubers or corms, not a coralloid mass)

12. Leaves a single opposite pair, definitely cauline, not at all sheathing the stem . . . . 10. **Listera**

12. Leaves solitary, alternate, absent, or basal (or almost basal, with sheathing bases)

13. Stem leafy, with 4 or more conspicuous broadly ovate-lanceolate to elliptic leaves; perianth ca. 7–10 mm long; flowers greenish, at least the petals suffused with pink; upper part of stem and axis of inflorescence finely pubescent. . . . . 11. **Epipactis**

13. Stem with the leaves fewer, narrower, and/or basal (or absent); perianth various, but if pinkish then 10 mm or more long and the vegetative parts completely glabrous

14. Perianth 10–12 mm long, white or creamy, the flowers in a dense spike-like inflorescence . . . . 18. **Spiranthes**

14. Perianth longer or shorter, or not whitish and the inflorescence not spike-like

15. Perianth 10 mm or more long, at least in part normally with some shade or pink or purple (yellowish in a form of *Aplectrum*)

16. Flowers ca. 2–3 cm or more broad, the lip uppermost, bearded with a tuft of yellow-tipped hairs; leaf solitary (rarely 2), several times as long as broad . . . . 12. **Calopogon**

16. Flowers less than 1.5 cm broad, the lip lowermost and not bearded; blade of leaf not over 3.5 times as long as broad

17. Leaves cauline, sessile and clasping, very seldom over 1.5 cm long . . . . 13. **Triphora**

17. Leaves basal or absent at flowering time, sheathing at base or petioled, much larger (usually 7–15 cm long)

18. Leaf solitary, petioled, developing in fall and overwintering, usually withered before plant flowers . . . . 14. **Aplectrum**

18. Leaves 2, sheathing at base, developing in current season and present at flowering. . . . . 15. **Liparis**

15. Perianth less than 10 mm long, greenish, white, or yellowish, with no trace of pink or purple

19. Leaves 1 or 2, sheathing at the base, the scape naked to the inflorescence; flowers on short pedicels, the raceme glabrous and not 1-sided nor noticeably twisted

20. Leaf 1 (very rarely 2); perianth less than 4 mm long . . . . 16. **Malaxis**

20. Leaves 2; perianth over 4 mm long . . . . 15. **Liparis**

19. Leaves 3 or more (or withering at flowering time), the stem above them bearing small bracts or scales; flowers sessile or almost so in a narrow spike-like inflorescence which is 1-sided or spirally twisted, or pubescent (or both)

21. Leaves ovate to elliptic, basal or nearly so, present and firm at flowering time, the midvein and/or other veins margined in white or pale green (not always visible in dry plants); lip pouched or saccate at the base . . . . 17. **Goodyera**
21. Leaves ovate-elliptic to linear and grass-like, sometimes cauline, often withering at flowering time (in wider-leaved species), not marked with whitish; lip not pouched or saccate . . . . 18. **Spiranthes**

1. Cypripedium   Lady-slipper

KEY TO THE SPECIES

1. Leaves all basal or nearly so; lip split down the middle above . . . . 1. **C.** acaule
1. Leaves cauline; lip closed above (open at base only)
 2. Lip pale to deep yellow . . . . 2. **C.** calceolus
 2. Lip variously white and purple
  3. The 3 sepals all distinct nearly to base; lip prolonged below to a blunt conical pouch, strongly marked throughout with purplish . . . . 3. **C.** arietinum
  3. The 2 lateral (or lower) sepals fused most or all of their length; lip smoothly rounded with no conical prolongation below, entirely white outside or with extensive white areas
   4. Lip entirely white outside, up to 2.5 cm long; sepals and lateral petals greenish yellow; plants small, with 3–4 leaves crowded near middle of stem, sparingly puberulent . . . . 4. **C.** candidum
   4. Lip white, marked with pink or purple outside, 2.5–5 cm long; lateral petals and sepals white; plants tall, with more leaves, the stem and leaves copiously hirsute . . . . 5. **C.** reginae

1. **C. acaule** Aiton   Moccasin Flower; Pink or Stemless Lady-slipper
Map 673. Sandy woodlands under red and jack pine, mixed oak and pine, or aspens; also in low ground, especially on hummocks in sphagnum bogs and under cedar, tamarack, or spruce in coniferous swamps and bogs.

Plants with the lip white [f. *albiflorum* Rand & Redf.] occasionally are found; the normal pink flowers, however, tend to be much paler when just opening. Some flowers are much deeper pink than the average.

2. **C. calceolus** L.   Plate 7-D   Yellow Lady-slipper
Map 674. In a great diversity of habitats except the driest: damp woods (coniferous, mixed, deciduous), bogs, meadows, borders of woods and clearings, often under cedar.

Correll refers all of our plants to var. *pubescens* (Willd.) Correll. Many of them have traditionally been called var. *parviflorum* (Salisb.) Fern., a problematical taxon which ordinarily differs in its smaller flowers (lip under 3 cm long), sepals more purple, and tendency to prefer wet, boggy or swampy habitats rather than the more mesophytic situations in which var. *pubescens* (sens. str.) usually occurs. Case, who undoubtedly knows the plants best in the field in Michigan, treats var. *parviflorum* in a narrow sense and indicates it almost solely in the southern half of the Lower Peninsula.

Dry specimens in which the original color is no longer apparent (and for which the collector neglected to note color on the label) may sometimes superficially resemble *C. reginae*. However, the sepals of *C. calceolus* are sharply acute, and the plant is less hirsute than in *C. reginae*, which is a hirsute plant with obtuse or rounded sepals. See also comments under *C. candidum*.

3. **C. arietinum** R. Br.   Plate 7-E   Ram's Head Lady-slipper
    Map 675. In Michigan, thrives best on low dunes, in partial shade of fringing conifers, along the northern shores of Lakes Michigan and Huron and on Lake Superior (where thin soil over rock is the rule westward); inland, under jack pine and oak and also in coniferous swamps (cedar, tamarack, spruce, fir). Reports from Gratiot County are apparently based on collections by C. A. Davis labeled "about Alma"; according to Davis' notes and more complete labels, however, he collected this species several times in Isabella County, not actually in Gratiot County.
    Albinos sometimes occur, as in the Grand Marais area (fide F. W. Case).

4. **C. candidum** Willd.   White Lady-slipper
    Map 676. Open marly bogs and swampy meadows, usually with tamarack and shrubby cinquefoil; very local. An old specimen labeled as collected in Keweenaw County (*Farwell 763* in 1884 or 1890, BLH) is highly suspicious and is not mapped.
    Dry herbarium specimens for which the collector failed to note flower color on the label sometimes resemble dry material of small-flowered plants of *C. calceolus*. The 3–4 leaves of *C. candidum* are crowded toward the middle of the stem, their bases overlapping and concealing the internodes; in *C. calceolus*, the leaves tend to be less crowded, with the internodes visible. The lip of dry *C. candidum* specimens ranges from 1.6 to 2.5 cm long; the dorsal sepal, from 20 to 30 (33) mm long; and the lateral petals are less than 37 (40) mm long. In *C. calceolus*, the lip is often over 2.5 cm long, the dorsal sepal often over 30 mm long, and the lateral petals almost always over 37 mm long — but smaller flowers occasionally are found.

673. Cypripedium acaule   674. Cypripedium   675. Cypripedium
                                  calceolus   arietinum

Hybrids with *C. calceolus* [*C.* × *andrewsii* Fuller] have been found with the parent species on the Baker Sanctuary of the Michigan Audubon Society, near Battle Creek, Calhoun County; and in Jackson, Oakland, and Washtenaw counties. They are intermediate in appearance.

5. **C. reginae** Walter   Showy Lady-slipper

Map 677. Bogs, especially in shrubby areas, and coniferous swamps, often associated with tamarack and cedar, less often with spruce and fir; not tolerant of dense shade, but thriving in open glades, clearings, old roads through boggy ground, etc.; occasionally in other swampy situations (e. g., with red maple) and along calcarous ridges and dunes, with conifers, along the northern shores of Lakes Michigan and Huron.

The lip is rarely pure white [f. *albolabium* Fern. & Schub.] . The hairs of this species, and of *C. calceolus*, are irritating to some people, producing a rash resembling poison-ivy.

2. Calypso

1. **C. bulbosa** (L.) Oakes   Plate 8-A   Calypso; Fairy-slipper

Map 678. Mixed damp woods of conifers and hardwoods (e. g., balsam fir and paper birch) or, more often, mostly coniferous woods (fir and cedar, spruce

676. Cypripedium
candidum

677. Cypripedium reginae

678. Calypso bulbosa

679. Isotria medeoloides

680. Isotria verticillata

681. Arethusa bulbosa

Page 438

and fir, hemlock), especially characteristic of old beach ridges under conifers near the shores of the Great Lakes; usually ± well shaded.

This is one of our most beautiful wildflowers, very local in occurrence (colonies of hundreds, as on Isle Royale, a truly handsome sight!). It deserves all of the protection which can be given to it (and the places where it grows). The sepals and upper petals are normally pink or magenta, but occasionally pure white or apricot; the lip is strongly lined with deep purplish, spotted toward the white "apron," which bears a yellow beard apically.

3. Isotria

Plants of this genus resemble *Medeola virginiana* when in a vegetative state. They can be distinguished by the glabrous, hollow stem and absence of a white tuberous rhizome (although one should never dig up a plant suspected of being one of these rare orchids).

REFERENCE

Case, Frederick W., Jr., with William Schwab. 1971. Isotria medeoloides, the Smaller Whorled Pogonia, in Michigan. Mich. Bot. 10: 39–43.

KEY TO THE SPECIES

1. Flower sessile or nearly so (though the slender elongate ovary may resemble a pedicel); sepals much less than twice as long as petals, clear green; lip whitish to pale green; stem and leaves distinctly glaucous, the plant with a decided gray or hoary aspect . . . . 1. **I. medeoloides**
1. Flower pedicelled; sepals about twice as long as petals or longer, purplish; lip greenish yellow, streaked with purple; stem and leaves not glaucous (or the latter somewhat so beneath) . . . . 2. **I. verticillata**

1. **I. medeoloides** (Pursh) Raf.   Plate 8-B   Smaller Whorled Pogonia
Map 679. First discovered in Michigan in 1968 by William Schwab. The plants were in a low second-growth woods of red maple and other trees at the site of an old orchard near Lake Michigan in southern Berrien County. This is the only station for this rare and very local species west of the Appalachians and north of Missouri.

2. **I. verticillata** (Willd.) Raf.   Whorled Pogonia
Map 680. Local in older tamarack or black spruce zones of sphagnum bogs; much less often in moist mixed second-growth woods on sandy (acid) soil.

4. Arethusa

1. **A. bulbosa** L.   Plate 8-C   Arethusa; Dragon's Mouth
Map 681. Open bog mats, generally in sphagnum, in coniferous (cedar, black

spruce, tamarack) swamps, and in similar peaty situations.

One of the handsomest of our native orchids, becoming quite rare and local, thanks to destruction of habitats as well as to vandalism. Blooms a little earlier than *Calopogon tuberosus* and *Pogonia ophioglossoides*, which are usually in the same bogs. This is the only species of the genus in North America; there is another in Japan.

5. Pogonia

1. **P. ophioglossoides** (L.) Ker   Fig. 318   Rose Pogonia
Map 682. Sphagnum bogs (often in the wettest parts), boggy beach pools, wet meadows, even on mossy logs in boggy places; often in ± marly situations.

Commoner than the preceding species, usually a paler pink, and flowering a little later. The white f. *albiflora* Rand & Redf. occasionally occurs with the typical pink one. An occasional plant bears two flowers.

6. Tipularia

1. **T. discolor** (Pursh) Nutt.   Fig. 329   Cranefly Orchid
Map 683. Despite old reports, the first collection from Michigan seems to have been made by F. W. Case in 1970 (MICH, BLH), at a station where it had been discovered by William Schwab. It grows in Berrien County in a deciduous woods (*Quercus*, *Fagus*, *Liriodendron*, *Acer*, *Sassafras*, *Betula alleghaniensis*, etc.) on old dunes near Bridgman and has also been found near Stevensville by V. G. Soukup (see Mich. Bot. 11: 38. 1972).

A very distinctive orchid, the winter-green leaf (as in *Aplectrum*) withering in spring, before the somewhat asymmetrical greenish or yellowish (marked with purplish) flowers mature in the summer, the rhizome with a series of tubers. The persistent old fruiting stalk may be found with a well developed leaf at the base. This is another genus of eastern North America and eastern Asia.

7. Orchis

KEY TO THE SPECIES

1. Leaf 1; lip under 1 cm long, spotted, notched at apex and with a lateral lobe on
each side . . . . 1. **O. rotundifolia**
1. Leaves normally 2; lip over 1 cm long, unspotted, not lobed (or somewhat crenate)
. . . . 2. **O. spectabilis**

1. **O. rotundifolia** Pursh   Fig. 327   Round-leaved Orchis
Map 684. A northern species, very local as far south as Michigan and here found only in bogs with cedar, tamarack, spruce, and/or fir, often with underlying marl.

2. **O. spectabilis** L.   Showy Orchis

Map 685. Low rich woods, often along creeks and in moist areas.

8. Habenaria   "Rein Orchid"

KEY TO THE SPECIES

1. Lip prominently ciliate or fringed
  2. Lip simple, not deeply divided (except for fringe)
    3. Flowers white; longest cilia (usually lateral) of fringe about half as long as the undivided portion of the lip, or even shorter (fig. 319) . . . . . . **1. H. blephariglottis**
    3. Flowers yellow to orange; longest cilia of fringe distinctly more than half the length of the undivided portion of the lip (fig. 320) . . . . **2. H. ciliaris**
  2. Lip deeply 3-parted in addition to fringe
    4. Flowers pink-purple; divisions of the lip broadly fan-shaped, copiously lacerate-fringed, but the fringe usually cut less than half the distance to the base of the division of the lip (fig. 321) . . . . **3. H. psycodes**
    4. Flowers yellowish, cream, or greenish; at least the lateral divisions of the lip more narrowly cuneate, mostly cut into a long fringe more than half their length
      5. Sepals (6) 6.5–9 (10) mm long; lateral petals broadly obovate or cuneate, erose or denticulate at apex . . . . **4. H. leucophaea**
      5. Sepals 3.5–5 (6) mm long; lateral petals linear-oblong or almost lanceolate, usually essentially entire . . . . **5. H. lacera**
1. Lip entire or toothed, but not fringed
  6. Leaves all basal, the stem at most with reduced bracts
    7. Leaves about twice as long as wide, or longer; spur less than 12 mm long
      8. Lip ± cuneate, with truncate 3-toothed or crenate apex; spur 7–11 mm long, much exceeding the lip . . . . **11. H. clavellata**
      8. Lip tapered to pointed or rounded apex, untoothed; spur about equalling lip or at most ca. 2 mm longer
        9. Leaf solitary, present through anthesis; ovary short-pedicelled (evident on older flowers or fruit), the inflorescence a raceme usually shorter than 10 (12) cm; lateral sepals 3.5–6 mm long . . . . **6. H. obtusata**
        9. Leaves usually 2–3, withering during or before anthesis; ovary sessile, the inflorescence a spike (5) 10 cm or more long; sepals 1.5–3 (3.5) mm long . . . . **7. H. unalascensis**
    7. Leaves orbicular or almost so, less than twice as long as broad; spur 16–40 mm long

682. Pogonia ophioglossoides   683. Tipularia discolor   684. Orchis rotundifolia

10. Scape naked (rarely with a bract); spur (16−24 mm long) tapered ± evenly to rounded apex; lip yellowish green, tending to turn upward near the end . . . . 8. **H. hookeri**

10. Scape with 1−6 bracts between leaves and inflorescence; spur parallel-sided or even somewhat club-shaped toward apex; lip whitish green, tending to turn downward . . . . 9. **H. orbiculata**

6. Leaves cauline

11. Lip truncate and 2−3-toothed or -lobed at apex

12. Spur a thick pouch 2−3 mm long, much shorter than the lip; leaves 3 or more; lower bracts very long, much exceeding the flowers . . . . 10. **H. viridis**

12. Spur slender, ± clavate, 7−11 mm long, much longer than the lip; leaves only 1−2; bracts shorter than the flowers . . . . 11. **H. clavellata**

11. Lip tapered, rounded (or almost truncate and obscurely crenulate) but not 2−3-toothed at apex

13. Lip much shorter than the spur, broadly rounded (or almost truncate) at apex, with an erect tubercle near the base and a lateral tooth or projection on each side near the base (fig. 325) . . . . 12. **H. flava**

13. Lip 1−2 mm shorter than, about equalling, or slightly longer than the spur, tapered to narrow apex, with neither a tubercle nor lateral teeth (at most, broadly dilated basally)

14. Lip cuneate to strap-shaped, not dilated basally; flowers greenish or greenish yellow . . . . 13. **H. hyperborea**

14. Lip dilated or expanded basally (fig. 326); flowers usually white or greenish white . . . . 14. **H. dilatata**

1. **H. blephariglottis** (Willd.) Hooker   Fig. 319   White Fringed Orchid
Map 686. Only in sphagnum bogs, seldom much shaded though often associated with black spruce and tamarack.

A plant with the lip entire rather than fringed has been seen in Mud Lake Bog near the University of Michigan Biological Station, in Cheboygan County [f. *holopetala* (Lindley) Boivin]. This species is often rather difficult to distinguish from *H. ciliaris* when the specimens are dry, especially if there is no note as to the color of the fresh flowers. Hybrids, with all intermediates between the two, including lemon-yellow flowers, are reported as found in Allegan, Berrien, and Van Buren counties [*H.* ×*bicolor* (Raf.) Beckner].

685. Orchis spectabilis

686. Habenaria blephariglottis

687. Habenaria ciliaris

2. **H. ciliaris** (L.) R. Br.   Fig. 320   Orange Fringed Orchid
Map 687. In bogs with *H. blephariglottis* and there often hybridizing with it; also in damp sandy meadows.

3. **H. psycodes** (L.) Sprengel   Fig. 321   Purple Fringed Orchid
Map 688. Damp shores at borders of woods, marshes and wet meadows, creek borders and open wet ground generally; less often in sedge mats and tamarack bogs, depressions in deciduous woods, and swamp forests; rock ledges in the Lake Superior region.
Variant forms occur, the type locality for two recognized in *Gray's Manual* being Round Lake, Emmet County, where discovered by Dr. Charles H. Swift: f. *ecalcarata* (Bryan) Dole, with the spur lacking and the flowers more or less regular; and f. *varians* (Bryan) Fern., with the middle division of the lip obsolete. Specimens with the lip not fringed and the spur very short have been collected at the same place. The white-flowered f. *albiflora* R. Hoffm. is found occasionally with typical pink-flowered plants. Large-flowered plants have sometimes been reported as var. *grandiflora* (Bigelow) Gray [*H. fimbriata* (Aiton) R. Br.] , but this taxon, if anything other than a large extreme, probably occurs only east and south of Michigan.
Hybrids with *H. lacera* [*H.* × *andrewsii* Niles] have occasionally been reported from Michigan, but perhaps not always correctly. Apparently authentic collections seen include one from east of Ojibway, Keweenaw County (*Hermann 7931* in 1936, NY); and one from Round Lake, Emmet County (*Jones* in 1903, ALBC).

4. **H. leucophaea** (Nutt.) Gray   Prairie Fringed Orchid
Map 689. Open bogs (on floating mats) as well as wet prairie-marshes and other marshy sites.
A plant with peloric flowers (regular, rather than bilaterally symmetrical) was collected on Belle Isle in 1884 (*A. B. Lyons*, MICH).

5. **H. lacera** (Michaux) Lodd.   Ragged Fringed Orchid
Map 690. In bogs, both open and under tamarack or cedars, but also in wet

688. Habenaria psycodes   689. Habenaria leucophaea   690. Habenaria lacera

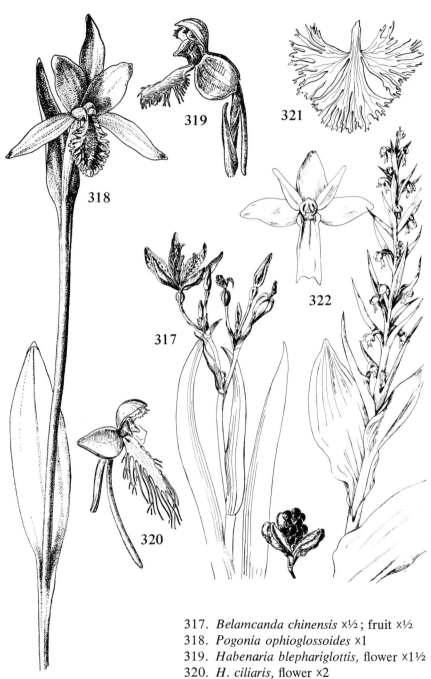

317. *Belamcanda chinensis* ×½ ; fruit ×½
318. *Pogonia ophioglossoides* ×1
319. *Habenaria blephariglottis,* flower ×1½
320. *H. ciliaris,* flower ×2
321. *H. psycodes,* lip ×3
322. *H. viridis* ×½ ; flower ×8

ground generally: damp grassy meadows, ditches, edges of moist woods, swampy thickets.

6. **H. obtusata** (Pursh) Richardson   Blunt-leaf Orchid
   Map 691. Usually in coniferous swamps and woods, boggy spots in mixed woods, and coniferous bog borders, often on mossy hummocks.

7. **H. unalascensis** (Sprengel) S. Watson   Alaska Orchid
   Map 692. A western species ranging from Alaska to northern Mexico, isolated on Anticosti Island, Quebec, and in the Niagara Escarpment region from the Bruce Peninsula of Ontario through Manitoulin Island to the eastern Upper Peninsula of Michigan. Here found in thin soil of open woods (aspen, birch), grassy borders of thickets, balsam—cedar woods, and clearings on or close to the underlying rock (dolomite).

8. **H. hookeri** Gray   Fig. 323   Hooker's Orchid
   Map 693. Coniferous or mixed woods, thickets, and borders, especially on wooded dunes and sandy soil near the Great Lakes; less often in deciduous forest or hemlock—hardwoods.
   Rarely a specimen occurs with one or both of the leaves about twice as long as broad, although the leaves are usually almost as orbicular as those of the next species. Fully developed spurs are 16—24 mm long, directed downward. This species blooms as much as two or three weeks earlier than *H. orbiculata* when the two are found in the same area.

9. **H. orbiculata** (Pursh) Torrey   Plate 8-D   Round-leaved Orchid
   Map 694. Rich deciduous or mixed woods; coniferous forests from virgin pine (Hartwick Pines) to swamps (or bogs) of cedar, balsam fir, hemlock, etc.
   The spurs are ordinarily 16—23 mm long, often ± horizontal in position; plants with larger spurs (25—43 mm) and very large leaves (to 22 cm across) have been called *H. macrophylla* Goldie, but this is probably only a large extreme of *H. orbiculata* [var. *macrophylla* (Goldie) Boivin]. This larger form is known from several counties in Michigan, including Charlevoix (Beaver Island),

691. Habenaria obtusata

692. Habenaria unalascensis

693. Habenaria hookeri

Cheboygan, Emmet, Houghton, Leelanau (South Fox Island), Mackinac (Round Island), and St. Clair.

10. **H. viridis** (L.) R. Br.   Fig. 322   Bracted Orchid
Map 695. Usually in beech–maple or northern hemlock–hardwoods, thriving particularly well on sandy soils of wooded dunes; less often in coniferous woods or (southward) mixed oak or oak–hickory.
American plants of this species are referred to var. *bracteata* (Willd.) Gray.

11. **H. clavellata** (Michaux) Sprengel   Fig. 324   Club-spur Orchid
Map 696. A rather inconspicuous yellow-green orchid in bogs (open mats to older cedar or tamarack swamps) and coniferous woods, mossy stream banks, boggy beach pools and shore meadows; often thriving along roadside ditches and in shallow sandy excavations, especially at the borders of coniferous thickets.

12. **H. flava** (L.) Sprengel   Fig. 325   Tubercled Orchid
Map 697. Usually in swampy woods, at seasonal pools, in sandy alder thickets, etc., but sometimes in open moist ground.
Plants from our region are generally referred to var. *herbiola* (R. Br.) Ames & Correll.

13. **H. hyperborea** (L.) R. Br.   Tall Northern Bog Orchid
Map 698. In almost all moist to wet situations: low woods, cedar swamps; coniferous, deciduous, and mixed woods; riverbanks, stream borders, and springy places; beach pools and meadows, and sometimes in bogs; ditches and wet grassy places.

694. Habenaria orbiculata   695. Habenaria viridis   696. Habenaria clavellata

323. *Habenaria hookeri* ×½
324. *H. clavellata,* flower ×8
325. *H. flava,* flower ×3
326. *H. dilatata,* flower ×3
327. *Orchis rotundifolia* ×½
328. *Listera cordata,* flower ×5

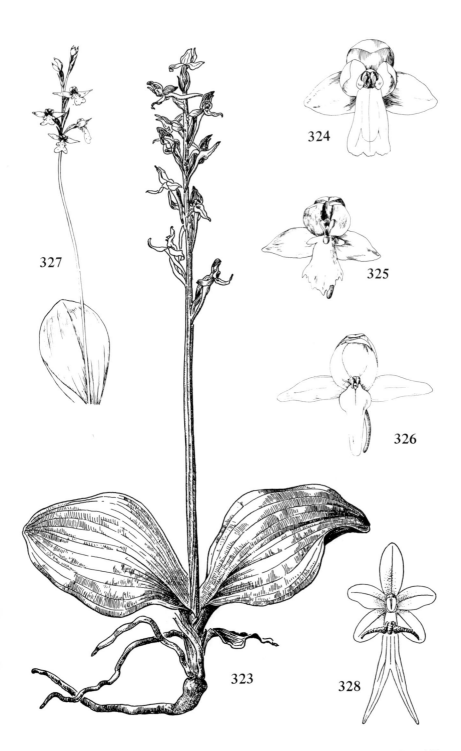

324

325

327

326

323

328

Extremely variable in size, the most vigorous clumps of large plants typical of marshy meadows and boggy interdunal swales along the northern shores of Lakes Michigan and Huron. Intermediates with *H. dilatata* are rather frequent (see comments under that species and in Case, 1964). Typical var. *hyperborea* is often considered to be a far-northern plant (south to Newfoundland and Hudson Bay); our specimens are then referred to var. *huronensis* (Nutt.) Farw. (TL: islands of Lakes Huron and Michigan [Mackinac County?]).

14. **H. dilatata** (Pursh) Hooker   Fig. 326   Tall White Bog Orchid; Bog-candle
    Map 699. In many kinds of moist, often marly ground: bogs, cedar and tamarack swamps (especially in openings); boggy meadows, pools, and marshes along the northern Great Lakes shores; springy banks and creek borders.

The flowers frequently have a spicy fragrance, suggestive of cloves. Plants with the lip dilated basally and greenish flowers are intermediate between *H. hyperborea* and typical white-flowered *H. dilatata*. Ames and some other authors treat these plants essentially as a green-flowered color form of *H. dilatata*. They are frequently impossible to distinguish from white-flowered plants when dried. The plants here mapped as *H. dilatata* include all those with definitely and ± abruptly dilated lips, regardless of flower color. Specimens mapped as *H. hyperborea* have narrow or somewhat narrowly triangular or cuneate lips, including some with a whitish and more fleshy petaloid texture, a tendency characteristic of *H. dilatata*. Not included in either map are some other intermediates of the *H. dilatata-hyperborea* complex, with broadly triangular or only slightly dilated lips and various flower textures. Some intermediate plants, at least, are presumably hybrids [*H.* ×*media* (Rydb.) Niles]. Intermediate or ambiguous plants seem to be most often encountered in moist or swampy woods and springy places, conifer swamps, and wet open ground along ditches and lake shores (not in bogs). In some plants, flowers from one part of the inflorescence have narrower and less dilated lips than those from another part. Hultén cites a cellular-papillose margin as characteristic of the lip of *H. dilatata* in Alaska, in contrast to a smooth margin in *H. hyperborea*, but in our area both types of margin occur in both species.

697. Habenaria flava

698. Habenaria hyperborea

699. Habenaria dilatata

9. Corallorhiza
Coral-root

The plants of this genus are "saprophytic," lacking in chlorophyll, and have characteristic coral-like masses of rhizomes (see Campbell, 1970). Fortunately, the species are almost always easily distinguished in the field, although the presence of uncommon color variations makes the key necessarily more complex. The color here called "purplish" varies somewhat, and has been variously termed by authors "reddish," "madder," "magenta-crimson," and "reddish purple."

REFERENCE

Campbell, Ella O. 1970. Morphology of the Fungal Association in Three Species of Corallorhiza in Michigan. Mich. Bot 9: 108–113.

KEY TO THE SPECIES

1. Lip with a small lobe or elongate tooth on each side near the base (sometimes difficult to see in dried specimens)
  2. Sepals and petals 3-nerved; summit of ovary with a low protuberance (like a rudimentary spur) usually visible below the base of the lip; lip 4.5–7 mm long   . . . . 1. **C. maculata**
  2. Sepals and petals 1-nerved (or the latter rarely weakly 3-nerved); summit of ovary with no visible protuberance; lip 2.5–4.5 (5) mm long. . . . . 2. **C. trifida**
1. Lip entire, or merely denticulate or erose
  3. Sepals and petals 3–5-nerved, 8–15 mm long, conspicuously striped with purplish, the lip solid purplish apically . . . . 3. **C. striata**
  3. Sepals and petals 1-nerved (or faintly 3-nerved), less than 6 mm long, not conspicuously striped
    4. Perianth 3–4.5 mm long, purplish; lip white, spotted with purplish; plant of southern Michigan, flowering August–September . . . . 4. **C. odontorhiza**
    4. Perianth 4–5.5 mm long, yellowish; lip unspotted white (or sometimes spotted in northern Michigan); plant found throughout the state, flowering in southern Michigan April–June (sometimes later northward) . . . . 2. **C. trifida**

1. **C. maculata** Raf.   Fig. 330
Spotted Coral-root
Map 700. In woods of all kinds: conifers, hardwoods, mixed; moist or dry; from sandy oak–hickory or red pine to low wet cedar–hemlock; often common in pine and spruce woods on old dunes as well as in beech–maple stands.

The stem is yellowish in f. *maculata* and the white lip is spotted with purplish. In f. *flavida* (Peck) Farw., the lip is plain white and the entire plant yellow. Plants of these forms have been found in both northern and southern Michigan, and may be distinguished from the yellowish *C. trifida* by the characters in the key. Two other forms recognized in current manuals were originally described from Michigan (both from Keweenaw County): f. *intermedia* (Farw.) Farw., with unspotted lip and the plant "purplish yellow" or "cinnamon drab"; and our common form, f. *punicea* (Bartlett) Weath. & Adams, in which the plants are reddish purple, with no trace of brown, and the lip is spotted.

2. **C. trifida** Chat.   Fig. 331   Early Coral-root
Map 701. Coniferous and mixed woods and swamps, especially under cedar, but also in beech—maple, aspen, and other woods; and less commonly in bogs and tamarack swamps; often in deep shade.

Usually easily recognized by its small stature (though rarely as tall as 28 cm), yellow or greenish yellow color, and unspotted lip (but cf. *C. maculata* f. *flavida*). These plants are var. *verna* (Nutt.) Fern. In the northern part of the state, ranging as far south as Midland and Clare counties, one occasionally finds plants with the white lip spotted with purplish; these are the typical var. *trifida*.

3. **C. striata** Lindley   Fig. 332   Striped Coral-root
Map 702. Coniferous and mixed woods and swamps, especially cedar thickets, often associated with limestone; also in beech—maple woods. May occur singly or in large handsome clumps of a dozen or more stems.

4. **C. odontorhiza** (Willd.) Nutt.   Fig. 333   Fall Coral-root
Map 703. Dry open oak woods, less often in beech—maple woods or under tamarack (Kalamazoo County).

The stem is ± bulbous or enlarged at the base in this species, unlike our others.

10. Listera   Twayblade

The flowers in *Listera* are purplish green or yellowish green, the deepest purplish flowers being found in *L. cordata*. In the other two species, the flowers are more often paler green.

REFERENCE

Case, Fred W., Jr. 1964. A Hybrid Twayblade and Its Rarer Parent, Listera auriculata, in Northern Michigan. Mich. Bot. 3: 67—70.

700. Corallorhiza maculata   701. Corallorhiza trifida   702. Corallorhiza striata

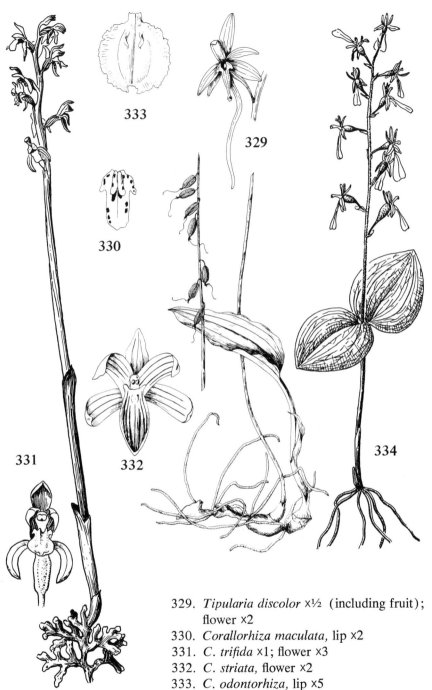

333

329

330

331

332

334

329. *Tipularia discolor* ×½ (including fruit); flower ×2
330. *Corallorhiza maculata*, lip ×2
331. *C. trifida* ×1; flower ×3
332. *C. striata*, flower ×2
333. *C. odontorhiza*, lip ×5
334. *Listera convallarioides* ×1

1. Lip deeply cleft about halfway to its base into two sharp-pointed, narrow lobes (fig. 328); axis of inflorescence glabrous (though the peduncle is puberulent above the leaves); leaves small, up to 2.5 cm long (often 1–1.5 cm), shorter than peduncle; column inconspicuous, ca. 0.5 mm long . . . . 1. **L. cordata**
1. Lip shallowly cleft into two ± rounded lobes; axis of inflorescence glandular-puberulent; leaves longer than peduncle, often over 2.5 cm long; column conspicuous, ca. 2–3 mm long
   2. Lip tapered to narrow base, at most with a weak lateral tooth on each side (not at base); pedicels ± finely glandular-puberulent and the ovary also somewhat so, at least on the main nerves . . . . 2. **L. convallarioides**
   2. Lip auriculate at base, with distinct rounded lobes; pedicels and ovary glabrous .
   . . . . 3. **L. auriculata**

1. **L. cordata** (L.) R. Br.   Fig. 328   Heartleaf Twayblade
Map 704. Usually in sphagnum of bogs and coniferous swamps, with cedar, tamarack, spruce, and/or fir; less commonly in hemlock groves and moist woods (e. g., along Lake Superior). Often common where found, but very delicate and inconspicuous. Plants in a colony may vary considerably in flower color.

2. **L. convallarioides** (Sw.) Torrey   Fig. 334   Broad-leaved Twayblade
Map 705. Especially characteristic of mossy or springy areas, seeping slopes, etc., in coniferous woods, hemlock–hardwoods, cedar swamps; often in wet sandy soil along stream borders.
Plants bearing flowers with the lip lobed but scarcely auriculate at the base and intermediate in pubescence, angle of the flowers, and leaves are presumably hybrids with *L. auriculata* (see below).

3. **L. auriculata** Wieg.   Auricled Twayblade
Map 706. Most characteristic of sandy soil in open alder thickets along rivers and Lake Superior; occasionally in mixed woods, or spruce–fir on Isle Royale.
Presumed hybrids with *L. convallarioides* are known from Alger and Luce counties and have been named *L. × veltmanii* Case (TL: Sable Falls, Alger Co.). The hybrids are intermediate in all respects between the parents and should be sought wherever the two species are in proximity (Case, 1964).

11. Epipactis

REFERENCES

Drew, W. B., & R. A. Giles. 1951. Epipactis helleborine (L.) Crantz in Michigan, and Its General Range in North America. Rhodora 53: 240–242.
Voss, Edward G. 1965. A Weedy Orchid? Mich. Audubon Newsl. 13(4): 4.

1. **E. helleborine** (L.) Crantz   Fig. 335   Helleborine
Map 707. Rich deciduous woods, hemlock–hardwoods, and mixed woods

and thickets on dunes near Lake Michigan (Berrien and Benzie counties). Naturalized from Europe; sporadic and apparently spreading in Michigan. Hardly a "weed," but said to be aggressive when established, and often associated with some disturbance of the woods where it grows.

Roots and seeds from near Buffalo, N. Y., were sent in 1891 to Ralph Ballard, of Niles (Berrien County), who planted them. No apparent results were noted until 1919, when the species was discovered to be established in the Niles area. (See Asa Gray Bull. 4: 19. 1896; Fuller, 1933, p. 110.) It is not certainly known whether any plants subsequently discovered at other sites in Michigan were derived from the Niles population. The Genesee County record is a specimen (MSC) labeled as from Grand Blanc Tp., from the herbarium of Daniel Clarke. It was identified as *Habenaria leucophaea*, of which there is also a good collection with similar label in Clarke's material. These collections are not dated, but presumably were made in the 1860's or 1870's, when Clarke was most active (he died in 1884). This would be an unusually early date for *Epipactis* (the first North American collection thus far known was in New York state in 1879), and it is possible that the specimen became associated with the wrong label.

The flowers of *Epipactis* might be confused by the inexperienced with those of a *Habenaria*, but they lack spurs and the axis is pubescent; the leafy stems are suggestive of *Cypripedium*. The plant is probably to be found at many additional stations in Michigan.

703. Corallorhiza odontorhiza

704. Listera cordata

705. Listera convallarioides

706. Listera auriculata

707. Epipactis helleborine

708. Calopogon tuberosus

12. Calopogon

1. C. tuberosus (L.) BSP.   Plate 8-E   Grass-pink
   Map 708. Bogs (including marly tamarack bogs), coniferous swamps (especially in openings), beachpool bogs along the Great Lakes shores, and wet meadows.
   This is our only native orchid in which the flowers are *not* resupinate, statements in many manuals notwithstanding; in all of our other species the flower is turned 180° so that the lip, which is morphologically the upper petal, becomes the lowest in position (i. e., resupinate). The species has long been known as *C. pulchellus* (Salisb.) R. Br.

13. Triphora

1. T. trianthophora (Sw.) Rydb.   Fig. 336   Nodding Pogonia; Three Birds
                                                                        Orchid
   Map 709. An extremely rare orchid, fluctuating greatly in numbers from one year to the next, blooming in late summer in humus of beech, oak, maple, or mixed woods. The flowers remain fresh only a few hours.

14. Aplectrum

1. A. hyemale (Willd.) Torrey   Fig. 337   Putty-root; Adam-and-Eve
   Map 710. Rich woods, both upland beech—maple and more swampy woods in low ground. The Keeenaw County record may look suspicious (*Farwell 415*, BLH) but Farwell's notes say "collected in 1878 by James Eade & analysed in school at that time."
   The pale-flowered form lacking the usual purplish markings is known as f. *pallidum* House, and has been found at a few localities in the state.

709. Triphora   710. Aplectrum hyemale   711. Liparis liliifolia
    trianthophora

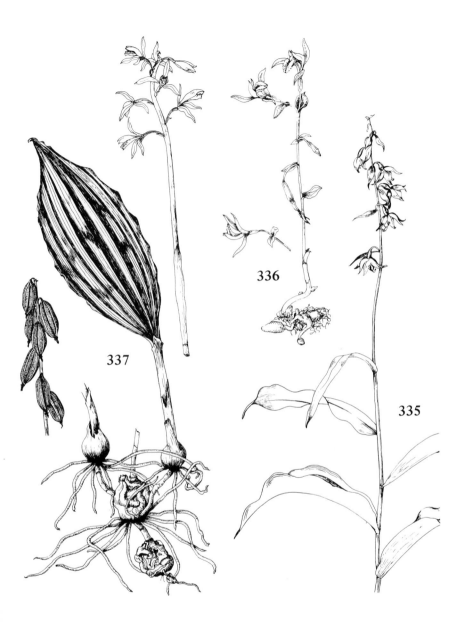

335. *Epipactis helleborine* ×½
336. *Triphora trianthophora* ×½
337. *Aplectrum hyemale* ×½

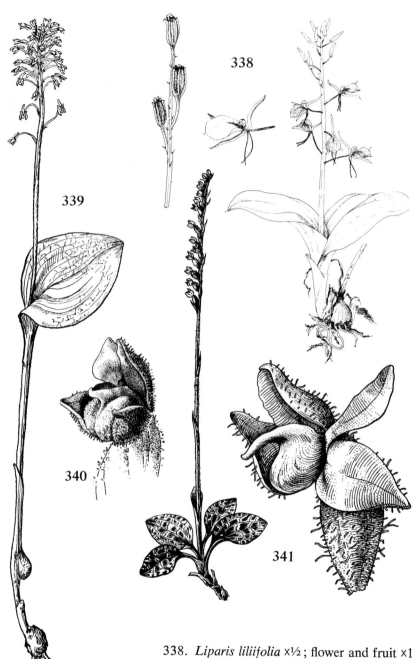

338. *Liparis liliifolia* ×½; flower and fruit ×1
339. *Malaxis unifolia* ×1
340. *Goodyera pubescens,* flower ×4
341. *G. repens* ×⅔; flower ×8

15. Liparis

KEY TO THE SPECIES

1. Lip ca. 10 mm long, purplish; capsules equalling or shorter than pedicels . . **1. L. liliifolia**
1. Lip 4–6 (6.5) mm long, yellow-green; capsules longer than pedicels . . . . . . **2. L. loeselii**

1. **L. liliifolia** (L.) Lindley   Fig. 338   Purple Twayblade
    Map 711. Local, but apparently somewhat aggressive and thriving in areas once cultivated or disturbed. Brushy second-growth thickets and mixed woods, floodplains, pine plantations.

2. **L. loeselii** (L.) Richard   Loesel's or Green Twayblade; Fen Orchid
    Map 712. An inconspicuous orchid, usually distinctly yellow-green throughout and generally ± hidden in grasses and sedges of bogs (open mat to older tamarack and cedar swamps), marshy shores, beach pools and interdunal swales, marly flats, roadside excavations, ditches, springs and creek beds. In swampy woods, sometimes very much larger (plants to 30 cm tall, with leaves to 6.5 cm broad and 22 cm long) than the small delicate plants of open sites.

16. Malaxis   Adder's-mouth

KEY TO THE SPECIES

1. Lip narrowly pointed at apex; raceme slender, elongate; pedicels 1–2 (3) mm long
. . . . **1. M. monophylla**
1. Lip 2-lobed at apex (with an indistinct central tooth); raceme thickish, usually ± condensed and flattened or rounded at top (fig. 339); pedicels at maturity 3.5–8 mm long . . . . **2. M. unifolia**

1. **M. monophylla** (L.) Sw.   White Adder's-mouth
    Map 713. Local, and seldom noticed, in mixed woods and swamp forests (often on hummocks or mossy logs), coniferous swamps and thickets by shores, jack pine and mixed woods on sandy soil along Lake Superior dunes, boggy places (seldom on sedge mats), especially along trails.

712. Liparis loeselii   713. Malaxis monophylla   714. Malaxis unifolia

The lip of plants from Europe and Alaska is uppermost; in those from eastern United States and Canada it is lowermost. Our plants have therefore been distinguished as var. *brachypoda* (Gray) Morris or as a distinct species, *M. brachypoda* (Gray) Fern. An aberrant plant with two leaves has been collected in Presque Isle County (*Case* in 1965, MICH).

2. M. unifolia Michaux   Fig. 339   Green Adder's-mouth
Map 714. Almost as local as the preceding species and likewise easily overlooked. Edges of woods, swamps, and clearings, often in sandy soil under bracken; bogs, thickets, and swamps of cedar, tamarack, and other conifers; also in rocky soil northward (including Isle Royale).

An aberrant plant with two leaves [f. *bifolia* Mousley] has been collected on Sugar Island, Chippewa County (*Hiltunen 845* in 1957, WUD).

17. Goodyera   Rattlesnake-plantain

In the flowers of *Goodyera*, the lateral petals and dorsal sepal are united into a hood; the lateral sepals are free. The lip, as usual, is lowermost. The species are often quite difficult to distinguish, especially when one deals with dry specimens. Correll (1950) states: ". . . the probable existence of a hybrid population (centered in the Lake States), involving *G. repens* var. *ophioides*, *G. tesselata* and *G. oblongifolia*, makes the determination of plants from that particular area a most unsatisfactory and perplexing task."

The key should work for the majority of specimens, but the unfortunate overlapping of measurements will make exact placement of a few intermediate specimens (extremes of variation and/or hybrids) impossible. The key is based in large part on specimens identified by Correll, but the measurements and descriptions are original, drawn from the Michigan specimens examined. In addition to the characters used here, there are technical ones involving the structure of the column but these are often extremely difficult to discern.

KEY TO THE SPECIES

1. Perianth 6—9 mm long; leaf blades with only the midvein outlined above in white or pale green, the largest usually 4—6 cm long; plants 20—50 cm tall . **1. G. oblongifolia**
1. Perianth 2.5—5.5 (rarely 6.5) mm long; leaf blades with white or pale green reticulation ± throughout (sometimes *not* on the midvein), the largest often less than 4 cm long; plants 5—32 cm tall (rarely to 36 cm)
   2. Stem with (6) 7—10 (14) cauline bracts (or undeveloped leaves); beak of lip (beyond the large pouch) 0.5—0.8 (1) mm long, about a fourth the total length of the lip or usually less (fig. 340); inflorescences ± densely flowered on all sides . .
   . . . . **2. G. pubescens**
   2. Stem with 2—5 (6) cauline bracts; beak of lip 1—2 mm long, about half the total length of the lip; pouch shallow or deep; inflorescence strongly one-sided or ± loosely flowered on all sides
      3. Lip shallowly saccate, the pouch longer than deep, the beak horizontal or slightly recurved (fig. 342); plants 13—32 (36) cm tall (usually 17—25); largest leaf blades 2—5 (6) cm long (usually 2.2—4 cm); cauline bracts 3—5 (6), usually

4–5, the uppermost glabrous or, more often, at least sparsely pubescent; perianth 3.5–5.5 (6) mm long . . . . 3. G. tesselata

3. Lip deeply saccate, the pouch about as deep as long, the beak often strongly turned downward at maturity (fig. 341); plants 5–20 (usually 10–17) cm tall; largest leaf blades 0.7–2.5 (usually 1–2) cm long; cauline bracts 2–4 (rarely 5, usually 3), the apical third to half of the uppermost one glabrous; perianth (2.5) 3–4 (4.5) mm long . . . . 4. G. repens

1. G. oblongifolia Raf.

Map 715. Moist to dry coniferous, mixed, or less commonly deciduous woods, on rich or sandy soils, often in beds of conifer needles or under cedars near Great Lakes shores.

2. G. pubescens (Willd.) R. Br.   Fig. 340

Map 716. Almost any kind of woods, though less often in beech–maple or oak–hickory stands than in coniferous or mixed woods; often associated with pine and hemlock, sometimes in moist ground and sometimes on hummocks and stumps in cedar or white pine–hardwood swamps.

3. G. tesselata Lodd.   Fig. 342

Map 717. Coniferous and mixed woods, ranging from cedar swamps to well wooded ravines and hollows on sand dunes. Often growing with G. oblongifolia and/or G. repens and apparently hybridizes with them, especially the latter; in fact, it has been suggested that G. tesselata itself may be of hybrid origin. The leaves are rather gray-green or dusty in shade, in contrast to those of G. repens which are deep green (apart from the paler reticulations).

4. G. repens (L.) R. Br.   Fig. 341

Map 718. Moist to dry coniferous or mixed woods, hummocks in coniferous swamps, etc.

Fresh plants have a strongly one-sided inflorescence, but this cannot always be seen in pressed specimens. European, northern Canadian, and western American plants tend to have plain green leaves and are typical var. repens. Our plants with strongly white-reticulate leaves are var. ophioides Fern.

715. Goodyera oblongifolia

716. Goodyera pubescens

717. Goodyera tesselata

18. **Spiranthes** Ladies'-tresses

KEY TO THE SPECIES

1. Leaves widely spreading, in a basal rosette, short-petioled, often withered or wilted at flowering time, their blades less than 4.5 (rarely 5.5) cm long, about two-fifths as broad as long or broader; flowers in a single row, the slender spike twisted or one-sided (fig. 343); perianth (2.5) 3.5–5.5 mm long . . . . 1. **S. lacera**
1. Leaves ascending, sheathing basally and not distinctly petioled, present at flowering time, their blades (non-sheathing portion) over 4.5 cm long and less than two-fifths as broad as long; flowers usually in 2 or more rows in a ± crowded spike (sometimes one-sided); perianth usually larger
  2. Largest leaves (non-sheathing portion) about 5–10 times as long as wide, lanceolate to oblanceolate; lip with yellow center; leaves all basal (rarely 1 cauline), 4.5–10 cm long, the stem bearing 1–2 cauline bracts (including reduced leaf if present); plant flowering in June and early July . . . . 2. **S. lucida**
  2. Largest leaves commonly over 10 times as long as wide; lip white or creamy throughout; leaves often present on lower portion of stem, the cauline bracts and leaves totaling 3–6; plant flowering in late summer and early fall
    3. Flowers in one row (but this may be obscured by twisted rachis); lip pubescent beneath, at least centrally, with prominent incurved basal auricles . . . . . 3. **S. vernalis**
    3. Flowers in 2 or more rows; lip (but not necessarily sepals which may overlap it) glabrous beneath except on the less prominent auricles
      4. Lip fiddle-shaped, strongly constricted behind expanded apex (fig. 344); lateral sepals united for at least half their length with dorsal sepal and lateral petals, forming a hood . . . . 4. **S. romanzoffiana**
      4. Lip ± oblong, often rather erose-margined but not strongly constricted (fig. 345); at least the lateral sepals free (or easily separated if connivent when young) . . . . 5. **S. cernua**

1. **S. lacera** (Raf.) Raf. Fig. 343 Slender Ladies'-tresses
Map 719. Characteristic of dry sandy soil of jack pine plains, often with blueberry and bracken; also under red pine and oak, in moist aspen woods and conifer thickets along shores and dunes, and in thin soil on rocks in the western Upper Peninsula; rarely in boggy places.

The perianth is 3.5–5.5 mm long (rarely smaller), the lip with a central green portion; the stem arises from a cluster of 2–several tuberous roots – which are not always collected by those who make specimens. A related species, *S. tuberosa* Raf. [*S. grayi* Ames], has been reported from Michigan by Correll on the basis of a few plants from jack pine woods in Cheboygan County; the flowers are smaller (perianth 2.5–3.5 mm). In *S. tuberosa* the lip is entirely white and the stem arises from a single tuberous root. The Michigan specimens I have seen which are the basis of this report (*F. C. Gates 16641* in 1931, UMBS) lack basal parts, and quite probably are merely forms of *S. lacera* with somewhat smaller flowers than usual, as the latter species is frequent in the jack pine stand whence the supposed *S. tuberosa* came. Specimens from an old field in Kalamazoo County (*B. Stergios 528* in 1969, MSC) also have small flowers, apparently without a green portion on the lip. The range of *S. tuberosa* is otherwise to the south and east of Michigan.

This species is often called *S. gracilis* (Bigelow) Beck, but the name used here is older. Another viewpoint, adopted in *Gray's Manual* but not generally accepted elsewhere, is that there are two species involved, for which both names are used; in this case, most if not all Michigan plants would still presumably be the more northern species, *S. lacera*.

2. S. lucida (H. H. Eaton) Ames   Shining Ladies'-tresses

Map 720. Stream banks, lake shores, wet hillsides; in sand or muck, often somewhat disturbed sites; very local.

This is a slender, rather delicate plant, the stems 0.5–1.5 mm in diameter just below the inflorescence, under 25 cm tall. The next three species tend to be more robust, often (though by no means always) taller and/or with thicker stems.

3. S. vernalis Engelm. & Gray

Map 721. This species is generally considered to range east and south of Michigan, and Michigan plants referred to it are quite possibly hybrids, *S. cernua* × *S. lacera* (see Case, 1964, pp. 86–87). Specimens very tentatively referred to *S. vernalis* have been found at a few places in sandy moist ground in Michigan, where it is rather variable vegetatively. One collection in particular (*Voss 509* in 1948, MICH, det. Correll), from Devereaux Lake, Cheboygan County, differs

718. Goodyera repens

719. Spiranthes lacera

720. Spiranthes lucida

721. Spiranthes vernalis

722. Spiranthes
romanzoffiana

723. Spiranthes cernua

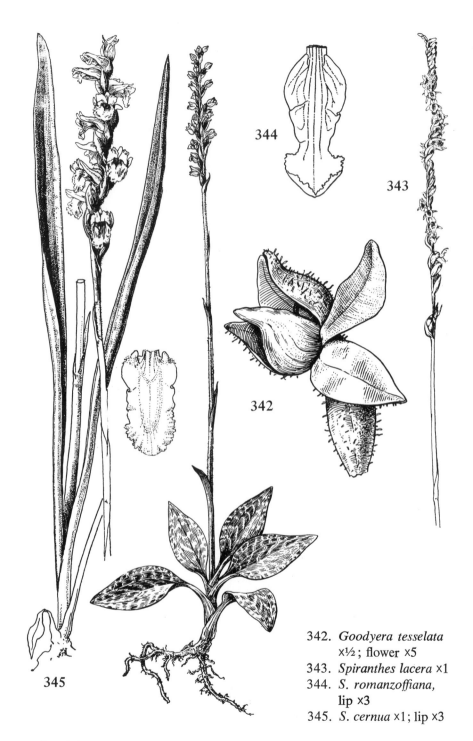

342. *Goodyera tesselata*
     x½; flower x5
343. *Spiranthes lacera* x1
344. *S. romanzoffiana,*
     lip x3
345. *S. cernua* x1; lip x3

from the usual appearance of the species in having relatively short, broad leaves, quite unlike those of *S. cernua* or *S. romanzoffiana*; the rachis is much twisted, obscuring the single-ranked nature of the inflorescence; the rachis and ovaries are densely pubescent with jointed hairs. Collections from Isabella County (*Parmelee 999* in 1949, MSC) and Genesee County (*Clarke*, n. d., MSC) with some pubescence on the lip beneath are also included here. A Marquette County collection (*W. P. Stoutamire 3254*, BLH) has a glabrous lip and appears as if it might be *S. cernua* × *S. lacera*. Ames (Rhodora 23: 79. 1921) suggested such a hybrid origin for northern plants referred to *S. vernalis*.

4. **S. romanzoffiana** Cham.   Fig. 344   Hooded Ladies'-tresses
    Map 722. In most kinds of wet ± open places: bogs, including springy marly areas; tamarack and cedar thickets and openings; sandy or mucky shores; moist roadsides, ditches, and sandy excavations; meadows, beach pools and marshes, interdunal swales.
    The flowers are very fragrant, and tend to be somewhat ascending. The rachis and bracts of the inflorescence are ordinarily much less densely glandular-pubescent than in the next species. Sometimes hybridizes with *S. cernua*, with which it may grow; hybrids [*S.* × *steigeri* Correll] have been discovered by Case in Barry and Chippewa counties and are to be expected elsewhere.

5. **S. cernua** (L.) Rich.   Fig. 345   Nodding Ladies'-tresses
    Map 723. Open moist places generally and often striking because of its late blooming season when little else may be in flower among the grasses and sedges. Shores, ditches, grassy roadsides and openings, sandy excavations, meadows, beach pools and interdunal swales, shaded rock ledges northward; often in sandy soil, sometimes marly or peaty, at bog borders. Tends to grow in somewhat less wet sites than *S. romanzoffiana*.
    The flowers are generally less fragrant than those of the preceding species, and the perianth tends to be horizontal or somewhat nodding at maturity. The inflorescence is usually rather densely glandular-pubescent. Dry specimens of these two species are often difficult to distinguish, especially until one is familiar with the shape of the lip and the tendencies not sufficiently clearcut to use in a key. Small or depauperate specimens may have the inflorescence appearing one-ranked. Hybridizes with *S. romanzoffiana* and perhaps with *S. lacera* (see comments under *S. vernalis*).

Glossary

Most terms relating to habitat are discussed on pp. 17–21.

Terms relating to abundance are discussed on pp. 23–24.

Some specialized terms are explained at the time of use, and are not repeated here.

See also the material on style, conventions, and abbreviations at the beginning of the Taxonomic Section, pp. 41–43.

When this glossary lists a noun, obvious adjectives derived from it are usually not listed separately: e. g., *whorled* means "in a whorl"; *petioled* or *petiolate* means "with a petiole"; *apiculate* means "with an apiculus"; *pistillate* means "possessing or pertaining to pistils"; *involucrate* means "having an involucre"; *involucral* means "pertaining to the involucre"; *stipitate* means "having a stipe"; and so forth.

Negatives are not usually defined: *imperfect* means "not perfect", etc.

Other terms not listed here are hardly specialized botanical ones; if they are unfamiliar, use a good dictionary.

For many terms, representative figures are cited where the application of the term may be seen. When it may not be evident from the definition or the figure legend itself to what part of a figure the term applies, a brief suggestion is added.

Abaxial. Away from the axis; e. g., the "lower" or dorsal surface of a leaf. Cf. adaxial.

Achene. A dry indehiscent fruit, strictly speaking one derived from a single superior carpel, but generally used for similar fruits ("nutlets") derived from more than one carpel. [Figs. 41, 241]

Acuminate. Prolonged into a very acute point (often slightly concave below the point). [Fig. 305, leaf tips]

Acute. With the sides or margins meeting at less than a 90° angle. [Figs. 278, 342, leaf tips]

Adaxial. Toward the axis; e. g., the "upper" or ventral surface of a leaf. Cf. abaxial.

Adnate. United (fused) to parts of a different kind; e. g., stamens to petals; stipule to blade. Cf. connate.

Adventive. Spreading from a native or naturalized source but not [yet] well established.

Albino. Lacking normal color; i. e., white – usually in reference to flowers, at least the perianth.

Alternate. Arranged singly at the nodes, as leaves on a stem or branches in an inflorescence; neither opposite nor whorled. [Fig. 306, leaf arrangement]

Annual. Living for one year; i. e., germinating, flowering, and setting seed in a single growing season (lacking perennial roots, rhizomes, or other such parts).

Anther. The pollen-bearing part of a stamen. [Figs. 35, 314; plate 6-E]

Anthesis. The time at which a flower is fully expanded and functional.

Antrorse. Directed toward the apex or "upward"; e. g., barbs on a bristle or awn. Cf. retrorse. [Figs. 138, barbs on bristles; 182, teeth on beak]

Apiculus. A very small sharp beak-like tip. [Fig. 228]

Appressed. Oriented in a parallel or nearly parallel manner to the surface or axis to which attached. [Fig. 73, pedicels and spikelets]

Articulate(d). With a definite point of separation or "joint."

Ascending. Directed strongly upward (or forward, in relation to the site of attachment). [Fig. 153, panicle branches]

Attenuate. Drawn out to a slender tapering apex or base. [Fig. 98, glumes]

Auricle. A lobe or appendage, often small and "ear-like," typically projecting at the base or summit of an organ (as on a blade or sheath). [Fig. 63]

Awn. A bristle, often a terminal appendage or elongation. [Figs. 51, 94, 98]

Axil. The angle where a leaf or branch joins a stem or main axis.

Basal. At the base; i. e., unless the context suggests otherwise, at the base of the plant, or ground level.

Beak. A comparatively slender prolongation on a broader organ. [Fig. 22, at summit of achene; figs. 173, 176, etc., at summit of perigynia]

Beard. A concentration of hairs; in wheat, the awns of "bearded" varieties. [Figs. 51, 117, at base of lemma; fig. 75]

Berry. A fleshy indehiscent several-seeded fruit derived from a single ovary. [Fig. 295]

Biconvex. Convex on both sides. Cf. plano-convex.

Bifid. Cleft in two. [Fig. 49, summit of lemma]

Bilaterally symmetrical. Capable of division into similar (mirror-image) halves on only one plane; zygomorphic. Cf. regular. [Fig. 340; plate 7-D]

Blade. The expanded portion of a leaf or other flat structure. [Fig. 40]

Bract. A reduced leaf-like, sometimes scale-like, structure, often subtending a flower, inflorescence, branch, etc. [Fig. 40, at base of pedicels; fig. 342, on stem]

Bracteole. A secondary bract; bractlet. [Fig. 290b]

Bulb. A short underground shoot which bears fleshy overlapping leaves (as in an onion).

Bulblet. A vegetative propagule, ± bulb-like. [Fig. 302]

Bulbous. With a bulb or bulb-like thickening.

Calcareous. Limy – rich in calcium carbonate, as from limestone or dolomite or marl.

Callus. The hard often enlarged area at the base of a grass floret. [Fig. 98]

Calyx. The outer series of perianth parts (or the only one); the sepals, collectively. [Fig. 296]

Capillary. Hair-like; i. e., extremely slender.

Capsule. A fruit which dehisces along 2 or more sutures (derived from 2 or more carpels), usually several- or many-seeded. [Fig. 317]

Carpel. The basic female structural unit of the flower, homologous to a sepal, petal, or stamen; in a compound pistil, the carpels are united (connate), but the number can often be determined from the number of styles, stigmas, lobes, or locules (compartments in the ovary).

Cauline. On or pertaining to the stem – often in contrast to basal.

Cespitose. Growing in tufts or dense clumps. [Fig. 262]

Cilia. Hairs along the margin or edge. [Fig. 287, leaf bases; plate 7-A, leaf margins]

Clavate. Club shaped; i. e., with a ± prolonged and narrow base. [Fig. 37, fruit]

Claw. A ± abruptly or strongly narrowed basal portion of some blades; e. g., of petals or tepals. [Fig. 287, base of petals]

Cleistogamous. Fertilized or setting seed without opening.

Column. In the orchids, the structure resulting from fusion (adnation) of stamen(s) and pistil. [Fig. 331, at center of flower]

Compound. Composed of more than one part, or branched; e. g., a leaf with 2 or more blades (leaflets), a pistil of more than 1 carpel; a branched inflorescence. Cf. simple.

Connate. United (fused) to other parts of the same kind; e. g., petals to petals; leaf margin to leaf margin. Cf. adnate.

Connivent. Coming into close contact but not actually fused.

Cordate. Broadly 2-lobed; heart-shaped. [Fig. 279, plate 7-B, leaf shape]

Coriaceous. Leathery in texture; firm.

Corm. A short thick underground stem without thickened leaves (cf. bulb). [Fig. 337]

Corolla. The inner series of perianth parts (when there are 2 series); the petals collectively. [Fig. 296]

Corymb. An inflorescence of the racemose type but with the lower pedicels longer than the upper, so that the inflorescence is ± flat-topped. [Fig. 314]

Crenate. With very rounded teeth; scalloped. [Fig. 324, apex of lip]

Culm. The stem of a grass or sedge.

Cuneate. Wedge-shaped; i. e., with straight but not parallel sides. [Fig. 324, outline of lip]

Cuspidate. With a sharp, firm point.

Cyme. A type of inflorescence in which the terminal (rather than the lower) flower matures first.

Decumbent. Prostrate basally but ascending toward the tip.

Decurrent. Extending downward and along, as a leaf base on a stem or a leaf blade on a petiole. [Plate 1-A, leaf bases]

Deflexed. = reflexed.

Dehiscent. Splitting open at maturity at 1 or more definite places. Cf. indehiscent.

Deltoid. Broadly triangular. [Fig. 180, body of perigynium]

Denticulate. With minute, usually ± remote, marginal teeth.

Depauperate. Stunted or otherwise poorly developed.

Dichotomous. Forking into two ± equal branches.

Dilated. Enlarged (widened or distended). [Fig. 326, base of lip]

Dimorphic. Of two forms.

Dioecious. Having the sexes on separate plants; i. e., all flowers on a single plant either staminate or pistillate. Cf. monoecious.

Diploid. See *n.*

Dissected. Very finely divided (as in some leaf blades) so that the blade tissue is more or less restricted to bordering the main veins.

Distal. Located toward the apex, i. e., toward the end opposite the end by which a structure is attached.

Divaricate. Strongly divergent; spreading or forking at about a 90° angle or more. [Fig. 67, awns]

Divergent. Spreading away from the surface or axis to which attached. [Fig. 155, panicle branches]

Dorsal. Pertaining to the surface (e. g., of a leaf, perigynium) away from the axis to which a structure is attached; abaxial; in relation to the ground, often the "lower" surface. Cf. ventral.

Eciliate. Without cilia.

Elliptic(al). Longer than wide, broadest at the middle, and tapering ± equally toward both ends. [Fig. 337, leaf blade]

Emersed. Normally extending above the water. Cf. submersed.

Entire. Without teeth; with a continuous margin.

Equitant. Folded lengthwise and straddling the structure beneath, as in the leaves of *Iris.* [Frontispiece]

Erose. Irregular (of a margin), as if chewed or gnawed. [Fig. 325, lip margin]

Excurrent. Running beyond, as a vein prolonged beyond the margin of a leaf or other structure. [Fig. 49, between teeth of lemma; fig. 272a]

Exserted. Protruding beyond the surrounding structure(s). Cf. included. [Fig. 86, awns; fig. 110, panicle]

Fertile. Normally reproductive; e. g. a fertile stamen produces pollen, a fertile flower bears seed (or at least reproductive parts); by extension, a structure associated with a fertile flower (as "a fertile lemma").

Filament. The stalk of a stamen, usually threadlike but sometimes flattened or expanded. [Figs. 35, 314]

Filamentose. Breaking into fibers, as in the sheaths of some species of *Carex.* [Fig. 239]

Filiform. Thread-like; very slender and approximately as broad as thick.

Flexuous. More or less loose and sinuous, bent or curved (usually several times in alternate directions). [Figs. 185, 193, inflorescences]

Floret. A reduced flower, as in a grass.

Follicle. A fruit which dehisces along a single suture (derived from a single carpel).

Fruit. A ripened ovary and any closely associated structures.

Glabrate. Nearly without hairs.

Glabrescent. Becoming hairless.

Glabrous. Without hairs.

Gland. A secreting structure; any protuberance (often of different texture, e. g., shiny or sticky in appearance) resembling such a structure.

Glaucous. Covered with a pale (gray to blue-green) waxy coating or "bloom."

Globose. Spherical. [Fig. 251, achene shape; fig. 294, head shape]

Glume. A bract or scale at the base of a grass spikelet. [Fig. 87]

Head. A compact inflorescence of sessile flowers crowded on a receptacle. [Plate 4-D]

Hispid. With stiff hairs.

Hispidulous. Minutely hispid.

Hyaline. Thin and translucent.

Illegitimate. Contrary to one or more Articles of the International Code of Botanical Nomenclature.

Imbricate. With the edges overlapping, like shingles on a roof. [Fig. 16, cone scales; fig. 286, floral bracts]

Included. Not protruding beyond the surrounding structure(s). Cf. exserted. [Fig. 111, base of panicle]

Indehiscent. Not splitting open naturally. Cf. dehiscent.

Indurate(d). Firm and hardened.

Inferior. (Of an ovary) below the perianth. Cf. superior. [Figs. 317, 342]

Inflorescence. An entire flower cluster, including pedicels and bracts.

Inserted. Attached to or on; appearing to arise from. [Fig. 92, awn at base of lemma]

Involucel. Bracts at the base of a unit in a compound inflorescence, in contrast to the involucre at the base of the entire inflorescence. [Fig. 274]

Involucre. The bract or bracts at the base of an inflorescence. [Fig. 274]

Involute. With the margins rolled in (i. e., adaxially). Cf. revolute.

Internode. The portion of a stem or rachis between nodes.

Keel. A ridge ± centrally located on the long axis of a structure. [Fig. 29, on fruit; fig. 49, on lemma]

Lacerate. Ragged, appearing as if torn. [Fig. 321, lip margin]

Lanceolate. Narrow and elongate, broadest below the middle. [Fig. 182]

Leaflet. One of the blades of a compound leaf.

Lemma. The lower or outer bract or scale at the base of a grass floret. [Figs. 49, 55, 87]

Lenticular. Lens-shaped; i. e., biconvex or at least 2-sided (rather than, e. g., 3-sided).

Ligule. An appendage (e. g., membranous collar or fringe of hairs) at the summit of a leaf sheath and on its adaxial side. [Figs. 81, 246]

Linear. Narrow and elongate with ± parallel sides. [Figs. 29, 291, leaf shape]

Lip. In the orchids, the odd petal which is specially modified and usually the lowest (through twisting of the ovary 180°). [Figs. 324–326] In many other bilaterally symmetrical flowers, one of a set of lobes.

Lobe. A projection or extension, usually ± rounded. [Fig. 279, at base of leaves]

-merous. -parted; i. e., with parts in the number cited or a multiple thereof.

Monoecious. Having the sexes in separate flowers but on the same individual. Cf. dioecious.

Mucro. A short, sharp, slender point. [Fig. 216, at tip of scale]

n. The haploid or gametic number of chromosomes; ordinary cells of a seed plant have this number of *pairs* of chromosomes. Many plants have more than a basic number of sets or complements of chromosomes and the number of these is indicated with the suffix -ploid: diploid = 2 sets; tetraploid = 4 sets; hexaploid or 6-ploid = six sets; etc.

Nerve. A vein or ridge, usually a relatively weak or less strong one.

Node. The point on a stem (extended to include the rachis of an inflorescence) at which a leaf or branch arises.

Nutlet. See achene.

Ob-. A prefix signifying inversion, usually with adjectives indicating shape; e. g., obovate or obpyramidal (with the small end basal).

Oblong. Longer than wide and ± parallel-sided (but not as elongate as "linear"). [Fig. 285, segments of the plant]

Obtuse. With the sides or margins meeting at more than a 90° angle. [Fig. 327, leaf apex]

Opposite. Two at a node (and ± 180° apart), as in some leaves. [Fig. 334] Centered upon (rather than alternating with), as stamens opposite the sepals.

Orbicular. Circular in outline, or nearly so. [Plate 8-D, leaves]

Ovary. The lower portion of a pistil, usually ± expanded, in which the seed or seeds are produced; ripens into a fruit. [Fig. 249]

Ovate. Shaped in general outline like a longitudinal section of an egg; i. e., broadest below the middle (but broader than lanceolate). [Fig. 186, body of perigynia; fig. 303, cauline leaves]

Ovoid. Egg-shaped.

Palea. The upper or inner bract or scale at the base of a grass floret. [Figs. 49, 55, 87]

Panicle. A "branched raceme"; i. e., an inflorescence in which the pedicels arise from a branched axis rather than a simple central axis and the lower flowers mature first. [Fig. 306]

Papilla. A minute blunt or rounded projection on a surface.

Pedicel. The stalk of an individual flower or spikelet.

Peduncle. The stalk of an entire inflorescence (or of a solitary flower when there is only one).

Peltate. With the stalk attached at or near the middle of a blade (rather than at the margin).

Perennial. Living 3 or more years.

Perfect. Containing both stamen(s) and pistil(s).

Perfoliate. With the stem (or other stalk) appearing to pass through the leaf (or other blade); i. e., the blade sessile and its tissue at the base surrounding the stem. [Plate 5-E]

Perianth. The "floral envelope" — all of the calyx and corolla together insofar as these are present, in contrast to the reproductive organs of the flower.

Perigynium. The flask-shaped or sac-like (sometimes flattened) structure surrounding the ovary (and later the achene) in *Carex.* [Figs. 190, 246, etc.]

Petal. One of the divisions of the corolla.

Petiole. The stalk portion of a leaf. [Fig. 40]

Phyllode. A bladeless petiole, often flattened or expanded and hence resembling a ribbon-like leaf, as in *Sagittaria cuneata.*

Pilose. With soft, usually long and ± straight, hairs.

Pinnate. With the parts arranged on opposite sides of an elongate rachis.

Pistil. One of the female or seed-producing structures of a flower, whether composed of a single carpel or two or more carpels; usually consisting of an ovary and one or more styles and stigmas. [Figs. 249, 296]

Plano-convex. Flat or flattish on one side and convex on the other.

Pollen. The grains (microspores, containing male gametes) produced in the anther.

-ploid. See *n.*

Propagule. Any part of a plant (whether seed or fruit and associated structures or a vegetative portion) by which dispersal occurs and from which a new individual may develop.

Puberulent. Minutely or finely pubescent.

Pubescent. With hairs (of whatever size or texture).

Pulverulent. Covered with a dust-like surface.

Pulvinus. A swelling at the base of a branch of the inflorescence (in some grasses) or at the base of a petiole or petiolule (the stalk of a leaflet).

Pustulate. With blister-like swellings.

Raceme. A type of inflorescence in which each flower is on an unbranched pedicel attached to a ± elongate unbranched central axis; the flowering sequence is from the base to the apex. [Figs. 37, 312]

Rachilla. The axis of a spikelet in the grasses and sedges.

Rachis. The central axis of an inflorescence or compound leaf.

-ranked. -rowed. Two-ranked structures are in two rows on opposite sides of an axis and 3-ranked structures are in 3 rows (best seen by examining from above the apex of the axis).

Receptacle. The surface on which the parts of a flower are inserted, or on which the flowers in a head are inserted.

Reflexed. Bent backward or downwards; deflexed. [Fig. 304, lower panicle branches; fig. 307, flowers; fig. 329, fruit]

Regular. With radial symmetry (capable of division into similar halves on more than 1 plane); actinomorphic. Cf. bilaterally symmetrical. [Fig. 309, flower]

Remote. Relatively far apart.

Reticulate. Having the appearance of a net. [Fig. 342, pattern on leaves]

Retrorse. Directed toward the base or "downward"; e. g., barbs on a bristle or awn. Cf. antrorse. [Fig. 264, barbs on bristles]

Revolute. With the margins rolled back or under (i. e., abaxially). Cf. involute.

Rhizome. An underground stem, usually ± elongate and growing horizontally (distinguishable from a root by the presence of nodes). [Figs. 169, 297]

Rosette. A ± dense and circular cluster of leaves. [Fig. 308]

Rugose. Wrinkled in appearance. [Fig. 138, lemma]

Sagittate. Arrowhead-shaped; with ± elongate and acute basal lobes. [Fig. 40]

Scabrous. Rough (to the touch).

Scale. A small bract, especially the one which subtends an individual flower in the sedges.

Scape. A peduncle arising from the base of a plant (directly from the root, rhizome, bulb, etc.) – a "leafless stem". [Figs. 300, 301]

Scarious. Thin and dry, papery in texture.

Sepal. One of the divisions of the calyx.

Septate. With cross-partitions. *Septate-nodulose,* with swollen cross-partitions as in cross-veins of some leaves. [Fig. 246]

Serrate. With sharp, ± forward-pointing marginal teeth. [Fig. 23]

Serrulate. Minutely serrate. [Fig. 185, margins of perigynia]

Sessile. Attached without a stalk. [Fig. 297, leaves]

Setaceous. Bristle-like.

Setulose. With minute bristles (or stiff cilia).

Simple. Composed of a single or unbranched part; e. g., a leaf with one blade, a pistil of one carpel, an unbranched inflorescence. Cf. compound.

Sinuous. With one or more bends or "kinks," as in the style of some *Carex.* [Fig. 240]

Sinus. The space or cleft between 2 lobes.

Spadix. An inflorescence consisting of small sessile flowers on a ± elongate fleshy axis. [Figs. 278, 279]

Spathe. A single bract or occasionally more at the base of an inflorescence (± equivalent to an involucre, but used only in the monocots). [Figs. 278, 279, 316]

Spicule. A minute sharp slender point, as on the margins of some leaves.

Spike. An elongate unbranched inflorescence in which the flowers are sessile; loosely, a dense elongate spike-like inflorescence. [Figs. 26, 343]

Spikelet. The unit of the inflorescence in a grass or sedge; i. e., a small spike with reduced flowers on a central axis. [Figs. 50, 199]

Spinulose. With minute spines. [Fig. 34, leaf margin]

Stamen. One of the male or pollen-producing structures of a flower, usually consisting of a filament and an anther. [Figs. 35, 314; plate 6-E]

Sterile. Not fertile; lacking reproductive parts or flowers.

Stigma. The part of a pistil which is receptive to pollen, usually distinguished by a sticky, papillose, or hairy surface. [Figs. 36, 243]

Stipe. A stalk (generally used when no more precise term, such as petiole, is applicable); e. g., the short stalk of some *Carex* perigynia or achenes. [Fig. 22, at base of achene; figs. 176 and 217, at base of perigynium]

Stipule. An appendage at the base of a leaf, in *Potamogeton* appearing solitary and axillary. [Figs. 9, 291]

Stolon. An elongate above-ground stem, growing ± horizontally and rooting at the nodes and/or apex. [Fig. 170]

Strigose. With short, straight, strongly appressed hairs.

Style. The portion of a pistil between the ovary and the stigma(s) − often narrow and elongate. [Fig. 243]

Sub-. A prefix meaning almost, not quite, just below; e. g., subterminal, just below the end; subglobose, almost spherical.

Submersed. Normally occurring under water. Cf. emersed.

Subtend. Occur immediately below, as a bract below a flower or pedicel.

Superior. (Of an ovary) with the perianth inserted beneath it. Cf. inferior. [Fig. 296]

Taxon. Any taxonomically recognized unit, regardless of rank; e. g., species, genus, variety.

Tendril. A slender coiling or twining organ, as on some vines.

Tepal. One of the divisions of a perianth when the sepals and petals are similar in size, color, and texture (though usually distinguishable by position, the sepals being the outer series and the petals the inner one). [Figs. 290b, 298; plate 6-E]

Terete. Round in cross-section.

Tetraploid. See *n.*

Truncate. Ending abruptly, as if cut off. [Fig. 241, body of perigynium; fig. 324, end of lip]

Tuber. A thickened portion of rhizome or root, usually a starch-storing organ.

Tubercle. A distinct enlargement or appendage, as at the summit of the achene in some sedges [Figs. 259, 261] or on the base of the lip in *Habenaria flava* [Fig. 325].

Umbel. An inflorescence in which the pedicels arise from the same point or nearly so. [Figs. 289, 300; plate 6-A]

Undulate. Wavy. [Fig. 26, leaf margin; fig. 66, awn]

Unisexual. (Of a flower) containing only stamen(s) or pistil(s); imperfect. Cf. perfect.

Vein. A bundle of vascular tissue; the external ridge marking the location of an underlying vein.

Ventral. Pertaining to the surface (e. g., of a leaf, perigynium, etc.) toward the axis to which a structure is attached; adaxial; in relation to the ground, often the "upper" surface. Cf. dorsal.

Villous. With soft, not necessarily straight, hairs.

Viviparous. Producing propagules which may sprout or germinate where borne on the parent plant. [Fig. 273, right]

Whorl. A ring of 3 or more similar structures around a stem or other axis (i. e., at the same node). [Fig. 46, leaves]

Index

Names of species followed by a map number (M-000) in parentheses are accepted ones for those recognized as part of our flora (i.e., keyed and mapped). An asterisk indicates that the species is illustrated. References to the color plates (between pages 40 and 41) and the figures (numbered consecutively throughout the volume) may be found at the main entry for each species – which is the first page cited. Additional page numbers refer to discussion in the Introductory Section or significant comments given under some other genus. Page numbers where species are compared within a genus are not indexed, nor are pages where taxa appear in the keys; species and genera can usually be found easily in the keys as they are numbered in sequence.

Names of species not followed by a map number represent synonyms, hybrids, or species mentioned incidentally, including those not considered established, those erroneously reported from the state, and dicots mentioned in the Introductory Section. (Species mentioned in habitat descriptions are generally not indexed unless there is definite phytogeographic or other information.)

Varieties and other infraspecific taxa are not mentioned in the index. Accepted names for families, tribes of Gramineae, and groups of *Carex* are in capitals. When only one species is mentioned in a genus the name of the genus is not separately indexed. Common names consisting of two elements are generally indexed under the second one (e.g., Kentucky *Bluegrass,* Indian *Grass,* Painted *Trillium*).

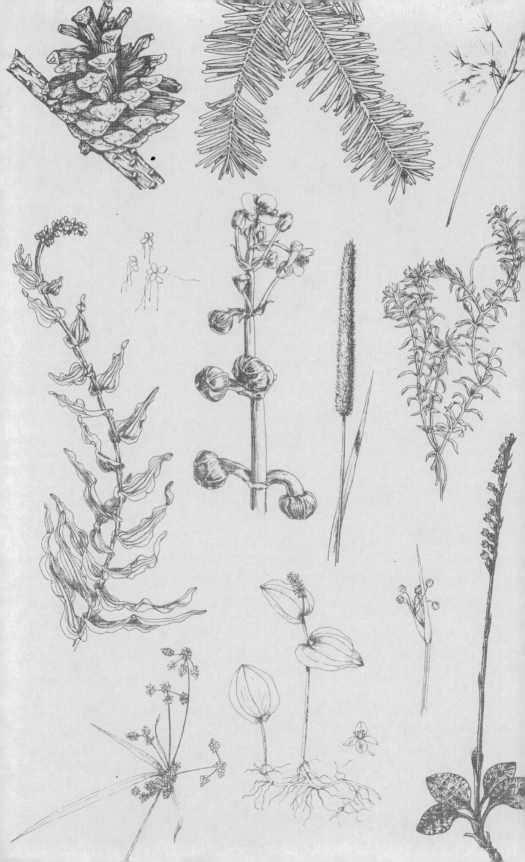